T5-CUL-547

DFG

TIAFT

Thin-Layer Chromatographic R_f Values

Distribution:
VCH, P.O. Box 10 11 61, D-6940 Weinheim (Federal Republic of Germany)
Switzerland: VCH, P.O. Box, CH-4020 Basel (Switzerland)
United Kingdom and Ireland: VCH (UK) Ldt., 8 Wellington Court, Cambridge CB1 1HZ (England)
USA and Canada: VCH, 220 East 23rd Street, New York, NY 10010–4606 (USA)

ISBN 3-527-27397-2 (VCH, Weinheim)
ISBN 1-56081-183-8 (VCH, New York)

ISSN 0930-7958

DFG Deutsche Forschungsgemeinschaft

TIAFT The International Association of
Forensic Toxicologists

Thin-Layer Chromatographic R_f Values of Toxicologically Relevant Substances on Standardized Systems

Second, Revised and Enlarged Edition

Prepared under the Auspices of the Committee for Systematic
Toxicological Analysis of TIAFT by
Rokus A. de Zeeuw (Groningen), Jan Piet Franke (Groningen),
Fritz Degel (Nürnberg), Günther Machbert (Erlangen),
Harald Schütz (Gießen) and Jaap Wijsbeek (Groningen).

Report XVII
of the DFG Commission for
Clinical-Toxicological Analysis
Special Issue
of the TIAFT Bulletin

Deutsche Forschungsgemeinschaft
Kennedyallee 40
D-5300 Bonn 2
Telephone (+49) 228 8851
FAX (+49) 228 8852221

The International Association of
Forensic Toxicologists
Dr. Vina Spiehler, Secretary
c/o Diagnostic Products Corporation
Los Angeles, CA 90045, USA
Telephone (+1) 213 776-0180
FAX (+1) 213 776-0204

Published jointly by

VCH Verlagsgesellschaft mbH, Weinheim (Federal Republic of Germany)
VCH Publishers Inc., New York, NY (USA)

Deutsche Bibliothek Cataloguing-in-Publication-Data

Thin layer chromatographic Rf values of toxicologically relevant
substances on standardized systems / Deutsche Forschungsgemeinschaft ;
The International Association of Forensic Toxicologists. –
2., revised and enl. ed. / prepared by Rokus A. de Zeeuw ... –
Weinheim ; Basel (Switzerland) ; Cambridge ; New York, NY :
VCH, 1992
(Mitteilung ... der Senatskommission für Klinisch-Toxikologische
Analytik / DFG, Deutsche Forschungsgemeinschaft ; (Report ... of the
DFG Commission for Clinical Toxicological Analysis ; 17) (TIAFT
bulletin : Special issue)
ISBN 3-527-27361-1 (Weinheim ...)
ISBN 0-89573-665-9 (New York)
NE: Zeeuw, Rokus A. de; Deutsche Forschungsgemeinschaft; Deutsche
Forschungsgemeinschaft / Kommission für Klinisch-Toxikologische
Analytik: Mitteilung ...

Printed on acid-free and low-chlorine paper

Production Manager: L & J Publikations-Service GmbH, D-6940 Weinheim
Composition: Satzrechenzentrum Kühn & Weyh Software GmbH, D-7800 Freiburg
Printing: Hans Rappold Offsetdruck GmbH, D-6720 Speyer
Bookbinding: Wilh. Osswald + Co., D-6730 Neustadt/Weinstr.

Printed in the Federal Republic of Germany

Contents

1 Preface

Systematic toxicological analysis (STA) comprises the logical chemical-analytical search for a harmful substance whose presence is unsuspected and whose identity is unknown. Thus, it is an essential part of all toxicological cases, even those where at least one poison is already known to be present. On the other hand, because of its complexity, STA is also a very difficult task for the analyst. How can he make sure that all substances with toxic potential are indeed detected and, once a substance is detected, how can he properly identify it against a background of thousands of others?

Alan Curry, in the 1976 edition of his unsurpassed book "Poison Detection in Human Organs", expressed his disappointment about the fact that very little was being published in the area of STA. This issue was discussed on a broad basis at the 1977 Conference of The International Association of Forensic Toxicologists (TIAFT) in Leipzig, Germany. The expression of deep concern resulted in the formation of a special TIAFT committee, the STA-Committee, at the 1978 TIAFT Conference in Wichita, Kansas, USA. The two major tasks of the Committee were:

- to provide adequate approaches for the detection of toxicologically relevant substances;
- to provide compilations of reliable analytical reference data on toxicologically relevant substances to facilitate correct identification.

Although there seemed to be a substantial literature on screening approaches for "general unknowns", as well as on data compilations, it soon became apparent that many publications were unsuitable, mainly for two reasons: a) insufficiently described or verified methods and the use of uncommon materials, b) insufficient agreement between data from different sources for the same substance. This prompted the Committee to establish interlaboratory procedures for the evaluation of methods and the verification of analytical data which turned out to be so time-consuming and laborious that it posed an almost unsurmountable task. In the early nineteeneighties, the STA-Committee was very fortunate to start a collaboration with the Senate Commission for Clinical-Toxicological Analysis of the Deutsche Forschungsgemeinschaft (DFG).

Among the tasks of the Senate Commission were the collection, evaluation and development of detection methods for substances frequently involved in human intoxications. Such methods should be reliable, simple and applicable close to the patient or victim. In particular the Commission was devoted to:

- the development and improvement of qualitative detection methods and quantitative assay methods for toxicologically relevant substances in body fluids, in order to identify the poison(s), and to assess the degree of intoxication and the efficacy of detoxification methods;
- the development and improvement of methods for drug monitoring in cases of risk-bearing drug therapy;
- the collection of substance-orientated analytical and clinical data on a broad, and possibly international scale.

The common interests of the STA-Committee and the Senate Commission provided an extra stimulus for both groups. The joint efforts resulted in two publications on gas chromatographic retention indices of toxicologically relevant substances (1982 and 1985) and one publication on thin-layer chromatographic R_f values (1987). All three were very well received by the toxicological community and sold out rapidly. Very recently, a separate publication was dedicated to the gas chromatographic retention indices of solvents and other volatile substances (1992). Also, a new, third edition of the gas chromatographic retention indices of toxicologically relevant substances was published in 1992.

The present book, the second on thin-layer chromatography, is the result of a further fruitful cooperation, providing reliable and comparable R_f values on toxicologically relevant substances. In comparison to the first edition, the number of substances covered has been increased from 1100 to some 1600 for the 10 systems that can be used for general screening purposes. These systems are known to work well for a large variety of drugs. However, it was also realized that for specific chemical or pharmacological groups, additional, special TLC systems may be useful to provide adequate separations within a given group. Therefore, we have added two special systems for pesticides and one special system to analyze benzodiazepine hydrolysis products (benzophenones) in urine. Finally, the book contains data for the Toxi-Lab screening system for basic and neutral drugs (Toxi-Lab, Inc., Irvine, CA, USA). This system utilizes a special form of sorbent material and the identification of spots is pursued by using the R_f value as well as the colors that can be observed in four different visualization stages.

Although the mandate of the Senate Commission officially ended in 1990, facilities were made available by the DFG to complete the present work. It is hoped that this compilation of data will serve a useful purpose in the daily practice of toxicological analysis. As with the previous data bases, the values taken up are the result of extensive interlaboratory investigations. The authors will be grateful for comments and critique from their colleagues, as well as for new data on additional substances, so that updates can be prepared in due time. Also, we would appreciate to be informed about any erroneous or misarranged data.

2 Acknowledgements

The authors thank the following colleagues on whose efforts much of this publication is based:

- A.H. Stead, R. Gill, T. Wright, J.P. Gibbs and A.C. Moffat, who generated the first collection of data, previously published in The Analyst 107 (1982) 1106-1168.
- Members and guests of the TIAFT STA-Committee and of the Task Groups 'Analysis' and 'Documentation' of the DFG Senate Commission.
- S.P. Sobol, US Drug Enforcement Administration, R.C. Hawkes, US National Institute of Drug Abuse, and K. Szendrei, UN Division of Narcotic Drugs, for providing reference samples of narcotic and other scheduled samples.
- J. Meindertsma and M. Kupsky, for preparing the manuscript.

TIAFT also acknowledges Bayer AG (Berlin, FRG), Hoffmann-La Roche AG (Grenzach, FRG), National Medical Services (Philadelphia, PA, USA) and Toxi-Lab, Inc. (Irvine, CA, USA) for their support in organizing various meetings of the STA-Committee.

3 Introduction

Thin-layer chromatography (TLC) is a popular technique for substance identification in analytical toxicology because of its speed, reliability and low cost. TLC is also straightforward to use with the standardized systems now available, although care must still be taken in order to obtain accurate results. In addition to its ability to separate substances of interest, the ease of applying general and/or specific chemical spot tests by judicious use of different spray or dip reagents has enhanced its value in identification procedures. It also has the advantage that once a drug has been placed on the plate, it cannot go undetected in a solvent front or by non-elution.

The technique has enormous flexibility in that both the stationary phase and the mobile phase can be changed to enable separations from the most polar to the most non-polar of materials. This flexibility has meant that many different systems have been developed for the analysis of drugs and has made the adoption of standard systems slow to be achieved. The Committee for Systematic Toxicological Analysis has studied this problem and has coordinated the work of Dr. A.C. Moffat and his co-workers at the Central Research Establishment, Home Office Forensic Science Service, Aldermaston, U.K., and Professor de Zeeuw and his co-workers at the University of Groningen, The Netherlands, to optimize and standardize TLC systems. The Committee can now recommend standardized TLC systems for use in general screening procedures for the identification of toxicologically relevant substances, be they present in pharmaceutical formulations, illicit products or in biological materials. Those systems that have been proven to be reliable and reproducible in practice throughout the world over many years of use, have been optimized and standardized in terms of their discrimination and reproducibility. In addition, some special systems are being suggested for use with particular substance classes (pesticides and benzodiazepines), or that utilize special materials (Toxi-Lab).

This publication outlines the principles to choose proper TLC systems for screening purposes, the methods of using these systems, and collections of data obtained in the laboratories of the authors and many collaborating members of TIAFT and DFG Task Groups.

4 Choice of Systems

4.1 General Systems

To select the best general TLC systems for screening purposes, the most important features for a good system were considered, viz (a) distribution of R_f values across the plate, (b) reproducibility of the measurement of those values, and (c) correlation of chromatographic properties between systems. There are three other features of importance, namely sensitivity, time for analysis and costs, but in general they do not vary much between systems and were therefore not considered when comparing systems. Although the steps needed to choose and standardize TLC systems have been amply laid down by Stead et al. (1982) and Schepers et al. (1983), it is worthwile outlining the main principles here.

A mathematical technique is required to compare the separating power of the TLC systems that brings together the features of distribution of R_f values, the reproducibility of those values and the correlation of chromatographic properties when more than one system is used. Five mathematical parameters have been used for this purpose, viz discriminating power (Moffat and Smalldon, 1974; Moffat and Clare, 1974; Owen et al., 1978a, 1978b), identification power, mean list length (de Zeeuw et al., 1978; Schepers et al., 1983), information content (Massart, 1973; de Clerq and Massart, 1975), quotient of distribution equality (Müller et al., 1976; Müller, 1980), and principal components analysis (Musumarra et al., 1983, 1984). Although each of these has its own advantages and disadvantages, the measurement of discriminating power or identification power was used to select the systems included here.

Discriminating Power

Discriminating power is defined as the probability that two drugs selected at random can be separated by a TLC system. It is calculated by comparing the measured R_f values for each drug in turn with R_f values for each of all the other drugs. If the R_f values are within a predetermined error window (E), they are considered to be undiscriminated (or matched) and the total number of matches from all the drugs is designated M. The total number of possible matches is N(N-1)/2 (where N = total number of drugs), and therefore the probability of two drugs selected at random being matched is 2M/N(N-1). Therefore,

$$\text{Discriminating power} = 1 - \frac{2M}{N(N-1)}$$

Measurements of discriminating power for combinations of systems are made by re-defining a match as follows: two drugs are matched if they cannot be separated in any of the systems used. The value of N is changed accordingly, depending on the number of systems examined. The theory of discriminating power measurements has been described by Smalldon and Moffat (1973) and its application to chromatography by Moffat et al. (1974).

Identification Power

For the calculation of the identification power (IP) the assumption is made that the data obtained in an analytical procedure for a given substance show fluctuations, from day to day as well as from laboratory to laboratory, and that these fluctuations follow a certain, known distribution pattern. In TLC, it is assumed that the R_f value of a given substance varies, following a normal, gaussian distribution with the listed value (in the data base) as the mean value and with a given, system-dependent, standard deviation. For each substance in the data base the probability can be calculated that the substance is a candidate for identification for a given R_f value. The larger the difference between the listed R_f value of a substance and the given R_f value the lower the probability will become. After normalization in such a way that the sum of the probabilities becomes 1, the substances can be ranked in decreasing order of probability. The top of the list then gives the substances with the highest probability and the end of the list of substances is reached at an arbitrarily set limit; for example, a cumulative probability of 95%. The list length (LL) then indicates the number of substances that qualify for identification at the 95% confidence level. This procedure can be repeated for all other R_f values in a given data base, after which a mean list length (MLL) can be calculated. This can be done for a single system as well as for combinations of systems (i.e. after using two or three different TLC systems, or using one TLC system and one GLC system).

Thus, the MLL is inversely proportional to the identification power of a system or combination of systems. That is, the lower the MLL, the better the system(s).

The ultimate is reached with a MLL of 1, which means that each substance in the data base can be unequivocally identified. The theory of MLL calculations has been described by Akkerboom et al. (1980) and its application to TLC by Schepers et al. (1983).

Since acidic and neutral drugs are often extracted together in analytical toxicology, TLC systems were selected to enable them to be chromatographed together. Four systems were chosen (systems 1 to 4, Table 5.1) which can be used for general screening purposes as well as for the separation of drugs in the same chemical or pharmacological class (Stead et al., 1982). It is worth noting that Clarke's Isolation and Identification of Drugs (Moffat, 1986) recommends the use of three of these systems (systems 1, 2 and 4) for the general screening of acidic and neutral drugs.

From the discriminating power and identification power measurements five TLC systems were chosen for basic drugs (systems 5, 6, 8, 9 and 10, Table 5.1) To these were added the methanol-concentrated ammonia system (system 7) because there was such a large existing data base and the ethyl acetate-methanol-concentrated ammonia system (system 4) because it is widely used for the analysis of basic drugs. It is worth noting that system 4 (ethyl acetate-methanol-concentrated ammonia) can be used for acidic, neutral and basic drugs. Also, systems 3 and 9 use the same mobile phase (chloroform-methanol), so that acidic, neutral and basic drugs can be run in the same tank, although on separate plates. Clarke's Isolation and Identification of Drugs (Moffat, 1986) advocates the use of three of these systems (systems 7, 8 and 9) for the general screening of basic drugs. It should be noted that the above systems for basic drugs are also able to handle neutral drugs if the latter are present in the sample or in the basic extract thereof.

When there is a need to separate only a few drugs from one another, the alphabetical listing of drugs and R_f data in Table 8.1 can be used to choose an appropriate system. If a large number of drugs is to be separated, or a general screening system is needed, then the TLC system should be chosen from Table 5.1 on the basis of high discriminating power (and consequently low mean list length). Thus, the best system for acidic and neutral drugs is the ethyl acetate system (system 2) while that for basic drugs is the methanol system (system 5).

When more than one TLC system is to be used, the correlation of R_f data between the systems must be taken into consideration, otherwise effort will be wasted with no further gain of information. Table 5.2 gives the correlation coefficients of R_f data for pairs of TLC systems and should be used in conjunction with Table 5.1 when more than one system is to be used. For example, if the ethyl acetate system (system 2) is to be used first for the analysis of acidic and neutral drugs, the next system to be used should be the ethyl acetate-methanol-concentrated ammonia system (system 4) since the correlation coefficient between these systems is only 0.464. Conversely, it would not be a good idea to run the chloroform-acetone system (system 1) and chloroform-methanol system (system 3) together since they give highly correlated R_f data (correlation coefficient 0.890).

4.2 Special Systems

Although the general systems described in the previous section work well in the broad screening for a large variety of substances, specific chemical or pharmacological classes of substances may require additional, special TLC systems to provide adequate differentiation between individual representatives within a class or to obtain adequately shaped spots. This is due to the fact that the general systems are geared to deal with a great variety of substances exhibiting large polarity differences. Hence, substances from a particular class with closely resembling polarities may give very similar R_f values in these systems. Two classes of substances for which special TLC systems appeared useful are pesticides and benzodiazepines. Both classes contain large numbers of representatives with very similar structures and are highly relevant in analytical toxicology. Based on extensive investigations, two TLC systems were selected for pesticides, whereas for benzodiazepines a TLC system was selected that proved to be very useful to analyze benzophenones resulting from hydrolyzing benzodiazepines and/or their metabolites in urine. When the general systems indicate that pesticides or benzodiazepines are present in a given case, the special TLC systems can be used subsequently to further differentiate between substances within these classes. As with the 10 general systems, the special systems were extensively evaluated to ascertain their interlaboratory accuracy and reproducibility.

Another special TLC system deemed worthwhile to be taken up is the so-called Toxi-Lab A system. It has been developed for the screening of basic and neutral substances by Toxi-Lab, Inc., Irvine, CA, USA, and is being used extensively, particularly in North America. An international Compendium with data for drugs available on the European market, has recently become available (Degel and Paulus, 1991).

Data on other special TLC systems for various classes of substances have been collected by Moffat (1986) and Müller (1991) and may be used when the systems given in this publication require additional information.

Pesticides (Systems P1 and P2)

As can be seen in the various Tables in Chapter 8, the general screening systems are not very useful to separate pesticides. Due to their rather lipophylic character they tend to migrate rather high on the plate, with hR_f values usually between 80-99. Therefore, a number of non-polar solvent systems were evaluated for a large selection of pesticides, representing various sub-classes, such as organophosphates, phenylurea's, carbamates, triazines and halogenated hydrocarbons. Based on identification power and correlation coefficients, two systems were finally selected (P1 and P2). They are given in Table 5.3 and a data base for 170 commonly used pesticides was generated (Erdmann et al., 1990).

Benzodiazepine Hydrolysis Products (System B)

Although the general systems are usually quite helpful in the identification of a variety of benzodiazepines and their metabolites, it is now well realized that important information can be gained from the analysis of hydrolized urine (Schütz, 1989). Most benzodiazepines are extensively metabolized in man before being excreted in the urine. The resulting metabolites, including those in conjugated form, can be hydrolyzed to yield aminobenzophenone derivatives, which are then extracted, separated by TLC and photolytically dealkylated. The resulting products are then diazotised and coupled with azo-dyes by means of the Bratton-Marshall reagent, which allows very sensitive detection.

It was found that toluene was a very suitable solvent system for the analysis of the benzophenones (see Table 5.3, system B). Since the latter can also be run relatively well in the 10 general systems, it is advised to utilize the information of both the special system and the general systems when trying to identify the benzophenones found. On the other hand, it must be realized that identification of a certain benzophenone may not be sufficient to draw conclusions on which benzodiazepine was ingested, because different benzodiazepines may yield the same benzophenone.

Some benzodiazepines and their metabolites (e.g. bromazepam) do not yield benzophenones, but benzoylpyridines. However, the latter behave similarly.

Toxi-Lab A (System T)

This system utilizes a special sorbent material, namely silica embedded in a glass fiber matrix, together with special development equipment in a standardized procedure. Therefore the results usually show good reproducibility. The system (see Table 5.3, system T) is geared towards the broad screening for basic and neutral drugs and is particularly useful for the analysis of biofluids, such as plasma and urine, since data for a large number of metabolites are available. Moreover, a special feature of the Toxi-Lab system is that it utilizes a well defined detection system, of four color stages, obtained by subsequent dipping of the plate in various solutions as follows:

Color stage 1: The plate is put in a jar containing formaldehyde vapors. After vapor adsorption for two minutes, the plate is dipped in concentrated sulfuric acid. Since Toxi-Gram A plates come pre-impregnated with ammonium metavanadate, the above sequence results in a combined Marquis-Mandelin reaction, to which many substances are susceptible.

Color stage 2: The wet plate from stage 1 is dipped once in water. The dilution of the remaining sulfuric acid with the water will develop heat, resulting in a change of colors for many spots.

Color stage 3: The plate is redipped several times in water to remove all remaining sulfuric acid and then viewed in the dark under long wave UV light of 366 nm. As a result of the treatment in stages 1 and 2, many basic and neutral substances will now fluoresce.

Color stage 4: The plate is dipped into modified Dragendorff's reagent. Most substances containing nitrogen will react. After each stage the resulting colors are observed. and compared with a color coding chart (see Chapter 5).

Table 8.12 contains the hR_f values and the various color codes for the Toxi-Lab A system. The combined information gained from the R_f value and the four color stages greatly enhances the identification power of TLC (Hegge et al., 1991).

5 Running the Systems

5.1 General Systems

The authors favour silica gel with fluorescence indicator e.g. $60F_{254}$ (Merck, Darmstadt, FRG), on TLC plates of 0.25 mm layer thickness. For systems 7 to 10 the plates are either sprayed, or dipped, in methanolic potassium hydroxide solution (0.1 mol/l) and then dried. For systems 1 to 6, the plates may be used as received from the manufacturer. All the systems except system 5 (methanol) and 6 (methanol-n-butanol-NaBr) utilize saturated tanks for development. For normal 20 x 20 cm plates a saturation time of 30 min is recommended; for smaller plates in smaller tanks saturation times can be reduced. For each of the 10 systems suggested for use, there is a mixture of reference compounds to be applied to the plates before running (2 µl of a solution containing 2mg/ml of each drug). There are four reference compounds for each system, chosen to give a good spread of R_f values across the plate. They are used as internal references to correct the R_f values observed (see Chapter 6). These solutions of reference compounds should be applied at three separate positions along the base line of each plate to detect any non-uniform migration of the solvent front (Stead et al., 1982; Bogusz et al., 1985b). All systems require ascending development.

Most of the data contained in the Tables were obtained using 10 cm distance from origin to solvent front. However, shorter development distances can be used. A 7 cm distance has been found to give a good balance between saving time and maintaining resolution (Franke et al., 1982; Franke et al., 1983).

5.2 Special Systems

Most of the remarks under Chapter 5.1 also apply to the systems P1, P2 and B, for pesticides and benzodiazepines, respectively. P1 and P2 are to be run in saturated tanks, whereas B is run in an unsaturated tank (i.e. one without filter paper lining and where the plate is introduced immediately after putting in the solvent).

The application of system B for benzodiazepines requires hydrolysis of the urine and extraction of the resulting benzophenones from the matrix. After chromatography detection can be performed by photolytical N-dealkylation, diazotation of the amine function and coupling with azo-dyes. Details for the use of system B can be found in Appendix D.

The T-system requires special materials. These and the standardized procedures for development of the plate and for the color reactions are to be obtained from Toxi-Lab, Inc., Irvine, CA, USA. For R_f correction a mixture of three substances is recommended. The latter comes pre-impregnated on a disk that can be directly inserted into a starting position on the plate. In order to allow fast and reliable assignment of colors a system has been developed by which the observed colors are given a numerical code, based on a series of reference colors. These reference colors and their corresponding numerical codes are given on the Color Wheel depicted in Fig. 5.1. A loose Color Wheel can be found in the inside back cover. When encoding the colors of the spots, the following should be observed:

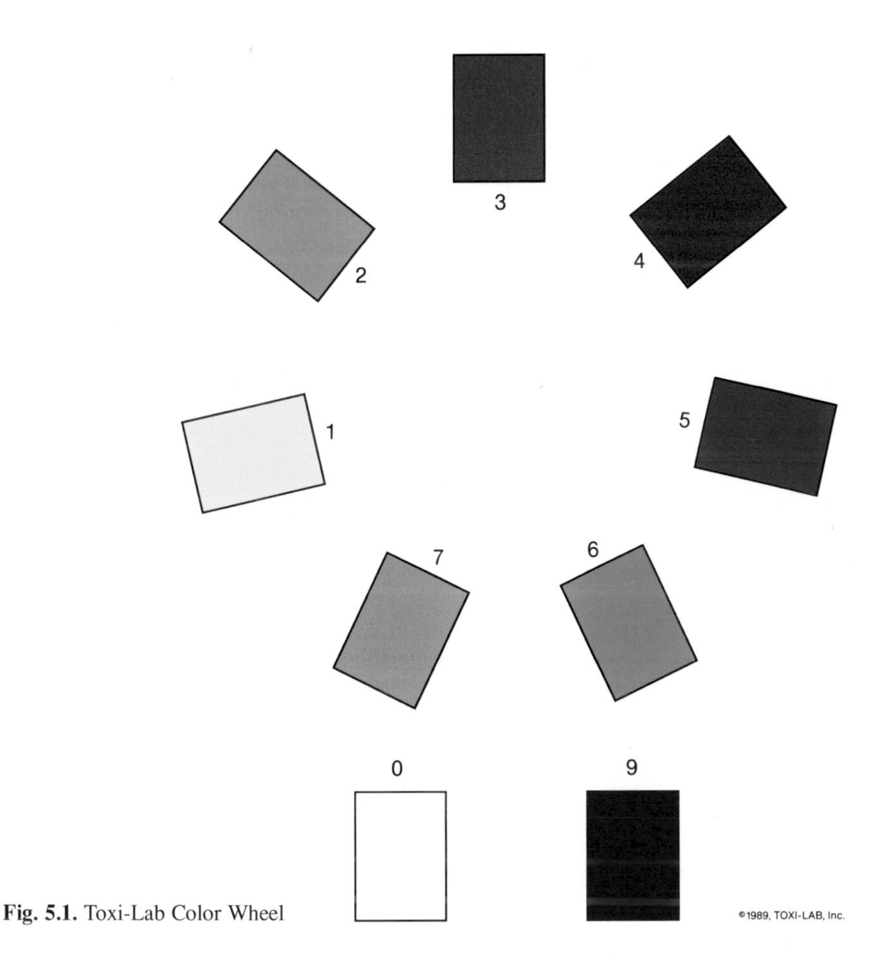

Fig. 5.1. Toxi-Lab Color Wheel

- When no spot is observed in one or more of the color stages, the value according to the background is given, namely white (0) in color stages 1, 2 and 4 and black (9) in color stage 3.
- When the observed color on the plate deviates somewhat from the colors on the wheel, the best match is sought and coded as such.
- One spot may show more than one color; for example the center may differ from the rim. In these cases, the predominating color is noted, which is usually present in the center of the spot.
- Colors may be dependent on the substance concentration. If it is just a matter of color intensity, coding will be relatively simple in that the spot is coded towards the more intense color. For example, a grey spot is coded as black; a pink spot is coded as red. However, when different concentrations of a given substance result in distinctly different colors, multiple entries for that substance are given that relate to low concentrations (suffix L) or high concentrations (suffix H).

For reasons of clarity, the color codes have only been taken up in Table 8.12 which lists the substances in the Toxi-Lab A system in alphabetical order, together with their hR_f values, color codes and CAS registry numbers.

When using system P1 and/or P2 for pesticides detection may be performed on plates containing fluorescence indicator by observing the plate under UV light of 254 nm. Many of the pesticides listed will show fluorescence quenching. In addition, Schütz et al. (1992) recommend six different detection reagents which can produce a variety of colors to facilitate differentiation. Although most pesticides react with one or more of these reagents, it should be noted that they cannot be oversprayed and that each detection system requires a separate plate. Appendix C gives details on the application of these detection reactions, abbreviated RG1 to RG6 and the resulting colors are given in Tables 8.13 and 8.14 for use with the R_f values in systems P1 and P2, respectively. For reasons of convenience, the colors have been encoded in the same way as described above for system T so that the Toxi-Lab Color Wheel can be utilized. Tables 8.13 and 8.14 also contain information on fluorescence quenching under the column UV.

Table 5.1. General TLC Systems.

Solvent[1]	Adsorbent	Reference compounds[2]	hR_f^c	Error window[3]	DP[4]	IP[5]
(1) Chloroform-Acetone (80 + 20)	Silica	Paracetamol	15	7	0.83	14
		Clonazepam	35			
		Secobarbital	55			
		Methylphenobarbital	70			
(2) Ethyl acetate	Silica	Sulfathiazole	20	8	0.88	10
		Phenacetin	38			
		Salicylamide	55			
		Secobarbital	68			
(3) Chloroform-Methanol (90 + 10)	Silica	Hydrochlorothiazide	11	8	0.78	17
		Sulfafurazole	33			
		Phenacetin	52			
		Prazepam	72			
(4a) Ethyl acetate-Methanol-conc. Ammonia (85 + 10 + 5)	Silica	Sulfadimidine	13	11	0.76	19
		Hydrochlorothiazide	34			
		Temazepam	63			
		Prazepam	81			
(4b) Ethyl acetate-Methanol-conc. Ammonia (85 + 10 + 5)	Silica	Morphine	20	10	0.71	21
		Codeine	35			
		Hydroxyzine	53			
		Trimipramine	80			
(5) Methanol	Silica	Codeine	20	8	0.83	17
		Trimipramine	36			
		Hydroxyzine	56			
		Diazepam	82			
(6) Methanol-n-Butanol (60 + 40);0.1 mol/1 NaBr	Silica	Codeine	22	9	0.78	19
		Diphenhydramine	48			
		Quinine	65			
		Diazepam	85			
(7) Methanol-conc. Ammonia (100 + 1.5)	Silica impregnated with 0.1 mol/1 KOH and dried	Atropine	18	9	0.77	18
		Codeine	33			
		Chlorprothixene	56			
		Diazepam	75			
(8) Cyclohexane-Toluene-Diethylamine (75 + 15 + 10)	Silica impregnated with 0.1 mol/1 KOH and dried	Codeine	6	8	0.75	19
		Desipramine	20			
		Prazepam	36			
		Trimipramine	62			
(9) Chloroform-Methanol (90 + 10)	Silica impregnated with 0.1 mol/1 KOH and dried	Desipramine	11	11	0.76	18
		Physostigmine	36			
		Trimipramine	54			
		Lidocaine	71			
(10) Acetone	Silica impregnated with 0.1 mol/1 KOH and dried	Amitriptyline	15	9	0.74	20
		Procaine	30			
		Papaverine	47			
		Cinnarizine	65			

[1] Eluent composition: volume + volume; Saturated systems are used except for systems 5 and 6 which are used with unsaturated solvent tanks. System 4 is split: 4a for acidic and neutral substances and 4b for basic and neutral substances.

[2] Solutions of the four reference compounds at a concentration of approximately 2 mg/ml of each substance.

[3] The error window for each system is based on multiplying by three the interlaboratory standard deviation of measurement of hR_f values.

[4] Discriminating power calculated using the error window in the fifth column.

[5] Identification power calculated using the error window in the fifth column and expressed as mean list length.

Table 5.2. Correlation coefficients of R_f data for pairs of general TLC systems.

	1	2	3	4	5	6	7	8	9	10
1.	–									
2.	0.820									
3.	0.890	–								
4.	0.530	0.748	–	–						
5.		0.464	0.593	0.460	–					
6.				0.436	0.614	–				
7.				0.700	0.745	0.552	–			
8.				0.593	–0.128	–0.045	0.228	–		
9.				0.723	0.748	0.472	0.728	0.342	–	
10.				0.710	0.750	0.655	0.771	0.206	0.820	–

Table 5.3. Special TLC systems.

Solvent[1]	Adsorbent	References compounds[2]	hR_f^c	Error window[3]
Pesticides				
(P1) n-Hexane-Acetone (80 + 20)	Silica	Triazophos	20	7
		Parathion-methyl	30	
		Pirimiphos-methyl	49	
		Quintozen	84	
(P2) Toluene-Acetone (95 + 5)	Silica	Carbofuran	20	7
		Azinphos-methyl	46	
		Methidathion	60	
		Parathion-ethyl	85	
Benzodiazepines				
(B) Toluene	Silica	ANB	15	5
		ACB	27	
		MACB	52	
		CCB	68	
Toxi-Lab A[4]				
(T) Ethyl acetate-Methanol-Water (87+3+1.5), 3 ml + 15 µl conc. Ammonia	Silica in glass fiber matrix	Strychnine	12	9
		Amitriptyline	58	
		Methaqualone	81	

[1] Eluent composition: volume + volume. Saturated tanks are used for systems 1 and 2; systems 3 and 4 are used in unsaturated tanks.

[2] Solutions of the reference compounds at a concentration of approximately 2 mg/ml of each substance. Abbreviations explained in Chapter 7.3.

[3] The error window for each system is based on multiplying by three the interlaboratory standard deviation of measurement of hR_f values.

[4] Requires special Toxi-Lab materials.

6 Measurement and Correction of hR_f Values

The hR_f value of a substance is defined as

$$hR_f = \frac{\text{Distance the substance travels from the origin}}{\text{Distance the solvent front travels from the origin}} \times 100$$

Its reproducibility is governed by many factors such as the amount of drug applied to the plate, running distance, state of saturation of the tank, etc. However, the effect of these factors can be reduced by the use of reference compounds and using corrected hR_f values (hR_f^c). The accurately known hR_f^c values for the reference compounds (and in each of the systems) are given in Tables 5.1 and 5.3. These should be plotted against the practically obtained values to give a six point correction graph, including the start point (0,0) and solvent front (100,100) point as depicted in Fig. 1. The experimentally determined Rf value can then simply be corrected by interpolation, which can be carried out graphically (see Fig. 1)* or by calculation (de Zeeuw et al., 1978; Galanos and Kapoulas, 1964). In the latter procedure, the corrected hR_f value of the unknown compound, $hR_f^c(X)$, can be calculated using the following equation:

$$hR_f^c(X) = hR_f^c(A) + \frac{\Delta^c}{\Delta}\,[hR_f(X) - hR_f(A)], \text{ where}$$

$$\Delta^c = hR_f^c(B) - hR_f^c(A) \text{ and}$$

$$\Delta = hR_f(B) - hR_f(A)$$

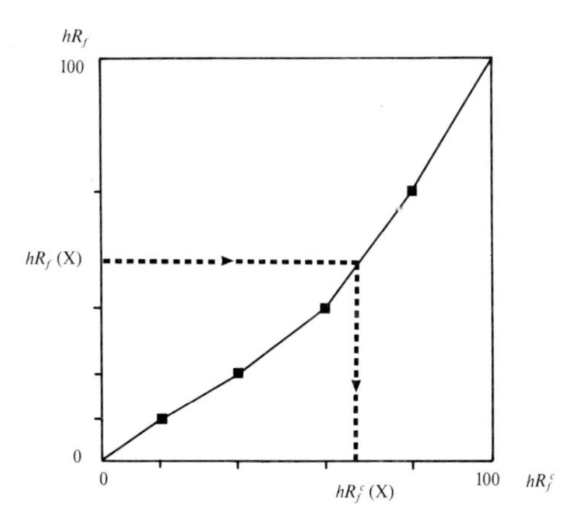

Fig. 6.1. Six-point correction graph.

---

* System T in Table 5.3 uses a mixture of three reference substances. This will give a five point correction graph.

The R_f value of the unknown compound, $R_f(X)$, is found between the bracketing reference substances A and B, where A can also be the starting point $[hR_f^c(A) = hR_f(A) = 0]$, and where B can also be the solvent front $[hR_f^c(B) = hR_f(B) = 100]$. This correction procedure reduces the standard deviation of measurement of hR_f values by about 50% (Moffat, 1975).

The following example will illustrate how the latter correction procedure works in practice: Using system 4b, the chromatogram, as depicted in Fig. 2, is obtained. For the unknown substance X, $hR_f(X)$ is found to be 14, while for the reference substances morphine, codeine, hydroxyzine and trimipramine hR_f values of 19, 36, 63 and 81 are found, respectively. From Table 5.1 we can see that the respective hR_f^c values for these drugs are 20, 35, 53 and 80. For $hR_f(X) = 14$, the bracketing reference points A and B are the starting point and morphine. Thus, substituting the relevant values in the above equation yields:

$$\Delta^c = 20 - 0 = 20 \text{ and}$$
$$\Delta = 19 - 0 = 19$$

$$hR_f^c(X) = 0 + \frac{20}{19}[14 - 0] = 15$$

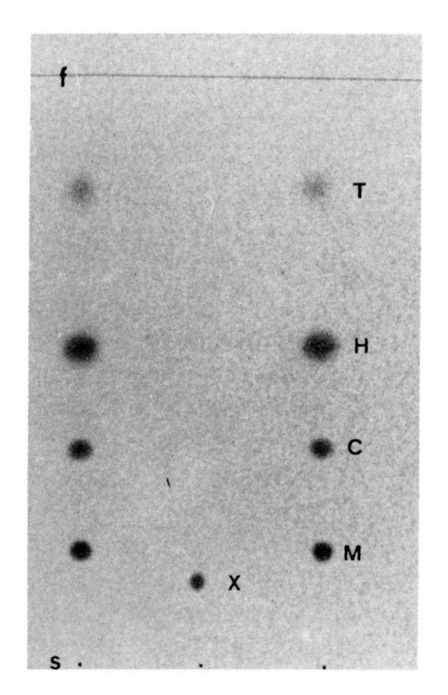

s = starting point;
f = front;
M = morphine;
C = codeine;
H = hydroxyzine;
T = trimipramine;
X = unknown substance.

Fig. 6.2. Thin-layer chromatograms as an example for R_f correction.

Apart from the above two procedures, correction of R_f values can also be carried out conveniently by programmable calculators or personal computers. To this end, programs in BASIC and in TURBO PASCAL are given in Appendices A and B, respectively.

7 R_f Data

7.1 Comments on the Tables

7.1.1 Substance Names

a) Names of substances were chosen according to the following rules: Preference was given to International Non-proprietary Names (INN), as recommended or proposed for drugs by the World Health Organization.
b) If such names did not exist, the names found in Chemical Abstracts were used for simple substances.
c) If a suitable name still remained unavailable, trivial names from Martindale, Merck and Nanogen Index (1975) were considered and a choice was made based on what was assumed to be the most common practice in toxicology.
d) Metabolites are indicated by the suffix M and entered under the name of the parent drug, unless they are also marketed as a parent drug, or available as a pure compound.

In case of doubt, it is recommended to seek confirmation from the CAS registry numbers provided in Table 8.16. This Table should also allow rapid cross checking with other data bases.

For some substances well known abbreviations exist which have been added for convenience. For example, o,p'-clofenotane is also known as o,p'-DDT. Arrows are being used in the Tables to notify the reader that a second name exists.

Data for chiral substances can usually be found under the name of the racemate. However, if data were available for individual enantiomers, the latter are being given as well and an arrow after the name of the enantiomer then points to the corresponding racemate.

7.1.2 Reported Data, General Systems

In the Tables the TLC systems are arranged in such a way that systems 1 to 4a are applicable for acidic and neutral substances and the systems 4b to 10 for basic substances. This subdivision was made on practical grounds and based on the common practice that separate extracts are made for acidic and neutral substances on the one hand and for basic substances on the other. Yet, when available, data for neutral substances have also been taken up in systems for basic substances. This applies, for example, to many of the benzodiazepines. Due to visualization problems the TLC systems are not arranged for a single pH-extraction.

The R_f data in the Tables are standardized hR_f values in that the values found were corrected using a mixture of reference substances as described in Chapter 6. The R_f data for acidic and neutral substances in the systems 1-4a and the R_f data for basic and neutral substances are from a publication by Stead et al. (1982) and from interlaboratory studies initiated by the DFG Senate Commission on Clinical-Toxicological Analysis. The data for systems 4b, 5 and 6 are also from interlaboratory studies initiated by the DFG Commission. Accuracy and precision was assessed by running selected substances in a number of laboratories. For the large majority of the remaining substances mean values were obtained in at least two laboratories. In general, the spots of the substances were visualized under short wave UV-light (254 nm) or by using spray or dip reagents such as Dragendorff's reagent, potassium iodoplatinate reagent or furfural reagent. Recommendations regarding the use of these and other visualization reagents are given by Moffat (1986) and Stead et al. (1982).

7.1.3 Reported Data, Special Systems

The systems for pesticides (P1 and P2) and for benzophenones (B) were developed by Schütz and co-workers, who also developed the initial data bases. The latter were then checked, corrected and expanded after interlaboratory experiments by the authors. Most pesticides can be detected under UV-light 254 nm on F_{254} plates. Special detection methods can be found in Appendix C. Detection of the benzophenones after diazotation and spraying with the Bratton-Marshall reagent (Schütz, 1989) results in purple to violet spots. The data base for Toxi-Lab plates (system T) was developed by Toxi-Lab Inc. and checked, corrected and expanded after interlaboratory experiments by the authors. The recommended visualization procedures have been summarized in Chapter 4.2.

7.2 Using the Data

When an unknown compound is run in any of the TLC systems, its likely identity may be established by consulting the relevant Table of ascending hR_f values (Tables 8.2 – 8.15). A suggested error window is given for each system (Tables 5.1 and 5.3) which is based on multiplying by three the interlaboratory standard deviation of measurement of hR_f values. These were estimated from running hundreds of compounds many times over a period of years in different laboratories (unpublished work from the Task Groups "Analysis" and "Documentation" of the DFG Senate Commission for Clinical-Toxicological Analysis).

It should be noted that the hR_f values in the Tables are for pure substances. Extracted substances from biological materials may necessitate the need to use larger error windows. Co-extracted materials generally decrease R_f values and increase their standard deviation of measurement regardless of whether it is extracted from blood, plasma or liver (Bogusz et al., 1984). One obvious way to compensate for this is to extract the relevant substances from appropriate tissues and use these extracts as solutions of reference compounds for TLC. These extracted reference compounds should be stored at -20 °C in methanol and discarded every 4-6 weeks (Bogusz et al., 1985a).

When more than one TLC system is being used for the identification of an unknown substance and when information from color reactions is utilized also, it quickly becomes difficult to search all the data by hand. Because of this, computer search systems have been developed to enable the experimental data for the unknown substance to be matched with those in the various data bases, also when other techniques than TLC are employed (Schepers et al., 1983; Franke et al., 1985; Gill et al., 1985; Degel et al., 1990). Presently, there are two such systems commercially available which are able to handle R_f information and color reactions on the plate, namely the Toxi-Search System (Toxi-Lab, Inc., Irvine, CA, USA) and the Merck Tox Screening System (Merck, Darmstadt, FRG). Finally, it should be noted that – as with other techniques – it is virtually impossible to unequivocally identify a substance on the basis of TLC R_f values and color reactions alone, also if the substance is run in more than one TLC system. The latter usually requires the application of additional techniques, particularly mass spectrometry. In this regard, the computerized search systems are very valuable, because they give a list of substances that come into consideration, based on how close their data base values match the experimental values for the unknown. (Schepers et al., 1983; Franke et al., 1985). Obviously, the best possible result is obtained when only one substance is listed. This means that from all substances present in the data bases searched, only one matched the values found for the unknown. Whilst this may appear pretty straightforward, it should be remembered that the validity of any identification is greatly dependent on the size of the data basis used: A substance not present in the data bases will never be found or it will be misidentified.

7.3 Abbreviations

The following abbreviations are used in the Tables:

M — Metabolite of the parent substance to which it is added as suffix. When more than one metabolite of a substance is listed, they are numbered M#1, M#2, M#3, etc. usually with increasing polarity. If the structure of M is known, additional information is given in parentheses.

H — As suffix to a substance: Color codes relate to high concentrations.

L — As suffix to a substance: Color codes relate to low concentrations.

Hy — Hydrolysis product obtained after acid hydrolysis used for the conversion of benzodiazepines into benzophenones. The resulting benzophenones are abbreviated as follows:

ABFB — 2-amino-5-bromo-2′-fluorobenzophenone (e.g. from haloxazolam)

ABP — (2-amino-5-bromophenyl)(2-pyridyl)methanone (e.g. from bromazepam)

ACB — 2-amino-5-chlorobenzophenone (e.g. from oxazepam)

ACFB — 2-amino-5-chloro-2′-fluorobenzophenone (e.g. from desalkylflurazepam)

ADB — 2-amino-2′,5-dichlorobenzophenone (e.g. from lorazepam)

ANB — 2-amino-5-nitrobenzophenone (e.g. from nitrazepam)

ANCB — 2-amino-2′-chloro-5-nitrobenzophenone (e.g. from clonazepam)

ANFB — 2-amino-2′-fluoro-5-nitrobenzophenone (e.g. from 1-desmethylflunitrazepam)

CCB — 5-chloro-2-[(cyclopropylmethyl)amino]benzophenone (e.g. from prazepam)

CFMB — 5-chloro-2′-fluoro-2-(methylamino)benzophenone (e.g. from fludiazepam)

CFTB — 5-chloro-2′-fluoro-2-(2,2,2-trifluoroethylamino)-benzophenone (e.g. from quazepam)

CPB	5-chloro-2-(2-propinylamino)-benzophenone (e.g. from pinazepam)
DAB	2,5-diaminobenzophenone (e.g. from aminonitrazepam)
DCB	2′-chloro-2,5-diaminobenzophenone (e.g. from aminoclonazepam)
DCFB	5-chloro-2-[2-(diethylamino)-ethylamino]-2′-fluorobenzophenone (e.g. from flurazepam)
HCFB	5-chloro-2′-fluoro-2-(2-hydroxyethylamino)-benzophenone (e.g. from 2-hydroxyethylflurazepam)
MACB	5-chloro-2-(methylamino)benzophenone (e.g. from diazepam)
MDB	2′,5-dichloro-2-(methylamino)benzophenone (e.g. from lormetazepam)
MNB	2-(methylamino)-5-nitrobenzophenone (e.g. from nimetazepam)
MNFB	2′-fluoro-2-(methylamino)-5-nitrobenzophenone (e.g. from flunitrazepam)
TCB	5-chloro-2-(2,2,2-trifluoroethylamino)benzophenone (e.g. from halazepam).

8 Tables

8.1 Substance Names in Alphabetical Order vs. hR$_f$ Data and CAS Registry Numbers

Substance Name	TLC System														CAS No.
	1	2	3	4	5	6	7	8	9	10	T	P1	P2	B	
ABFB → Haloxazolam Hy	70	69	74	83	81	93	83	11	71	69	–	–	–	32	
ABP → Bromazepam Hy	64	61	66	85	86	88	77	9	68	69	–	–	–	2	1563-56-0
ACB → Oxazepam Hy	76	69	76	86	90	91	78	12	76	71	–	–	–	27	719-59-5
ACFB → Flurazepam M Hy	76	68	74	86	89	93	78	11	75	70	–	–	–	31	784-38-3
ADB → Lorazepam Hy	77	70	76	87	89	93	78	11	76	70	–	–	–	33	2958-30-3
ANB → Nitrazepam Hy	71	65	68	85	90	91	77	5	69	69	–	–	–	15	1775-95-7
ANCB → Clonazepam Hy	72	66	70	86	87	93	77	4	69	70	–	–	–	16	2011-66-7
ANFB → Flunitrazepam M Hy	71	66	68	86	87	91	77	4	69	70	–	–	–	16	344-80-9
Acebutolol	–	–	–	33	13	71	47	0	3	6	14	–	–	–	37517-30-9
Acecarbromal	49	48	60	59	84	84	–	–	–	–	–	–	–	–	77-66-7
Acefylline Piperazine	–	–	–	–	–	–	4	1	1	1	–	–	–	–	18833-13-1
Acemetacine	–	–	–	–	–	–	–	–	–	–	3	–	–	–	53164-05-9
Acenocoumarol	52	48	60	16	92	92	–	0	–	–	–	–	–	–	152-72-7
Acephate	–	–	–	–	–	–	–	–	–	–	–	0	0	–	30560-19-1
Acepromazine	–	–	–	63	28	–	48	26	24	12	–	–	–	–	61-00-7
Aceprometazine	–	–	–	65	34	–	–	–	–	–	49	–	–	–	13461-01-3
Aceprometazine M	–	–	–	45	24	–	–	–	–	–	–	–	–	–	
Acetaminophenazone M (Bis-nor-) AC	–	–	–	–	–	–	–	–	–	–	26	–	–	–	
Acetanilide	45	45	52	66	80	84	–	–	–	–	–	–	–	–	103-84-4
Acetazolamide	4	31	18	5	84	–	85	–	–	–	–	–	–	–	59-66-5
Acetohexamide	39	43	53	12	–	–	–	–	–	–	–	–	–	–	968-81-0
Acetophenazine	–	–	–	–	33	32	53	3	25	3	–	–	–	–	2751-68-0
Acetylcodeine	–	–	–	54	26	25	44	23	43	12	–	–	–	–	6703-27-1
Acetylsalicylic Acid	18	30	31	9	78	–	–	–	–	–	–	–	–	–	50-78-2
Aconitine	–	–	–	75	42	68	68	11	39	58	–	–	–	–	302-27-2
Adinazolam	9	2	51	57	68	66	69	6	55	22	–	–	–	–	37115-32-5
Adinazolam M (Nor-)	1	0	28	40	49	39	63	1	36	2	–	–	–	–	37115-33-6
Adiphenine	–	–	–	83	49	–	64	55	60	51	–	–	–	–	64-95-9
Ajmalicine	–	–	–	81	81	–	–	–	–	–	–	–	–	–	483-04-5
Ajmaline	–	–	–	56	22	–	62	–	–	–	43	–	–	–	4360-12-7
Alachlor	–	–	–	–	–	–	–	–	–	–	–	40	45	–	15972-60-8
Albuterol → Salbutamol	–	0	–	20	16	74	46	1	1	4	11	–	–	–	18559-94-9
Alclofenac	18	28	33	4	–	–	–	–	–	–	–	–	–	–	22131-79-9
Aldrin	–	–	–	–	–	–	–	–	–	–	–	89	98	–	309-00-2
Aletamine	–	–	–	–	–	–	59	37	40	42	–	–	–	–	4255-23-6
Alfadolone Acetate	35	40	62	71	–	–	–	–	–	–	–	–	–	–	23930-37-2
Alimemazine → Trimeprazine	–	–	–	77	32	46	58	54	39	31	73	–	–	–	84-96-8

Table 8.1. (continued)

Substance Name	TLC System														CAS No.
	1	2	3	4	5	6	7	8	9	10	T	P1	P2	B	
Allobarbital	50	66	56	34	87	–	–	–	–	–	–	–	–	–	52-43-7
Allopurinol	–	–	–	21	75	–	–	–	–	–	–	–	–	–	315-30-0
Allyl-5-butylbarbituric Acid,5- → Idobutal	55	69	59	41	–	–	–	–	–	–	–	–	–	–	3146-66-5
Allyl-5-ethylbarbituric Acid,5–	–	–	–	32	–	–	–	–	–	–	–	–	–	–	2373-84-4
Allylbarbituric Acid,5–	54	67	59	35	–	–	–	–	–	–	–	–	–	–	2565-43-7
Allylprodine	–	–	–	–	–	–	58	35	49	24	–	–	–	–	25384-17-2
Aloe-emodin	–	–	–	25	–	–	–	–	–	–	–	–	–	–	481-72-1
Aloxiprin	4	10	22	9	–	–	–	–	–	–	–	–	–	–	9014-67-9
Alphacetylmethadol	–	–	–	84	38	62	57	62	45	50	–	–	–	–	17199-58-5
Alphameprodine	–	–	–	–	–	–	1	33	40	16	–	–	–	–	468-51-9
Alphamethadol	–	–	–	–	–	–	58	56	43	38	–	–	–	–	17199-54-1
Alphamethylfentanyl	–	–	–	84	64	77	–	45	75	64	–	–	–	–	79704-88-4
Alphaprodine	–	–	–	–	–	–	50	30	35	11	–	–	–	–	77-20-3
Alprazolam	7	2	40	47	67	66	67	1	57	14	39	–	–	–	28981-97-7
Alprazolam M (α-Hydroxy-)	4	4	37	37	76	76	72	1	44	18	–	–	–	–	37115-43-8
Alprazolam M (4-Hydroxy-)	3	2	26	26	73	69	67	1	25	7	–	–	–	–	30896-57-2
Alprenolol	–	–	–	49	22	–	52	11	12	11	39	–	–	–	13655-52-2
Alverine	–	–	–	–	–	–	66	65	39	38	–	–	–	–	150-59-4
Amantadine	–	–	–	35	7	77	23	17	7	4	20	–	–	–	768-94-5
Ambazone	–	–	0	38	70	–	55	0	4	8	–	–	–	–	539-21-9
Ambroxol → Bromhexine M	–	–	–	67	59	–	–	6	–	–	78	–	–	–	18683-91-5
Ambucetamide	–	–	–	76	76	88	73	5	68	61	–	–	–	–	519-88-0
Amethocaine	–	–	–	64	43	39	57	15	32	16	–	–	–	–	94-24-6
Ametryne	–	–	–	–	–	–	–	–	–	–	–	26	23	–	834-12-8
Amfepramone	–	–	–	85	55	56	76	62	63	64	95	–	–	–	90-84-6
Amidefrine	–	–	–	0	1	–	15	0	1	2	–	–	–	–	3354-67-4
Amidopyrine → Aminophenazone	25	10	58	62	70	68	66	21	–	–	51	–	–	–	58-15-1
Amikacin	–	–	–	–	0	–	–	–	–	–	–	–	–	–	37517-28-5
Amiloride	–	–	0	24	6	74	24	0	1	2	–	–	–	–	2609-46-3
Amino-2′,5-dichlorobenzophenone,2- → Lorazepam Hy	77	70	76	87	89	93	78	11	76	70	–	–	–	33	2958-30-3
Amino-2′-fluoro-5-nitrobenzophenone,2- → Flunitrazepam M Hy	71	66	68	86	87	91	77	4	69	70	–	–	–	16	344-80-9
Amino-2′-chloro-5-nitrobenzophenone,2- → Clonazepam Hy	72	66	70	86	87	93	77	4	69	70	–	–	–	16	2011-66-7
Amino-5-bromo-2′-fluorobenzophenone,2- → Haloxazolam Hy	70	69	74	83	81	93	83	11	71	69	–	–	–	32	
Amino-5-bromo-3-hydroxyphenyl)(2-pyridyl)ketone, (2–	–	–	–	23	86	–	–	–	–	–	–	–	–	–	40951-53-9
Amino-5-bromo-benzoyl)pyridine,2-(2- → Bromazepam Hy	64	61	66	85	86	88	77	9	68	69	–	–	–	2	1563-56-0

Table 8.1. (continued)

Substance Name	TLC System 1	2	3	4	5	6	7	8	9	10	T	P1	P2	B	CAS No.
Amino-5-chloro-(2,2,2-trifluoro-ethyl)benzophenone,2- → Halazepam Hy	81	74	81	89	94	95	82	49	81	75	–	–	–	74	
Amino-5-chloro-2′-fluorobenzo-phenone,2- → Flurazepam Hy	76	68	74	86	89	93	78	11	75	70	–	–	–	31	784-38-3
Amino-5-chlorobenzophenone,2- → Oxazepam Hy	76	69	76	86	90	91	78	12	76	71	–	–	–	27	719-59-5
Amino-5-nitrobenzophenone,2- → Nitrazepam Hy	71	65	68	85	90	91	77	5	69	69	–	–	–	15	1775-95-7
Aminoacridine	–	–	–	56	9	–	–	–	–	–	–	–	–	–	90-45-9
Aminobenzoic Acid,p–	19	43	31	1	–	–	58	–	–	–	–	–	–	–	150-13-0
Aminoglutethimide	32	47	53	65	–	–	–	–	–	–	–	–	–	–	125-84-8
Aminophenazone	25	10	58	62	70	68	66	21	–	–	51	–	–	–	58-15-1
Aminophenazone M (Bis-nor-)	–	–	–	–	–	–	–	–	–	–	52	–	–	–	
Aminophenol,p- → Paracet-amol M	21	40	30	59	75	–	–	3	–	–	–	–	–	–	123-30-8
Aminophylline	–	–	–	–	–	–	2	0	0	0	–	–	–	–	317-34-0
Aminopromazine	–	–	–	60	12	–	50	42	19	8	–	–	–	–	58-37-7
Aminosalicylic Acid,p–	5	24	15	5	–	–	70	–	–	–	–	–	–	–	65-49-6
Aminoxytriphene	–	–	–	–	–	–	70	51	65	64	–	–	–	–	5585-64-8
Amiodarone	–	–	–	82	54	64	72	62	68	55	88	–	–	–	1951-25-3
Amiphenazole	–	–	–	62	–	–	61	2	33	57	–	–	–	–	490-55-1
Amitriptyline	–	–	–	69	27	51	51	50	32	15	58	–	–	–	50-48-6
Amitriptyline M #1	–	–	–	–	–	–	–	–	–	–	39	–	–	–	
Amitriptyline M #2	–	–	–	–	–	–	–	–	–	–	36	–	–	–	
Amitriptyline M/Nortriptyline	–	–	–	46	87	71	34	28	16	4	30	–	–	–	72-69-5
Amitriptyline M/Nortriptyline M #1	–	–	–	–	–	–	–	–	–	–	17	–	–	–	
Amitriptyline M/Nortriptyline M #2	–	–	–	–	–	–	–	–	–	–	14	–	–	–	
Amitriptyline-oxide	–	–	–	–	–	–	–				5	–	–	–	4317-14-0
Amitrole	–	–	–	–	–	–	–	–	–	–	–	0	0	–	61-82-5
Amobarbital	52	66	58	44	88	–	–	–	–	–	–	–	–	–	57-43-2
Amodiaquine	–	–	–	–	–	–	62	8	40	37	–	–	–	–	86-42-0
Amolanone	–	–	–	–	–	–	66	57	67	64	–	–	–	–	76-65-3
Amopyroquine	–	–	–	–	–	–	52	5	24	21	–	–	–	–	550-81-2
Amoxapine	–	–	–	–	–	–	–	–	–	–	39	–	–	–	14028-44-5
Amphetamine	–	–	–	43	12	75	43	20	9	18	32	–	–	–	300-62-9
Amphetamine M → Hydroxy-amphetamine	–	–	–	–	–	–	35	2	2	11	–	–	–	–	1518-86-1
Amphetaminil M artifact → Amphetamine	–	–	–	43	12	75	43	20	9	18	32	–	–	–	300-62-9
Amprotropine	–	–	–	–	–	–	69	26	45	67	–	–	–	–	148-32-3
Ampyrone	–	–	–	47	61	–	–	–	–	–	–	–	–	–	83-07-8
Amylocaine	–	–	–	–	–	–	73	60	67	63	–	–	–	–	644-26-8
Androsterone	–	–	–	77	–	–	–	–	–	–	–	–	–	–	53-41-8
Anhydrotetracycline	–	–	–	–	88	–	–	–	–	–	–	–	–	–	

Table 8.1. (continued)

Substance Name	TLC System 1	2	3	4	5	6	7	8	9	10	T	P1	P2	B	CAS No.
Anileridine	–	–	–	79	60	66	73	12	56	51	–	–	–	–	144-14-9
Antazoline	–	–	–	47	5	66	31	6	7	3	–	–	–	–	91-75-8
Antipyrine → Phenazone	18	14	50	45	66	66	65	4	–	–	49	–	–	–	60-80-0
Apoatropine	–	–	–	–	–	–	22	18	7	2	–	–	–	–	500-55-0
Apomorphine	–	–	–	58	54	–	83	0	21	27	–	–	–	–	58-00-4
Aprindine	–	–	–	76	20	–	–	63	–	–	62	–	–	–	37640-71-4
Aprindine M (Desethyl-)	–	–	–	55	12	–	–	42	–	–	–	–	–	–	
Aprobarbital	48	65	57	40	86	–	–	–	–	–	–	–	–	–	77-02-1
Apronal	–	–	–	62	–	–	–	–	–	–	–	–	–	–	528-92-7
Arecoline	–	–	–	62	43	–	53	–	–	–	–	–	–	–	63-75-2
Articaine	–	–	–	76	66	80	75	36	64	56	–	–	–	–	23964-58-1
Artifact (lipid)	–	–	–	–	–	–	–	–	–	–	99	–	–	–	
Artifact (stopper)	–	–	–	–	–	–	–	–	–	–	85	–	–	–	
Artifact (syringe)	–	–	–	–	–	–	–	–	–	–	90	–	–	–	
Astemizole	–	–	–	64	56	–	–	1	–	–	49	–	–	–	68844-77-9
Astemizole M (Nor-)	–	–	–	51	54	–	–	0	–	–	–	–	–	–	
Atenolol	–	–	–	22	14	–	45	0	2	2	10	–	–	–	29122-68-7
Atropine	–	–	–	24	5	28	18	5	3	1	5	–	–	–	51-55-8
Atropine Methonitrate	–	–	–	0	0	–	–	0	–	–	–	–	–	–	52-88-0
Azacyclonol	–	–	–	14	3	–	10	0	3	1	–	–	–	–	115-46-8
Azaperone	–	–	–	76	65	–	–	–	–	–	–	–	–	–	1649-18-9
Azapetine	–	–	–	78	14	–	70	57	67	56	–	–	–	–	146-36-1
Azapropazone	–	–	–	8	88	–	68	53	5	67	8	–	–	–	13539-59-8
Azatadine	–	–	–	–	–	–	–	–	–	–	17	–	–	–	3964-81-6
Azathioprine	–	–	71	6	80	–	53	3	8	4	–	–	–	–	446-86-6
Azinphos-ethyl	–	–	–	–	–	–	–	–	–	–	–	24	48	–	2642-71-9
Azinphos-methyl	91	93	94	99	81	87	84	40	99	99	–	20	42	–	86-50-0
Azintamid	–	–	–	73	84	–	–	–	–	–	–	–	–	–	1830-32-6
Aziprotryne	–	–	–	–	–	–	–	–	–	–	–	42	50	–	4658-28-0
Azocyclotin	–	–	–	–	–	–	–	–	–	–	–	1	0	–	41083-11-8
Baclofen	1	0	0	0	–	–	–	–	–	–	–	–	–	–	1134-47-0
Bamethan	–	–	–	38	23	–	55	3	6	0	–	–	–	–	3703-79-5
Bamifylline	–	–	–	59	71	–	65	0	54	34	–	–	–	–	2016-63-9
Bamipine	–	–	–	59	24	–	49	40	43	13	48	–	–	–	4945-47-5
Barbital	41	61	57	32	84	–	–	–	–	–	–	–	–	–	57-44-3
Barbituric Acid	0	0	0	0	84	–	–	–	–	–	–	–	–	–	67-52-7
Barbituric Acid,5-(2-Bromo-allyl)-5-(1-methylbutyl)- → Sigmodal	59	71	63	50	86	–	–	–	–	–	–	–	–	–	1216-40-6
Beclamide	–	–	–	76	90	–	65	8	65	64	88	–	–	–	501-68-8
Bemegride	52	53	64	68	–	88	–	–	–	–	–	–	–	–	64-65-3
Benactyzine	–	–	–	77	52	57	66	43	53	53	–	–	–	–	302-40-9
Bencyclane	–	–	–	61	79	–	–	–	–	–	41	–	–	–	2179-37-5
Bendroflumethiazide	25	71	30	57	–	–	–	–	–	–	–	–	–	–	73-48-3
Benethamine	–	–	–	–	–	–	64	48	59	43	–	–	–	–	3647-71-0
Benfluralin	–	–	–	80	80	–	–	–	–	–	–	–	–	–	1861-40-1
Benorilate	–	–	–	70	86	–	67	0	51	62	–	–	–	–	5003-48-5

Table 8.1. (continued)

Substance Name	TLC System 1	2	3	4	5	6	7	8	9	10	T	P1	P2	B	CAS No.
Benperidol	–	3	–	60	62	69	78	3	51	32	61	–	–	–	2062-84-2
Benproperine	–	–	–	83	28	–	–	–	–	–	–	–	–	–	2156-27-6
Benserazide	–	–	–	–	–	–	1	0	1	3	–	–	–	–	322-35-0
Benzalkonium Chloride	–	–	–	2	4	66	2	0	4	0	–	–	–	–	8001-54-5
Benzatropine	–	–	–	36	6	–	13	26	6	2	11	–	–	–	86-13-5
Benzbromarone	–	–	–	24	94	–	–	–	–	–	–	–	–	–	3562-84-3
Benzethidine	–	–	–	–	–	–	75	44	72	57	–	–	–	–	3691-78-9
Benzetimide	–	–	–	69	51	–	–	–	–	–	–	–	–	–	14051-33-3
Benzhexol → Trihexyphenidyl	–	–	–	83	43	75	68	66	61	59	–	–	–	–	144-11-6
Benziodarone	62	58	65	23	–	–	–	–	–	–	–	–	–	–	68-90-6
Benzocaine	56	62	63	76	84	87	67	5	57	66	–	–	–	–	94-09-7
Benzoctamine	–	–	–	77	38	–	59	56	52	43	82	–	–	–	17243-39-9
Benzoic Acid	26	28	35	4	–	–	–	–	–	–	–	–	–	–	65-85-0
Benzoylecgonine,N- → Cocaine M	–	–	–	2	22	14	21	0	1	3	0	–	–	–	519-09-5
Benzphetamine	–	–	–	87	60	–	73	67	70	70	–	–	–	–	156-08-1
Benzquinamide	–	–	–	–	–	–	65	7	69	36	52	–	–	–	63-12-7
Benzquinamide M (Desacetyl-)	–	–	–	53	62	–	–	12	–	–	–	–	–	–	
Benzquinamide M (Desacetyl-, N-deethyl)	–	–	–	38	69	–	–	0	–	–	–	–	–	–	
Benzquinamide M (Nor-)	–	–	–	47	68	–	–	2	–	–	–	–	–	–	
Benzthiazide	14	51	30	9	–	–	–	–	–	–	–	–	–	–	91-33-8
Benzthiazuron	–	–	–	–	–	–	–	–	–	–	–	8	3	–	1929-88-0
Benzydamine	–	–	–	50	16	–	44	36	22	9	–	–	–	–	642-72-8
Benzyl Alcohol	–	–	–	86	86	–	–	–	–	–	95	–	–	–	100-51-6
Benzyl Nicotinate	–	–	–	–	–	–	63	42	71	60	–	–	–	–	94-44-0
Benzylmorphine	–	–	–	41	20	23	41	6	23	8	–	–	–	–	14297-87-1
Berberine	–	–	–	8	4	56	7	38	0	0	–	–	–	–	2086-83-1
Betacetylmethadol	–	–	–	–	–	–	64	55	38	36	–	–	–	–	17199-59-6
Betahistine	–	–	–	8	–	–	10	–	–	–	–	–	–	–	5638-76-6
Betaine	–	–	–	0	16	11	–	0	0	–	–	–	–	–	107-43-7
Betameprodine	–	–	–	–	–	–	58	34	57	37	–	–	–	–	468-50-8
Betanaltrexol,6–	–	–	–	–	–	–	–	–	–	–	55	–	–	–	
Betanidine	–	–	–	–	–	–	1	0	0	0	–	–	–	–	55-73-2
Betaprodine	–	–	–	–	–	–	50	29	25	11	–	–	–	–	468-59-7
Betaxolol	–	0	–	51	22	–	–	11	–	–	35	–	–	–	63659-18-7
Betazole	–	–	–	–	5	42	26	11	0	24	–	–	–	–	105-20-4
Bezafibrate	–	–	–	5	89	–	–	–	–	–	–	–	–	–	41859-67-0
Bezitramide	–	–	–	65	92	96	71	41	79	70	–	–	–	–	15301-48-1
Bialamicol	–	–	–	–	–	–	74	62	80	73	–	–	–	–	493-75-4
Bioallethrin	–	–	–	–	–	–	–	–	–	–	–	48	52	–	584-79-2
Biperiden	–	–	–	83	45	–	64	68	64	64	–	–	–	–	514-65-8
Bisacodyl	–	–	–	80	82	87	74	17	76	66	90	–	–	–	603-50-9
Bisacodyl M (-diphenol)	–	–	–	59	–	–	–	–	–	–	–	–	–	–	
Bisbentiamin	–	–	–	49	62	–	–	–	–	–	–	–	–	–	2667-89-2
Bisoprolol	–	0	–	48	22	–	–	8	–	–	28	–	–	–	66722-44-9
Bitertanol	–	–	–	–	–	–	–	–	–	–	–	10	3	–	55179-31-2

Table 8.1. (continued)

Substance Name	TLC System														CAS No.
	1	2	3	4	5	6	7	8	9	10	T	P1	P2	B	
Bornaprine	–	24	–	75	25	–	–	58	–	–	57	–	–	–	20448-86-6
Brallobarbital	52	68	57	29	87	–	–	–	–	–	–	–	–	–	561-86-4
Brocresine	–	–	–	–	–	–	68	1	35	65	–	–	–	–	555-65-7
Bromazepam	13	18	47	63	73	69	61	6	41	53	74	–	–	–	1812-30-2
Bromazepam Hy/ABP/2-(2-Amino-5-bromo-benzoyl) pyridine	64	61	66	85	86	88	77	9	68	69	–	–	–	2	1563-56-0
Bromazepam M (3-Hydroxy-)	0	6	26	39	73	61	61	0	24	28	–	–	–	–	13132-73-5
Bromazepam M (N(py)-Oxide)	1	1	25	31	62	50	64	0	–	–	–	–	–	–	67664-86-2
Bromazine	–	–	–	69	27	48	54	42	43	13	–	–	–	–	118-23-0
Bromdiphenhydramine → Bromazine	–	–	–	69	27	48	54	42	43	13	–	–	–	–	118-23-0
Bromhexine	–	–	–	88	84	–	75	69	79	71	–	–	–	–	3572-43-8
Bromhexine M/Ambroxol	–	–	–	67	59	–	–	6	–	–	78	–	–	–	18683-91-5
Bromisoval	35	51	52	68	84	–	–	6	–	–	–	–	–	–	496-67-3
Bromo-2,5-dimethoxy-4-methylamphetamine → Bromo-STP	–	–	–	–	–	–	38	13	11	17	–	–	–	–	87112-45-6
Bromo-2,5-dimethoxyamphetamine,4–	–	–	–	43	10	76	36	15	14	17	–	–	–	–	32156-26-6
Bromo-STP/Bromo-2,5-dimethoxy-4-methylamphetamine	–	–	–	–	–	–	38	13	11	17	–	–	–	–	87112-45-6
Bromocriptine	–	–	–	73	84	88	72	2	69	61	79	–	–	–	25614-03-3
Bromolysergide,2–	–	–	–	70	74	76	75	13	58	51	–	–	–	–	478-84-2
Bromophos	–	–	–	99	83	–	–	–	–	–	–	58	92	–	2104-96-3
Bromophos-ethyl	–	–	–	99	84	–	–	–	–	–	–	69	93	–	4824-78-6
Bromopride	–	–	–	50	13	–	–	–	–	–	–	–	–	–	4093-35-0
Bromoxynil	–	–	–	–	–	–	–	–	–	–	–	8	16	–	1689-84-5
Bromperidol	–	3	–	73	50	–	–	13	–	–	68	–	–	–	10457-90-6
Brompheniramine	–	–	–	49	12	–	45	33	16	6	24	–	–	–	86-22-6
Brotizolam	15	5	53	52	72	71	72	5	52	27	54	–	–	–	57801-81-7
Brotizolam M (α-Hydroxy-)	7	6	41	45	78	78	72	2	46	31	–	–	–	–	62551-41-1
Brotizolam M (6-Hydroxy-)	5	4	37	28	76	76	68	1	35	13	–	–	–	–	88883-43-6
Broxaldine	–	–	–	78	78	–	74	52	79	71	–	–	–	–	3684-46-6
Broxyquinoline	–	–	–	–	–	–	51	0	6	3	–	–	–	–	521-74-4
Brucine	–	–	–	25	5	7	16	0	17	1	–	–	–	–	357-57-3
Buclizine	–	–	–	–	–	–	75	61	83	72	–	–	–	–	82-95-1
Buclosamide	–	–	–	47	90	–	90	2	67	70	–	–	–	–	575-74-6
Budesonide	26	42	52	63	89	93	77	0	52	56	–	–	–	–	51333-22-3
Bufexamac	11	19	31	18	–	–	–	–	–	–	–	–	–	–	2438-72-4
Buformin	–	–	–	–	–	–	2	0	0	0	–	–	–	–	692-13-7
Bufotenine	–	–	–	33	10	34	35	0	1	1	–	–	–	–	487-93-4
Bumadizon	–	–	–	8	86	–	–	–	–	–	6	–	–	–	3583-64-0
Bumetanide	1	10	6	4	87	–	–	–	–	–	–	–	–	–	28395-03-1
Bunitrolol	–	–	–	40	16	–	–	–	–	–	25	–	–	–	34915-68-9
Buphenine	–	–	–	62	33	83	74	3	14	50	–	–	–	–	447-41-6
Bupivacaine	–	–	–	80	69	79	69	42	73	65	–	–	–	–	2180-92-9
Bupivacaine M (4-Hydroxy)	–	–	–	56	68	–	–	1	–	–	–	–	–	–	

Table 8.1. (continued)

Substance Name	TLC System														CAS No.
	1	2	3	4	5	6	7	8	9	10	T	P1	P2	B	
Bupivacaine M/Mepivacaine M	–	–	–	49	34	–	–	12	–	–	–	–	–	–	
Bupranolol	–	0	–	49	20	–	–	12	–	–	36	–	–	–	14556-46-8
Buprenorphine	–	–	–	80	80	–	76	9	68	69	–	–	–	–	52485-79-7
Buspirone	–	5	–	71	64	–	–	30	–	–	70	–	–	–	36505-84-7
Busulfan	–	–	–	–	0	–	–	–	–	–	–	–	–	–	55-98-1
Butacaine	–	–	–	83	44	76	71	7	30	64	–	–	–	–	149-16-6
Butalamine	11	29	46	86	–	–	68	–	–	–	–	–	–	–	22131-35-7
Butalbital	54	67	57	44	87	–	–	1	–	–	–	–	–	–	77-26-9
Butallylonal	53	68	57	45	88	–	–	–	–	–	–	–	–	–	1142-70-7
Butamirate	–	–	–	66	17	–	–	–	–	–	–	–	–	–	18109-80-3
Butanilicaine	–	–	–	75	65	–	76	14	54	61	–	–	–	–	3785-21-5
Butaperazine	–	–	–	52	26	–	53	28	37	5	–	–	–	–	653-03-2
Butaperazine M	–	–	–	–	19	–	–	–	–	–	–	–	–	–	
Butaxamine	–	–	–	–	–	–	48	28	11	18	–	–	–	–	2922-20-5
Butetamate	–	–	–	81	48	56	69	59	57	47	–	–	–	–	14007-64-8
Butethal → Butobarbital	50	65	58	41	86	–	–	–	–	–	–	–	–	–	77-28-1
Butethamine	–	–	–	–	45	55	63	5	26	31	–	–	–	–	2090-89-3
Butinoline	–	–	–	–	–	–	–	–	–	–	63	–	–	–	968-63-8
Butobarbital	50	65	58	41	86	–	–	–	–	–	–	–	–	–	77-28-1
Butocarboxim	–	–	–	–	–	–	–	–	–	–	–	15	11	–	34681-10-2
Butocarboxim M (Sulphoxide)	–	–	–	–	–	–	–	–	–	–	–	0	13	–	
Butoxycarboxim	–	–	–	–	–	–	–	–	–	–	–	3	2	–	34681-23-7
Butoxyethyl Nicotinate,2–	–	–	–	–	–	–	63	45	69	62	–	–	–	–	13912-80-6
Butriptyline	–	–	–	–	–	–	59	61	48	38	–	–	–	–	35941-65-2
Buturon	–	–	–	–	–	–	–	–	–	–	–	20	29	–	3766-60-7
Butyl Aminobenzoate	–	–	–	–	83	90	75	6	63	70	–	–	–	–	94-25-7
Butylscopolammonium Bromide,N–	–	–	–	2	3	41	8	0	0	0	–	–	–	–	149-64-4
Buzepide	–	–	–	6	3	–	–	–	–	–	–	–	–	–	3691-21-2
CCB → Prazepam Hy	83	74	82	88	92	95	80	58	82	74	–	–	–	68	2897-00-9
CFMB → Fludiazepam Hy	71	69	81	83	77	90	75	48	77	69	–	–	–	55	1548-36-3
CFTB → Quazepam Hy	74	73	80	85	77	93	79	44	77	73	–	–	–	74	
CPB → Pinazepam Hy	75	73	80	86	86	91	83	48	77	72	–	–	–	57	
CZ-74 → Psilocin-(eth)	–	–	–	65	15	67	–	9	15	25	–	–	–	–	22204-89-3
Cafedrine	–	–	–	47	50	–	–	–	–	–	–	–	–	–	58166-83-9
Caffeine	15	10	55	52	59	55	52	3	58	25	64	–	–	–	58-08-2
Camazepam	55	32	69	75	82	83	76	12	73	65	86	–	–	–	36104-80-0
Camazepam M → Oxazepam	22	35	42	45	82	82	56	0	40	51	–	–	–	–	604-75-1
Camazepam M → Temazepam	51	47	65	62	82	82	53	8	59	53	81	–	–	–	846-50-4
Camphor	–	–	–	–	–	93	–	–	–	–	–	–	–	–	76-22-2
Camylofine	–	–	–	70	21	–	–	–	–	–	–	–	–	–	54-30-8
Captodiame	–	–	–	77	47	–	66	49	–	–	–	–	–	–	486-17-9
Captopril	1	1	6	0	–	–	–	–	–	–	–	–	–	–	62571-86-2
Carazolol	–	0	–	47	21	–	–	1	–	–	39	–	–	–	57775-29-8
Carbachol	–	–	–	0	4	23	0	0	0	0	–	–	–	–	51-83-2
Carbamazepine	–	–	–	56	79	75	60	2	56	47	70	–	–	–	298-46-4
Carbamazepine M #1	–	–	–	–	–	–	–	–	–	–	82	–	–	–	

Table 8.1. (continued)

Substance Name	TLC System														CAS No.
	1	2	3	4	5	6	7	8	9	10	T	P1	P2	B	
Carbamazepine M #2	–	–	–	–	–	–	–	–	–	–	58	–	–	–	
Carbamazepine M #3	–	–	–	–	–	–	–	–	–	–	33	–	–	–	
Carbaryl	–	–	–	–	–	–	–	–	–	–	–	18	25	–	63-25-2
Carbazochrome	0	0	10	16	–	–	–	–	–	–	–	–	–	–	69-81-8
Carbenoxolone	7	17	28	0	–	–	–	–	–	–	–	–	–	–	5697-56-3
Carbetapentane → Pentoxyverine	–	–	–	68	16	49	48	47	22	14	48	–	–	–	77-23-6
Carbidopa	0	2	4	0	–	–	–	–	–	–	–	–	–	–	28860-95-9
Carbifene	–	–	–	–	–	–	71	48	76	62	–	–	–	–	15687-16-8
Carbimazole	63	47	68	42	75	75	72	1	–	–	–	–	–	–	22232-54-8
Carbinoxamine	–	–	–	50	13	16	48	26	19	4	27	–	–	–	486-16-8
Carbocromen	–	–	–	62	18	–	48	17	24	12	–	–	–	–	804-10-4
Carbofuran	–	–	–	79	–	–	–	–	–	–	–	17	20	–	1563-66-2
Carbosulfan	–	–	–	–	–	–	–	–	–	–	–	49	66	–	55285-14-8
Carboxypropyl)phenylpropionic acid,2-4′-(2- → Ibuprofen M	–	13	–	0	78	–	–	–	–	–	–	–	–	–	
Carbromal	53	55	64	75	85	87	–	12	–	–	85	–	–	–	77-65-6
Carbutamide	–	–	–	–	87	–	–	–	–	–	–	–	–	–	339-43-5
Carfenazine	–	–	–	39	39	–	54	5	27	7	–	–	–	–	2622-30-2
Carisoprodol	36	53	59	75	85	79	–	4	–	–	–	–	–	–	78-44-4
Carminic Acid	–	–	–	–	–	1	–	–	–	–	–	–	–	–	1260-17-9
Carnidazole	–	–	–	62	78	81	71	4	55	52	–	–	–	–	42116-76-7
Caroverine	–	–	–	80	57	–	–	–	–	–	–	–	–	–	23465-76-1
Carprofen	–	–	–	–	–	–	–	–	–	–	3	–	–	–	53716-49-7
Carteolol	–	–	–	33	14	–	–	–	–	–	19	–	–	–	51781-06-7
Cathine → Pseudoephedrine M (Nor-)	–	–	–	29	12	–	42	25	5	46	–	–	–	–	39393-56-3
Celiprolol	–	–	–	33	15	–	–	–	–	–	12	–	–	–	56980-93-9
Cephaeline	–	–	–	–	–	–	53	1	19	8	–	–	–	–	483-17-0
Cetrimide	–	–	–	0	0	29	0	0	1	0	–	–	–	–	8044-71-1
Cetylpyridinium	–	–	–	0	0	29	0	0	1	0	–	–	–	–	7773-52-6
Chlorambucil	33	40	50	6	84	–	–	–	–	–	–	–	–	–	305-03-3
Chloramphenicol	11	31	34	36	86	90	69	0	31	48	–	–	–	–	56-75-7
Chlorazanil	–	–	–	69	75	–	–	–	–	–	–	–	–	–	500-42-5
Chlorbenzoxamine	–	–	–	–	74	–	–	–	–	–	–	–	–	–	522-18-9
Chlorbromuron	–	–	–	–	–	–	–	–	–	–	–	22	30	–	13360-45-7
Chlorcyclizine	–	–	–	67	35	52	57	42	46	14	–	–	–	–	82-93-9
Chlordiazepoxide	10	10	53	52	76	77	62	2	50	22	58	–	–	–	58-25-3
Chlordiazepoxide M	–	–	–	–	–	–	–	–	–	–	65	–	–	–	
Chlordiazepoxide M (Nor-)	3	3	33	40	68	60	69	0	32	42	–	–	–	–	7722-15-8
Chlordiazepoxide M → Demoxepam	15	24	42	41	81	83	63	0	35	51	–	–	–	–	963-39-3
Chlordiazepoxide M → Nordazepam	34	45	57	69	82	83	62	3	55	60	81	–	–	–	1088-11-5
Chlordiazepoxide M → Oxazepam	22	35	42	45	82	82	56	0	40	51	–	–	–	–	604-75-1
Chlorfenvinphos	–	–	–	92	87	–	–	–	–	–	–	25	25	–	470-90-6

Table 8.1. (continued)

Substance Name	TLC System 1	2	3	4	5	6	7	8	9	10	T	P1	P2	B	CAS No.
Chlorguanide → Proguanil	–	–	–	18	7	79	3	0	1	1	–	–	–	–	500-92-5
Chlorhexidine	–	–	–	13	1	–	33	–	–	–	–	–	–	–	55-56-1
Chlorimiphenine → Imiclopazine	–	–	–	39	39	–	–	–	–	–	–	–	–	–	7224-08-0
Chlormephos	–	–	–	99	86	–	–	–	–	–	–	64	91	–	24934-91-6
Chlormezanone	–	–	–	68	84	80	66	1	63	57	–	–	–	–	80-77-3
Chloro-2′-fluoro-2-(2,2,2-trifluoroethyl)aminobenzophenone,5- →Quazepam Hy	74	73	80	85	77	93	79	44	77	73	–	–	–	74	
Chloro-2′-fluoro-2-(hydroxyethylamino)benzophenone,5- → Flurazepam Hy	48	50	57	82	87	91	77	7	60	66	2	–	–	–	35231-38-0
Chloro-2′-fluoro-2-methylaminobenzophenone,5- → Fludiazepam Hy	71	69	81	83	77	90	75	48	77	69	–	–	–	55	1548-36-3
Chloro-2-(2-diethylamino)-ethylamino-2′-fluoro-benzophenone,5- → Flurazepam Hy	9	8	41	83	45	51	67	53	60	51	0	–	–	–	36105-18-7
Chloro-2-(2-propinyl)benzophenone,5- → Pinazepam Hy	75	73	80	86	86	91	83	48	77	72	–	–	–	57	
Chloro-2-cyclopropyl-methylaminobenzophenone,5- → Prazepam Hy	83	74	82	88	92	95	80	58	82	74	–	–	–	68	2897-00-9
Chloro-2-methylaminobenzophenone,5- → Oxazepam Hy	82	72	82	88	89	93	79	53	81	71	–	–	–	–	33184-55-3
Chloro-o-tolyloxy)acetic Acid,(4- → MCPA	9	11	15	5	–	–	–	–	–	–	–	0	0	–	94-74-6
Chlorobenzoic Acid,m–	–	–	–	–	81	–	–	–	–	–	–	–	–	–	535-80-8
Chlorobenzoic Acid,o–	–	–	–	83	–	–	–	–	–	–	–	–	–	–	118-91-2
Chlorobenzoic Acid,p–	23	26	32	5	–	–	–	–	–	–	–	–	–	–	74-11-3
Chlorophacinon	–	–	–	–	–	–	–	–	–	–	–	0	1		3691-35-8
Chloroprocaine	–	–	–	–	–	–	59	5	23	37	–	–	–	–	133-16-4
Chloropyramine	–	–	0	63	22	–	52	41	28	17	–	–	–	–	59-32-5
Chloroquine	–	–	0	46	4	14	38	14	4	2	13	–	–	–	54-05-7
Chlorotheophylline,8–	–	–	–	14	85	–	–	–	–	–	–	–	–	–	85-18-7
Chlorothiazide	2	16	11	2	–	–	–	–	–	–	–	–	–	–	58-94-6
Chloroxuron	–	–	–	–	–	–	–	–	–	–	–	10	7	–	1982-47-4
Chlorphenamine	–	–	–	46	12	21	45	35	18	2	21	–	–	–	132-22-9
Chlorphenesin Carbamate	11	34	32	52	87	–	82	0	–	–	80	–	–	–	886-74-8
Chlorpheniramine → Chlorphenamine	–	–	–	46	12	21	45	35	18	2	21	–	–	–	132-22-9
Chlorphenol,p-/PCP	–	–	–	–	–	–	–	–	–	–	–	22	32	–	106-48-9
Chlorphenoxamine	–	–	–	70	29	–	53	47	36	17	61	–	–	–	77-38-3
Chlorphenprocoumon, p–	–	50	–	17	93	–	–	0	–	–	–	–	–	–	
Chlorphentermine	–	–	–	48	14	77	44	18	17	8	–	–	–	–	461-78-9
Chlorproguanil	–	–	–	–	–	–	3	0	1	1	–	–	–	–	537-21-3
Chlorpromazine	–	–	–	70	25	45	49	45	35	17	55	–	–	–	50-53-3

Table 8.1. (continued)

Substance Name	TLC System 1	2	3	4	5	6	7	8	9	10	T	P1	P2	B	CAS No.
Chlorpromazine M	–	–	–	42	16	–	–	–	–	–	–	–	–	–	
Chlorpropamide	38	43	49	10	87	88	72	0	–	–	–	–	–	–	94-20-2
Chlorprothixene	–	–	–	74	34	51	56	51	51	25	66	–	–	–	113-59-7
Chlorprothixene M	–	–	–	62	27	–	–	–	–	–	46	–	–	–	
Chlorpyriphos	–	–	–	99	85	–	–	–	–	–	–	64	95	–	2921-88-2
Chlorpyriphos-methyl	–	–	–	–	–	–	–	–	–	–	–	56	89	–	5598-13-0
Chlortalidone	4	40	23	42	88	83	–	–	–	–	–	–	–	–	77-36-1
Chlorthion	–	–	–	–	–	–	–	–	–	–	–	31	71	–	500-28-7
Chlorthiophos	–	–	–	99	86	–	–	–	–	–	–	55	91	–	60238-56-4
Chlortoluron	–	–	–	–	–	–	–	–	–	–	–	15	9	–	15545-48-9
Chlorzoxazone	54	61	56	33	88	90	85	0	51	47	–	–	–	–	95-25-0
Cholesterol	61	60	64	76	84	96	–	22	64	69	–	–	–	–	57-88-5
Chrysophanol	–	–	–	57	–	–	–	–	–	–	–	–	–	–	481-74-3
Cimetidine	–	–	–	27	53	55	54	0	9	12	22	–	–	–	51481-61-9
Cimetidine M (Sulfoxide)	–	–	–	9	48	–	–	–	–	–	–	–	–	–	
Cinchocaine	–	–	–	67	42	–	63	25	34	35	–	–	–	–	85-79-0
Cinchonidine	–	–	–	44	24	55	49	6	8	6	–	–	–	–	485-71-2
Cinchonine	–	–	–	44	19	61	49	6	12	5	–	–	–	–	118-10-5
Cinchophen	0	0	7	8	82	72	75	0	2	0	–	–	–	–	132-60-5
Cinnamoylcocaine	–	–	–	78	34	30	70	42	60	55	–	–	–	–	521-67-5
Cinnarizine	–	–	–	86	79	87	76	54	78	65	92	–	–	–	298-57-7
Cis-Ordinol → Clopenthixol	–	–	–	44	45	–	56	7	32	11	88	–	–	–	982-24-1
Cisapride	–	–	–	62	66	–	–	0	–	–	–	–	–	–	81098-60-4
Clamoxyquine	–	–	–	–	–	–	2	0	0	0	–	–	–	–	2545-39-3
Clefamide	–	–	–	–	–	–	69	0	56	68	–	–	–	–	3576-64-5
Clemastine	–	–	–	58	18	49	46	49	25	9	40	–	–	–	15686-51-8
Clemizole	–	–	–	78	76	73	78	31	69	52	–	–	–	–	442-52-4
Clenbuterol	–	–	–	58	22	–	–	13	–	–	–	–	–	–	37148-27-9
Clenbuterol M #1 (Hydroxy-)	–	–	–	43	30	–	–	1	–	–	–	–	–	–	
Clenbuterol M #2	–	0	–	2	87	–	–	0	–	–	–	–	–	–	
Clenbuterol M #3	–	19	–	6	87	–	–	0	–	–	–	–	–	–	
Clidinium Bromide	–	–	–	1	3	–	2	–	–	–	–	–	–	–	3485-62-9
Clindamycin	–	–	–	28	81	–	–	0	–	–	42	–	–	–	18323-44-9
Clindamycin Palmitate	–	–	–	–	–	–	–	–	–	–	92	–	–	–	36688-78-5
Clioquinol	–	–	70	30	40	88	56	0	5	9	–	–	–	–	130-26-7
Clobazam	53	47	70	75	84	85	62	8	70	62	87	–	–	–	22316-47-8
Clobazam M (Nor-)	32	43	54	64	84	90	75	0	52	61	–	–	–	–	22316-55-8
Clobutinol	–	–	–	78	20	–	–	–	–	–	–	–	–	–	14860-49-2
Clocinizine	–	–	–	86	85	90	83	53	77	66	–	–	–	–	298-55-5
Clofazimine	–	–	–	86	33	86	70	57	59	68	–	–	–	–	2030-63-9
Clofedanol	–	–	–	–	32	–	52	41	37	29	–	–	–	–	791-35-5
Clofenotane (o,p′-Isomer)/o,p′-DDT	–	–	–	87	–	–	–	–	–	–	–	–	–	–	789-02-6
Clofenotane (p,p′-Isomer)/p,p′-DDT	82	74	80	87	–	–	–	–	–	–	–	76	98	–	50-29-3
Clofexamide	–	–	–	–	22	–	–	–	–	–	–	–	–	–	1223-36-5
Clofibrate	75	66	71	84	81	–	–	–	–	–	–	–	–	–	637-07-0

Table 8.1. (continued)

Substance Name	1	2	3	4	5	6	7	8	9	10	T	P1	P2	B	CAS No.
Clomethiazole	–	–	–	76	80	85	64	44	69	58	90	–	–	–	533-45-9
Clomifene	–	–	–	82	30	–	60	56	52	35	–	–	–	–	911-45-5
Clomipramine	–	–	–	72	26	54	51	53	34	18	59	–	–	–	303-49-1
Clomipramine M #1	–	–	–	–	–	–	–	–	–	–	40	–	–	–	
Clomipramine M #2	–	–	–	–	–	–	–	–	–	–	16	–	–	–	
Clomipramine M (10-Hydroxy-)	–	–	–	54	25	42	49	15	19	9	–	–	–	–	
Clomipramine M (2-Hydroxy-)	–	–	–	45	26	51	70	1	24	11	–	–	–	–	
Clomipramine M (8-Hydroxy-)	–	–	–	51	26	51	51	4	17	10	–	–	–	–	
Clomipramine M (8-Hydroxy-nor-)	–	–	–	32	8	75	33	1	4	1	–	–	–	–	
Clomipramine M (N-Oxid)	–	–	–	16	27	20	41	0	9	0	–	–	–	–	
Clomipramine M (Nor-)	–	–	–	45	8	75	34	24	13	3	27	–	–	–	
Clonazepam	35	45	56	67	85	87	72	0	53	61	86	–	–	–	1622-61-3
Clonazepam Hy/ANCB/2-Amino-2′-chloro-5-nitrobenzophenone	72	66	70	86	87	93	77	4	69	70	–	–	–	16	2011-66-7
Clonazepam M Hy/DCB/2,5-Diamino-2′-chloro-benzophenone	41	46	55	82	80	80	74	5	59	61	–	–	–	0	58479-51-9
Clonazepam M (7-Acetamido-)	7	11	30	42	79	82	73	0	22	45	–	–	–	–	41993-30-0
Clonazepam M (7-Amino-)	11	20	40	52	76	77	73	0	30	47	–	–	–	–	4959-17-5
Clonazepam M (7-Amino-3-hydroxy-)	–	–	–	37	73	–	–	–	–	–	–	–	–	–	41993-29-7
Clonidine	–	–	–	70	44	76	62	8	31	53	80	–	–	–	4205-90-7
Clonitazene	–	–	–	–	–	–	66	8	69	60	–	–	–	–	3861-76-5
Clopamide	19	38	39	55	–	–	79	–	–	–	–	–	–	–	636-54-4
Clopenthixol	–	–	–	44	45	–	56	7	32	11	88	–	–	–	982-24-1
Clopenthixol M	–	–	–	33	32	–	–	–	–	–	–	–	–	–	
Clopenthixol Artifact	–	–	–	–	–	–	–	–	–	–	36	–	–	–	
Clopyralid	–	–	–	–	–	–	–	–	–	–	–	0	0	–	1702-17-6
Clorazepic Acid	34	46	57	68	83	87	84	3	56	60	87	–	–		23887-31-2
Clorazepic Acid → Nordazepam	34	45	57	69	82	83	62	3	55	60	81	–	–	–	1088-11-5
Clorazepic Acid M → Oxazepam	22	35	42	45	82	82	56	0	40	51	–	–	–	–	604-75-1
Clorexolone	31	51	47	60	79	–	76	–	–	–	–	–	–	–	2127-01-7
Clorgiline	–	–	–	–	–	–	67	42	70	59	–	–	–	–	17780-72-2
Clorprenaline	–	–	–	–	–	–	57	18	15	20	–	–	–	–	3811-25-4
Closantel	–	–	–	26	96	98	95	0	10	90	–	–	–	–	57808-65-8
Clotiapine	–	–	–	–	–	–	59	41	59	23	–	–	–	–	2058-52-8
Clotiazepam	55	48	66	80	84	87	78	31	69	66	91	–	–	–	33671-46-4
Clotrimazole	–	–	68	76	80	–	–	–	–	–	–	–	–	–	23593-75-1
Cloxazolam	39	45	63	75	78	84	73	0	66	59	–	–	–	–	24166-13-0
Clozapine	4	5	29	55	42	–	57	4	38	17	50	–	–	–	5786-21-0
Clozapine M	–	–	–	–	–	–	–	–	–	–	29	–	–	–	
Clozapine M (N-Oxyde)	–	–	–	8	19	–	–	0	–	–	–	–	–	–	
Clozapine M (Nor-)	–	–	–	41	19	–	–	1	–	–	–	–	–	–	
Cocaine	–	–	–	77	35	30	65	45	47	54	81	–	–	–	50-36-2

Table 8.1. (continued)

Substance Name	TLC System														CAS No.
	1	2	3	4	5	6	7	8	9	10	T	P1	P2	B	
Cocaine M (Nor-)	–	–	–	59	39	50	–	21	48	12	–	–	–	–	18717-72-1
Cocaine M/Ecgonine	–	–	–	0	16	11	17	0	0	0	–	–	–	–	481-37-8
Cocaine M/N-Benzoylecgonine	–	–	–	2	22	14	21	0	1	3	0	–	–	–	519-09-5
Codeine	–	–	–	35	21	22	33	6	18	3	24	–	–	–	76-57-3
Codeine M (-6-Glucuronide)	–	–	–	0	30	2	–	0	0	0	–	–	–	–	20736-11-2
Codeine M (N-Oxide)	–	–	–	–	–	–	18	1	3	0	–	–	–	–	3688-65-1
Codeine M (Nor-)	–	–	–	–	–	–	13	0	5	0	6	–	–	–	467-15-2
Colchicine	–	–	–	33	69	63	55	0	37	12	23	–	–	–	64-86-8
Conessine	–	–	–	–	–	–	28	49	3	3	–	–	–	–	546-06-5
Coniine	–	–	–	37	5	70	26	39	13	5	–	–	–	–	458-88-8
Cotarnine	–	–	–	26	1	22	2	38	1	0	–	–	–	–	82-54-2
Cotinine	–	–	–	41	53	–	–	4	–	–	33	–	–	–	486-56-6
Coumachlor	–	–	–	86	86	–	–	–	–	–	–	–	–	–	81-82-3
Coumaphos	–	–	–	–	–	–	–	–	–	–	–	27	61	–	56-72-4
Coumarine	–	–	–	74	82	–	–	–	–	–	–	–	–	–	91-64-5
Coumatetralyl	–	–	–	–	–	–	–	–	–	–	–	14	30	–	5836-29-3
Cromoglicic Acid	0	0	0	0	–	–	–	–	–	–	–	–	–	–	16110-51-3
Cropropamide	–	–	–	74	25	83	70	29	69	57	–	–	–	–	633-47-6
Crotamiton	–	–	–	83	84	–	83	–	–	–	–	–	–	–	483-63-6
Crotethamide	–	–	–	69	83	79	68	28	67	55	–	–	–	–	6168-76-9
Crotylbarbital	–	65	–	43	81	–	–	–	–	–	–	–	–	–	1952-67-6
Cryptopine	–	–	–	73	14	50	–	22	46	43	–	–	–	–	482-74-6
Cyanazine	–	–	–	–	–	–	–	–	–	–	–	16	12	–	21725-46-2
Cyclamic Acid	5	10	10	0	–	–	–	–	–	–	–	–	–	–	100-88-9
Cyclandelate	74	77	73	80	87	95	–	37	–	–	–	–	–	–	456-59-7
Cyclazocine	–	–	–	65	24	74	53	15	13	25	–	–	–	–	3572-80-3
Cyclizine	–	–	–	68	40	52	57	49	41	16	–	–	–	–	82-92-8
Cyclobarbital	50	64	58	40	88	–	–	–	–	–	–	–	–	–	52-31-3
Cyclobenzaprine	–	–	–	69	–	–	–	–	–	–	56	–	–	–	303-53-7
Cyclomethycaine	–	–	–	–	–	–	58	55	36	25	–	–	–	–	139-62-8
Cyclopentamine	–	–	–	66	6	68	20	32	10	2	–	–	–	–	102-45-4
Cyclopenthiazide	21	62	27	66	–	–	–	–	–	–	–	–	–	–	742-20-1
Cyclopentobarbital	50	65	59	39	90	–	–	–	–	–	–	–	–	–	76-68-6
Cyclopentolate	–	–	–	–	–	–	57	27	39	26	–	–	–	–	512-15-2
Cyclophosphamide	–	–	–	–	74	–	–	–	–	–	–	–	–	–	50-18-0
Cycloserine	–	–	–	–	–	–	44	1	1	2	–	–	–	–	68-41-7
Cyclothiazide	18	60	26	59	–	–	77	–	–	–	–	–	–	–	2259-96-3
Cycluron	–	–	–	–	–	–	–	–	–	–	–	13	5	–	2163-69-1
Cycrimine	–	–	–	–	–	–	66	67	61	60	–	–	–	–	77-39-4
Cyfluthrin	–	–	–	–	–	–	–	–	–	–	–	37	89	–	68359-37-5
Cyhexatin	–	–	–	–	–	–	–	–	–	–	–	0	0	–	13121-70-5
Cypermethrin	–	–	–	–	–	–	–	–	–	–	–	41	91	–	52315-07-8
Cyprazepam	–	–	–	74	86	–	–	–	–	–	–	–	–	–	15687-07-7
Cyprenorphine	–	–	–	–	–	–	72	5	59	61	–	–	–	–	4406-22-8
Cyproheptadine	–	–	–	64	30	50	51	45	44	13	58	–	–	–	129-03-3
Cytarabine	–	–	0	–	69	–	5	0	1	1	–	–	–	–	147-94-4
Cytisine	–	–	–	–	–	–	40	1	10	2	–	–	–	–	485-35-8

Table 8.1. (continued)

Substance Name	TLC System														CAS No.
	1	2	3	4	5	6	7	8	9	10	T	P1	P2	B	
D,2,4- → (2,4-Dichloro-phenoxy)acetic Acid	4	6	11	4	–	–	–	–	–	–	–	0	2	–	94-75-7
DAB → Nitrazepam M Hy	37	41	54	81	80	79	72	5	57	57	–	–	–	0	18330-94-4
DCB → Clonazepam Hy	41	46	55	82	80	80	74	5	59	61	–	–	–	0	58479-51-9
DCFB → Flurazepam Hy	9	8	41	83	45	51	67	53	60	51	–	–	–	0	36105-18-7
DDT,o,p-′ → Clofenotane (o,p′-Isomer)	–	–	–	87	–	–	–	–	–	–	–	–	–	–	789-02-6
DDT,p,p′- → Clofenotane (p,p′-Isomer)	82	74	80	87	–	–	–	–	–	–	–	76	98	–	50-29-3
DDVP → Dichlorvos	–	–	–	83	85	–	–	–	–	–	–	20	20	–	62-73-7
DET → Diethyltryptamine,N,N–	–	–	–	63	14	56	46	14	10	11	–	–	–	–	61-51-8
DMT → Dimethyltryptamine	–	–	–	50	14	39	40	9	9	6	–	–	–	–	61-50-7
DNOC → 4,6-Dinitro-o-cresol	–	–	–	–	–	–	–	–	–	–	–	6	38	–	534-52-1
DOB → 4-Bromo-2,5-dimethoxyamphetamine	–	–	–	43	10	76	36	15	14	17	–	–	–	–	32156-26-6
DOET → 2,5-Dimethoxy-4-ethylamphetamine	–	–	–	43	8	75	36	24	10	14	–	–	–	–	22004-32-6
DOM → 2,5-Dimethoxy-4-methylamphetamine	–	–	–	41	9	76	51	15	17	16	–	–	–	–	15588-95-1
Danthron	80	69	78	43	–	–	–	–	–	–	–	–	–	–	117-10-2
Dantrolene	19	36	50	9	–	–	–	–	–	–	–	–	–	–	7261-97-4
Debrisoquine	–	–	–	–	–	–	1	0	0	0	–	–	–	–	1131-64-2
Decamethonium	–	–	–	0	0	2	0	0	0	0	–	–	–	–	156-74-1
Decamethrin	–	–	–	–	–	–	–	–	–	–	–	40	90	–	52918-63-5
Deet → N,N-Diethyl-m-toluamide	–	–	–	–	–	–	73	–	–	–	–	–	–	–	134-62-3
Dehydroemetine	–	–	–	–	–	–	43	6	21	2	–	–	–	–	4914-30-1
Delorazepam	35	41	57	72	82	86	73	5	58	56	–	–	–	–	2894-67-9
Deltamethrin → Decamethrin	–	–	–	–	–	–	–	–	–	–	–	40	90	–	52918-63-5
Demecolcine	–	–	–	35	56	–	53	0	41	11	–	–	–	–	477-30-5
Demeton	–	–	–	–	–	–	–	–	–	–	–	17	17	–	8065-48-3
Demeton	–	–	–	–	–	–	–	–	–	–	–	83	81	–	8065-48-3
Demeton-S-methyl	–	–	–	81	86	–	–	–	–	–	–	18	13	–	919-86-8
Demeton-S-methylsulfone	–	–	–	–	–	–	–	–	–	–	–	1	3	–	17040-19-6
Demoxepam	15	24	42	41	81	83	63	0	35	51	–	–	–	–	963-39-3
Deoxyephedrine,(+/-)–	–	–	–	42	9	63	31	28	13	5	–	–	–	–	4846-07-5
Deptropine	–	–	–	–	–	–	13	24	4	1	–	–	–	–	604-51-3
Deserpidine	–	–	–	81	73	–	72	3	77	66	–	–	–	–	131-01-1
Desipramine	–	–	–	40	7	71	26	19	11	3	24	–	–	–	50-47-5
Desipramine M #1	–	–	–	–	–	–	–	–	–	–	15	–	–	–	
Desipramine M #2	–	–	–	–	–	–	–	–	–	–	5	–	–	–	
Desmetryn	–	–	–	–	–	–	–	–	–	–	–	21	16	–	1014-69-3
Desomorphine	–	–	–	–	–	–	33	18	17	8	–	–	–	–	427-00-9
Dexetimide	–	–	–	70	51	67	71	21	48	44	–	–	–	–	21888-48-2
Dextromethorphan	–	–	–	47	10	42	33	42	18	6	26	–	–	–	125-71-3
Dextromoramide	–	–	–	79	72	78	73	42	71	60	86	–	–	–	357-56-2

Table 8.1. (continued)

Substance Name	TLC System														CAS No.
	1	2	3	4	5	6	7	8	9	10	T	P1	P2	B	
Dextropropoxyphene → Propoxyphene	–	–	–	82	50	63	68	58	55	54	82	–	–	–	469-62-5
Dextrorphan/Levorphanol	–	–	–	42	10	49	35	14	4	3	–	–	–	–	125-73-5
Dextrorphan M/Levorphanol M (Nor-)	–	–	–	–	–	–	9	2	1	1	–	–	–	–	1531-12-0
Diamino-2′-chloro-benzophenone,2-5- → Clonazepam Hy	41	46	55	82	80	80	74	5	59	61	–	–	–	0	58479-51-9
Diaminobenzophenone,2,5- → Nitrazepam Hy	37	41	54	81	80	79	72	5	57	57	–	–	–	0	18330-94-4
Diamocaine	–	–	–	71	11	28	54	41	21	9	–	–	–	–	27112-37-4
Diamorphine	–	–	–	49	26	33	47	15	38	4	–	–	–	–	561-27-3
Diamorphine M → Morphine-6-acetate	–	–	–	44	25	27	46	6	26	8	30	–	–	–	2784-73-8
Diampromide	–	–	–	–	–	–	71	51	72	65	–	–	–	–	552-25-0
Diamthazole → Dimazole	–	–	–	–	–	–	52	30	30	17	–	–	–	–	95-27-2
Diazepam	58	49	72	76	82	85	75	27	73	59	90	–	–	–	439-14-5
Diazepam Hy/MACB/2-Methylamino-5-chlorobenzophenone	82	72	82	88	89	93	79	53	81	71	–	–	–	52	1022-13-5
Diazepam M → Nordazepam	34	45	57	69	82	83	62	3	55	60	81	–	–	–	1088-11-5
Diazepam M → Oxazepam	22	35	42	45	82	82	56	0	40	51	–	–	–	–	604-75-1
Diazepam M → Temazepam	51	47	65	62	82	82	53	8	59	53	81	–	–	–	846-50-4
Diazinon → Dimpylate	–	–	–	–	86	–	–	–	–	–	–	47	50	–	333-41-5
Diazoxide	–	–	–	–	–	–	82	1	28	41	–	–	–	–	364-98-7
Dibenzepin	–	–	–	55	38	22	54	22	35	14	45	–	–	–	4498-32-2
Dibenzepin M (5-Desmethyl-)	–	–	–	52	39	–	–	4	–	–	–	–	–	–	
Dibenzepin M (Dinor-)	–	–	–	39	21	–	–	3	–	–	–	–	–	–	
Dibenzepin M (Dinor-,5-desmethyl-)	–	–	–	38	20	–	–	0	–	–	–	–	–	–	
Dibenzepin M (Nor-)	–	–	–	40	19	–	–	6	–	–	–	–	–	–	
Dibenzepin M (Nor-,5-desmethyl-)	–	–	–	38	20	–	–	1	–	–	–	–	–	–	
Dibucaine → Cinchocaine	–	–	–	67	42	–	63	25	34	35	–	–	–	–	85-79-0
Dichlofluanid	–	–	–	–	–	–	–	–	–	–	–	30	64	–	1085-98-9
Dichloro-2-methylaminobenzophenone,2′-5- → Lormetazepam Hy	81	69	81	87	82	93	75	49	81	7	–	–	–	57	
Dichlorophen	58	67	57	34	91	–	–	–	–	–	–	12	16	–	97-23-4
Dichlorophenoxy)acetic Acid, (2,4-/2,4-D	4	6	11	4	–	–	–	–	–	–	–	0	2	–	94-75-7
Dichlorprop	–	–	–	–	–	–	–	–	–	–	–	3	2	–	120-36-5
Dichlorvos	–	–	–	83	85	–	–	–	–	–	–	20	20	–	62-73-7
Diclofenac	25	27	47	12	90	–	–	–	–	–	12	–	–	–	15307-86-5
Diclofenamide	14	64	23	33	–	–	82	–	–	–	–	–	–	–	120-97-8
Diclofop-methyl	–	–	–	–	–	–	–	–	–	–	–	43	76	–	51338-27-3
Dicofol	–	–	–	–	–	–	–	–	–	–	–	43	80	–	115-32-2
Dicoumarol	18	32	33	30	–	–	–	–	–	–	–	–	–	–	66-76-2
Dicyclomine → Dicycloverine	–	–	–	–	52	–	68	67	64	54	–	–	–	–	77-19-0

Table 8.1. (continued)

Substance Name	TLC System														CAS No.
	1	2	3	4	5	6	7	8	9	10	T	P1	P2	B	
Dicycloverine	–	–	–	–	52	–	68	67	64	54	–	–	–	–	77-19-0
Dieldrin	–	–	–	–	–	–	–	–	–	–	–	65	87	–	60-57-1
Diethazine	–	–	–	77	33	54	58	57	51	39	–	–	–	–	60-91-3
Diethyl-m-toluamide,N,N-/Deet	–	–	–	–	–	–	73	–	–	–	–	–	–	–	134-62-3
Diethylaminoethyl Diphenyl-propionate	–	–	–	–	–	–	67	58	62	49	–	–	–	–	3563-01-7
Diethylcarbamazine	–	–	–	–	–	–	52	17	26	5	–	–	–	–	90-89-1
Diethylpropion → Amfepramone	–	–	–	85	55	56	76	62	63	64	95	–	–	–	90-84-6
Diethylstilbestrol	–	–	–	73	92	92	–	–	–	–	–	–	–	–	56-53-1
Diethylthiambutene	–	–	–	–	–	–	70	61	43	43	–	–	–	–	86-14-6
Diethyltryptamine,N,N-/DET	–	–	–	63	14	56	46	14	10	11	–	–	–	–	61-51-8
Difenidol	–	–	–	–	–	–	61	56	45	51	–	–	–	–	972-02-1
Difenoxine	–	–	–	–	–	–	–	–	–	–	6	–	–	–	28782-42-5
Difenoxuron	–	–	–	–	–	–	–	–	–	–	–	8	5	–	14214-32-5
Diflubenzuron	–	–	–	–	–	–	–	–	–	–	–	18	24	–	35367-38-5
Diflunisal	8	5	18	16	89	–	–	–	–	–	12	–	–	–	22494-42-4
Digitoxin	3	10	42	36	88	–	–	–	–	–	–	–	–	–	71-63-6
Digoxin	1	5	28	33	85	–	–	–	–	–	–	–	–	–	20830-75-5
Dihydralazine	–	–	–	15	–	–	55	36	2	1	–	–	–	–	484-23-1
Dihydrocodeine	–	–	–	29	11	19	26	8	13	2	12	–	–	–	125-28-0
Dihydroergocristine	–	–	–	83	5	–	–	–	–	–	–	–	–	–	17479-19-5
Dihydroergotamine	–	–	–	42	58	–	60	1	28	16	33	–	–	–	511-12-6
Dihydroergotamine M (8′-OH)	–	–	–	40	60	–	–	0	–	–	–	–	–	–	
Dihydromorphine	–	–	–	18	12	–	25	2	3	1	–	–	–	–	509-60-4
Dihydrostreptomycin	–	–	–	–	0	–	–	–	–	–	–	–	–	–	128-46-1
Diiodohydroxyquinoline	17	19	72	28	–	–	–	–	–	–	–	–	–	–	83-73-8
Diisopromine	–	–	–	86	23	75	55	68	36	52	–	–	–	–	5966-41-6
Diloxanide	–	–	–	–	–	–	66	16	74	67	–	–	–	–	579-38-4
Diltiazem	–	–	–	56	47	–	–	–	–	–	50	–	–	–	42399-41-7
Dimazole	–	–	–	–	–	–	52	30	30	17	–	–	–	–	95-27-2
Dimefline	–	–	–	–	–	–	59	15	48	24	–	–	–	–	1165-48-6
Dimefos	–	–	–	99	89	–	–	–	–	–	–	–	–	–	
Dimenhydrinate	–	–	–	2	87	46	88	0	10	10	–	–	–	–	523-87-5
Dimenhydrinate	–	–	–	68	28	48	55	45	33	15	–	–	–	–	523-87-5
Dimetacrine	–	–	–	62	22	–	–	–	–	–	–	–	–	–	4757-55-5
Dimethoate	–	–	–	99	80	–	–	–	–	–	–	4	4	–	60-51-5
Dimethocaine	–	–	–	–	–	–	68	7	45	67	–	–	–	–	94-15-5
Dimethoxanate	–	–	–	–	21	38	39	18	24	6	–	–	–	–	477-93-0
Dimethoxy-4-ethylamphet-amine,2,5-/DOET	–	–	–	43	8	75	36	24	10	14	–	–	–	–	22004-32-6
Dimethoxy-4-methylamphet-amine,2,5-/STP/DOM	–	–	–	41	9	76	51	15	17	16	–	–	–	–	15588-95-1
Dimethoxyamphetamine,2,5–	–	–	–	45	8	75	36	19	10	17	–	–	–	–	2801-68-5
Dimethoxyphenethylamine,3,4–	–	–	–	31	6	61	26	5	11	14	–	–	–	–	120-20-7
Dimethylaminobenzaldehyde,p–	–	–	–	–	81	–	–	–	–	–	–	–	–	–	100-10-7

Table 8.1. (continued)

Substance Name	TLC System 1	2	3	4	5	6	7	8	9	10	T	P1	P2	B	CAS No.
Dimethylphenylthiazolanimin (3,4-Dimethyl-5-phenyl-2-thiazolidinimine)	–	–	–	–	–	–	49	26	14	14	–	–	–	–	14007-67-1
Dimethylthiambutene	–	–	–	–	–	–	62	50	41	21	–	–	–	–	524-84-5
Dimethyltryptamine,N,N-/DMT	–	–	–	50	14	39	40	9	9	6	–	–	–	–	61-50-7
Dimetindene	–	–	–	47	10	–	42	35	13	6	27	–	–	–	5636-83-9
Dimetiotazine	–	–	–	66	43	–	56	13	48	28	–	–	–	–	7456-24-8
Diminazene Aceturate	–	–	–	–	–	–	0	0	0	0	–	–	–	–	908-54-3
Dimoxyline	–	–	–	–	–	–	68	16	75	58	–	–	–	–	147-27-3
Dimpylate/Diazinon	–	–	–	–	86	–	–	–	–	–	–	47	50	–	333-41-5
Dinoseb	–	–	–	–	–	–	–	–	–	–	–	40	81	–	88-85-7
Dioxaphetyl Butyrate	–	–	–	–	–	–	80	46	74	62	–	–	–	–	467-86-7
Diperodon	–	–	–	–	–	–	70	15	58	66	–	–	–	–	101-08-6
Diphenadione	11	33	53	46	–	–	–	–	–	–	–	–	–	–	82-66-6
Diphenazoline	–	–	–	–	–	–	37	9	9	3	–	–	–	–	6703-39-5
Diphenhydramine	–	–	–	65	27	48	55	44	33	15	55	–	–	–	58-73-1
Diphenhydramine M	–	–	–	–	–	–	–	–	–	–	30	–	–	–	
Diphenoxylate	–	–	–	87	90	92	74	42	84	70	–	–	–	–	915-30-0
Diphenylamine	–	–	–	–	–	89	–	–	–	–	–	–	–	–	122-39-4
Diphenylpyraline	–	–	–	68	23	49	46	37	28	8	–	–	–	–	147-20-6
Dipipanone	–	–	–	87	27	72	66	67	33	70	–	–	–	–	467-83-4
Dipropetryn	–	–	–	–	–	–	–	–	–	–	–	35	35	–	4147-51-7
Diprophylline	–	–	–	25	70	59	48	0	12	12	–	–	–	–	479-18-5
Dipyridamole	–	–	–	44	82	87	68	0	37	42	58	–	–	–	58-32-2
Dipyrone → Metamizol	0	0	2	2	85	59	84	0	1	2	53	–	–	–	50567-35-6
Diquat Dibromide	–	–	–	0	0	0	0	0	0	0	–	–	–	–	85-00-7
Disopyramide	–	–	–	60	9	7	45	7	8	13	32	–	–	–	3737-09-5
Disopyramide M	–	–	–	–	–	–	–	–	–	–	10	–	–	–	
Distigmine Bromide	–	–	–	3	2	–	–	–	–	–	–	–	–	–	15876-67-2
Disulfiram	–	–	–	78	78	81	71	21	78	69	92	–	–	–	97-77-8
Disulfoton	–	–	–	–	–	–	–	–	–	–	–	58	89	–	298-04-4
Diuron	–	–	–	–	–	–	–	–	–	–	–	15	10	–	330-54-1
Dixyracine	–	–	–	49	47	–	–	–	–	–	36	–	–	–	2470-73-7
Dixyracine M	–	–	–	29	31	–	–	–	–	–	–	–	–	–	
Dobutamine	–	–	–	49	87	–	52	0	1	3	10	–	–	–	34368-04-2
Domperidone	–	–	–	41	69	66	76	0	41	8	–	–	–	–	57808-66-9
Dopamine	–	–	–	43	14	59	18	0	0	0	–	–	–	–	51-61-6
Dothiepin	–	–	–	70	27	41	51	49	42	16	55	–	–	–	113-53-1
Dothiepin M	–	–	–	–	–	–	–	–	–	–	21	–	–	–	
Dothiepin M (Nor-)	–	–	–	44	19	22	45	17	28	5	–	–	–	–	1154-09-2
Dothiepin M (Nor-,Sulfoxide)	–	–	–	28	6	37	27	2	8	1	–	–	–	–	
Dothiepin M (Sulfoxide)	–	–	–	46	20	22	45	11	28	5	–	–	–	–	
Doxapram	–	–	–	72	69	–	64	20	70	54	–	–	–	–	309-29-5
Doxazosin	–	–	–	81	71	–	–	0	–	–	–	–	–	–	74191-85-8
Doxepin	–	–	–	63	24	45	51	48	37	13	52	–	–	–	1668-19-5
Doxycycline	–	–	–	–	88	–	–	–	–	–	–	–	–	–	17086-28-1
Doxylamine	–	–	–	60	12	–	48	41	10	9	38	–	–	–	469-21-6

Table 8.1. (continued)

Substance Name	TLC System														CAS No.
	1	2	3	4	5	6	7	8	9	10	T	P1	P2	B	
Doxylamine M	–	–	–	–	–	–	–	–	–	–	10	–	–	–	
Drofenine	–	–	–	84	–	–	–	–	–	–	–	–	–	–	1679-76-1
Droperidol	–	–	–	58	71	73	67	2	48	36	59	–	–	–	548-73-2
Dropropizine	–	–	–	38	59	–	65	–	–	–	–	–	–	–	17692-31-8
Dyclonine	–	–	–	–	–	–	60	49	40	25	–	–	–	–	586-60-7
EPTC	–	–	–	–	–	–	–	–	–	–	–	69	63	–	759-94-4
Ecgonine → Cocaine M	–	–	–	0	16	11	17	0	0	0	–	–	–	–	481-37-8
Econazole	–	–	–	75	78	80	80	9	61	–	–	–	–	–	27220-47-9
Ectylurea	–	–	–	62	82	–	–	–	–	–	–	–	–	–	95-04-5
Embramine	–	–	–	–	–	–	54	50	32	17	–	–	–	–	3565-72-8
Emepronium Bromide	–	–	–	2	3	–	5	0	–	–	–	–	–	–	3614-30-0
Emetine	–	–	–	53	19	–	54	13	34	12	25	–	–	–	483-18-1
Enalapril	0	0	0	0	–	–	–	–	–	–	–	–	–	–	75847-73-3
Enallylpropymal	71	71	70	58	–	–	–	–	–	–	–	–	–	–	1861-21-8
Encainide	–	–	–	54	16	–	–	28	–	–	–	–	–	–	66778-36-7
Endosulfan	–	–	–	–	–	–	–	–	–	–	–	40	77	–	959-98-8
Endosulfan	–	–	–	–	–	–	–	–	–	–	–	77	95	–	959-98-8
Endralazine Mesilate	–	–	–	46	65	14	52	2	52	21	–	–	–	–	39715-02-1
Endrin	–	–	–	–	–	–	–	–	–	–	–	71	90	–	72-20-8
Enilconazole	–	–	–	72	78	–	–	–	–	–	–	5	2	–	35554-44-0
Enoxolone	21	46	47	7	–	–	–	–	–	–	–	–	–	–	471-53-4
Ephedrine	–	–	–	25	10	64	30	5	5	1	–	–	–	–	299-42-3
Ephedrine M/Phenylpropanol-amine	–	0	–	30	12	75	44	4	4	33	23	–	–	–	492-41-1
Epinephrine	–	–	–	13	3	–	0	0	1	0	–	–	–	–	51-43-4
Epithiazide	13	62	25	44	–	–	–	–	–	–	–	–	–	–	1764-85-8
Eprazinone	–	–	–	–	69	–	–	–	–	–	–	–	–	–	10402-90-1
Ergocristine	–	–	–	57	74	77	77	0	59	50	–	–	–	–	511-08-0
Ergocryptine	–	–	–	59	74	77	78	0	58	50	–	–	–	–	511-09-1
Ergoloid	–	–	–	58	69	–	66	1	48	38	–	–	–	–	11032-41-0
Ergometrine	–	–	–	33	62	60	57	0	12	8	–	–	–	–	60-79-7
Ergosine	–	–	–	64	64	67	75	0	47	29	–	–	–	–	561-94-4
Ergosterol	60	58	63	79	–	–	–	–	–	–	–	–	–	–	57-87-4
Ergotamine	–	–	–	44	68	64	63	1	34	22	40	–	–	–	113-15-5
Ergotoxine	–	–	–	–	–	–	66	1	62	48	–	–	–	–	8006-25-5
Erythromycin	–	–	–	46	24	–	–	11	–	–	30	–	–	–	114-07-8
Erythromycin ester	–	–	–	–	–	–	–	–	–	–	50	–	–	–	
Estazolam	7	5	44	50	71	60	71	2	53	25	–	–	–	–	29975-16-4
Estradiol	61	58	64	78	–	–	–	–	–	–	–	–	–	–	50-28-2
Etacrynic Acid	3	2	5	5	71	–	–	–	–	–	–	–	–	–	58-54-8
Etafedrine	–	–	–	56	14	–	44	35	9	15	–	–	–	–	48141-64-6
Etafenone	–	–	–	81	41	–	–	–	–	–	–	–	–	–	90-54-0
Etamiphyllin	–	–	–	–	–	–	54	12	39	17	–	–	–	–	314-35-2
Etamivan	38	35	59	41	85	–	–	–	–	–	–	–	–	–	304-84-7
Ethacridine	–	–	–	–	7	–	–	–	–	–	–	–	–	–	442-16-0
Ethambutol	–	–	0	76	12	–	30	3	2	2	–	–	–	–	74-55-5
Ethaverine	–	–	–	56	83	–	–	–	–	–	88	–	–	–	486-47-5

Table 8.1. (continued)

Substance Name	TLC System														CAS No.
	1	2	3	4	5	6	7	8	9	10	T	P1	P2	B	
Ethchlorvynol	81	74	82	86	87	–	–	0	–	–	–	–	–	–	113-18-8
Ethchlorvynol M	–	–	–	–	–	–	–	–	–	–	90	–	–	–	
Ethenzamide	–	–	–	76	87	–	64	3	59	55	85	–	–	–	938-73-8
Ethiazide	11	50	22	50	–	–	–	–	–	–	–	–	–	–	1824-58-4
Ethinamate	49	59	58	76	86	87	76	5	–	–	–	–	–	–	126-52-3
Ethinylestradiol	–	–	–	71	86	–	–	4	–	–	–	–	–	–	57-63-6
Ethiofencarb	–	–	–	–	–	–	–	–	–	–	–	21	25	–	29973-13-5
Ethionamide	–	–	–	–	–	–	65	0	36	55	–	–	–	–	536-33-4
Ethofumesate	–	–	–	–	–	–	–	–	–	–	–	25	50	–	26225-79-6
Ethoheptazine	–	–	–	55	12	41	40	45	19	4	–	–	–	–	77-15-6
Ethomoxane	–	–	–	–	–	–	60	34	47	36	–	–	–	–	16509-23-2
Ethoprofos	–	–	–	–	–	–	–	–	–	–	–	33	28	–	13194-48-4
Ethopropazine → Profenamine	–	–	–	83	31	55	67	64	47	66	–	–	–	–	522-00-9
Ethosuximide	50	53	59	66	84	–	70	5	–	–	–	–	–	–	77-67-8
Ethotoin	53	54	60	71	–	–	–	–	–	–	–	–	–	–	86-35-1
Ethoxazene	–	–	–	–	–	–	65	0	56	67	–	–	–	–	94-10-0
Ethoxzolamide	43	65	51	43	–	–	76	–	–	–	–	–	–	–	452-35-7
Ethyl Biscoumacetate	4	32	21	24	–	–	–	–	–	–	–	–	–	–	548-00-5
Ethyl Loflazepate	53	58	62	74	85	93	76	0	62	67	–	–	–	–	29177-84-2
Ethyl p-Hydroxybenzoate	64	66	57	46	88	–	–	–	–	–	–	–	–	–	120-47-8
Ethyl-1-phenylcyclohexyl-amine,N- → Eticyclidine	–	–	–	79	21	78	–	65	27	37	–	–	–	–	2201-15-2
Ethyl-2-(p-tolyl)malonamide,2–	–	–	–	52	84	–	–	–	–	–	–	–	–	–	68692-83-1
Ethylmethylthiambutene	–	–	–	–	–	–	59	57	42	28	–	–	–	–	441-61-2
Ethylmorphine	–	–	–	36	21	26	40	7	22	6	–	–	–	–	76-58-4
Ethylnorepinephrine	–	–	–	–	–	–	42	1	2	24	–	–	–	–	536-24-3
Ethylpiperidyl Benzilate/JB-318	–	–	–	78	59	54	75	44	63	55	–	–	–	–	3567-12-2
Eticyclidine	–	–	–	79	21	78	–	65	27	37	–	–	–	–	2201-15-2
Etidocaine	–	–	–	75	80	–	–	–	–	–	–	–	–	–	36637-18-0
Etifelmine	–	–	–	64	19	–	–	–	–	–	–	–	–	–	341-00-4
Etilamfetamine	–	–	–	59	16	74	47	47	19	8	–	–	–	–	457-87-4
Etilefrine	–	–	–	22	15	74	41	2	2	3	14	–	–	–	709-55-7
Etiroxate	–	–	–	34	83	–	–	–	–	–	–	–	–	–	17365-01-4
Etizolam	–	–	–	52	72	–	–	–	–	–	–	–	–	–	40054-69-1
Etodroxizine	–	–	–	49	52	–	–	–	–	–	–	–	–	–	17692-34-1
Etofenamate	–	–	–	78	89	–	–	–	–	–	–	–	–	–	30544-47-9
Etofylline	–	6	–	38	66	–	–	0	–	–	31	–	–	–	519-37-9
Etomidate	–	–	–	73	78	85	67	26	71	52	–	–	–	–	33125-97-2
Etonitazene	–	–	–	79	70	72	68	13	70	61	–	–	–	–	911-65-9
Etorphine	–	–	–	–	–	–	73	7	61	63	–	–	–	–	14521-96-1
Etoxeridine	–	–	–	–	–	–	60	9	42	18	–	–	–	–	469-82-9
Etozolin	–	–	–	86	78	–	–	–	–	–	–	–	–	–	73-09-6
Etrimfos	–	–	–	99	84	–	–	–	–	–	–	3	0	–	38260-54-7
Etrimfos	–	–	–	99	84	–	–	–	–	–	–	52	67	–	38260-54-7
Etryptamin	–	–	–	43	12	–	–	–	–	–	–	–	–	–	2235-90-7
Etymemazine	–	–	–	–	–	–	61	60	52	42	–	–	–	–	523-54-6
Eucaine,α–	–	–	–	82	55	57	–	54	60	62	–	–	–	–	470-68-8

Table 8.1. (continued)

Substance Name	TLC System														CAS No.
	1	2	3	4	5	6	7	8	9	10	T	P1	P2	B	
Eucaine,β–	–	–	–	66	27	74	54	42	27	17	–	–	–	–	500-34-5
Eucatropine	–	–	–	–	–	–	46	18	13	12	–	–	–	–	100-91-4
Famprofazone	–	–	–	87	90	–	72	37	74	67	–	–	–	–	22881-35-2
Fedrilate	–	–	–	–	69	–	–	–	–	–	–	–	–	–	23271-74-1
Fenarimol	–	–	–	–	–	–	–	–	–	–	–	16	12	–	60168-88-9
Fenbufen	18	30	39	4	–	–	–	–	–	–	–	–	–	–	36330-85-5
Fenbutrazate	–	–	–	86	88	–	72	47	78	67	–	–	–	–	4378-36-3
Fencamfamine	–	–	–	77	21	–	54	62	34	30	–	–	–	–	1209-98-9
Fendiline	–	–	–	85	43	–	–	–	–	–	92	–	–	–	13042-18-7
Fenetylline	–	–	–	54	44	–	55	3	45	14	41	–	–	–	3736-08-1
Fenfluramine	–	–	–	60	19	–	48	41	16	11	42	–	–	–	458-24-2
Fenimide	62	65	65	58	–	–	–	–	–	–	–	–	–	–	60-45-7
Fenitrothion	–	–	–	–	–	–	–	–	–	–	–	32	76	–	122-14-5
Fenoprofen	42	38	50	6	–	–	–	–	–	–	1	–	–	–	31879-05-7
Fenoprop	–	–	–	–	–	–	–	–	–	–	–	1	1	–	93-72-1
Fenoterol	–	0	–	25	38	81	76	0	1	4	–	–	–	–	13392-18-2
Fenpipramide	–	–	–	48	25	–	54	3	16	15	–	–	–	–	77-01-0
Fenpiverinium	–	–	–	5	4	39	2	0	4	0	–	–	–	–	125-60-0
Fenpropathrin	–	–	–	–	–	–	–	–	–	–	–	50	88	–	64257-84-7
Fenpropimorph	–	–	–	–	–	–	–	–	–	–	–	77	34	–	67306-03-0
Fenproporex	–	–	–	77	–	–	–	–	–	–	–	–	–	–	15686-61-0
Fentanyl	–	–	–	78	70	77	70	43	74	58	–	–	–	–	437-38-7
Fenthion	–	–	–	–	–	–	–	–	–	–	–	41	81	–	55-38-9
Fenticlor	–	–	–	–	91	–	–	–	–	–	–	–	–	–	97-24-5
Fenuron	–	–	–	–	–	–	–	–	–	–	–	11	7	–	101-42-8
Fenvalerate	–	–	–	–	–	–	–	–	–	–	–	41	92	–	51630-58-1
Fenyramidol	–	–	–	76	80	86	69	8	52	59	–	–	–	–	553-69-5
Feprazone	–	–	–	19	92	–	–	–	–	–	–	–	–	–	30748-29-9
Flamprop-isopropyl	–	–	–	–	–	–	–	–	–	–	–	29	35	–	52756-22-6
Flamprop-methyl	–	–	–	–	–	–	–	–	–	–	–	22	29	–	52756-25-9
Flavoxate	–	–	–	77	48	–	62	36	67	45	–	–	–	–	15301-69-6
Flecainide	–	–	–	49	28	–	–	6	–	–	37	–	–	–	54143-55-4
Floctafenine	–	–	–	85	85	–	–	–	–	–	–	–	–	–	23779-99-9
Fluanisone	–	23	–	82	67	75	73	39	68	60	88	–	–	–	1480-19-9
Fluazifop-butyl	–	–	–	–	–	–	–	–	–	–	–	46	66	–	69806-50-4
Flubendazole	–	–	–	58	83	85	80	0	55	6	–	–	–	–	31430-15-6
Fludiazepam	56	50	67	75	78	86	76	24	69	63	–	–	–	–	3900-31-0
Fludiazepam Hy/CFMB/5-Chloro-2′-fluoro-2-methylaminobenzophenone	71	69	81	83	77	90	75	48	77	69	–	–	–	55	1548-36-3
Flufenamic Acid	–	–	–	18	84	–	–	–	–	–	16	–	–	–	530-78-9
Flumazenil/RO 15-1788	30	14	61	61	76	72	71	3	63	44	–	–	–	–	78755-81-4
Flumazenil M/RO 15-3890	–	0	–	0	44	–	–	0	–	–	–	–	–	–	84378-44-9
Flunarizine	–	–	–	88	83	90	82	45	75	64	91	–	–	–	52468-60-7
Flunitrazepam	54	47	72	74	80	82	63	10	72	63	90	–	–	–	1622-62-4
Flunitrazepam M (3-Hydroxy-)	38	37	64	57	81	80	71	1	59	52	–	–	–	–	67739-71-3

Table 8.1. (continued)

Substance Name	TLC System														CAS No.
	1	2	3	4	5	6	7	8	9	10	T	P1	P2	B	
Flunitrazepam M (7-Acetamido-)	19	11	41	51	78	76	73	0	40	49	–	–	–	–	67739-72-4
Flunitrazepam M (7-Amino-)	21	21	52	63	77	74	74	1	55	52	–	–	–	–	34084-50-9
Flunitrazepam M (7-Amino-nor-)	11	19	35	54	78	75	72	0	34	47	–	–	–	–	894-76-8
Flunitrazepam M (Nor-)	35	46	56	68	84	85	76	0	58	59	–	–	–	–	2558-30-7
Flunitrazepam M Hy/ ANFB/ 2-Amino-2′-fluoro-5-nitrobenzophenone	71	66	68	86	87	91	77	4	69	70	–	–	–	16	344-80-9
Flunitrazepam Hy/ MNFB/ 2-Methylamino-2′-fluoro-5-nitrobenzophenone	7	63	80	86	89	90	77	20	79	71	–	–	–	29	735-06-8
Fluometuron	–	–	–	–	–	–	–	–	–	–	–	16	9	–	2164-17-2
Fluorofentanyl,p–	–	–	–	83	70	78	–	38	73	59	–	–	–	–	1506-86-1
Fluoxetine	–	–	–	–	–	–	–	–	–	–	30	–	–	–	54910-89-3
Flupentixol	–	–	–	46	50	–	62	6	33	–	36	–	–	–	2709-56-0
Flupentixol M	–	–	–	37	40	–	–	–	–	–	–	–	–	–	
Fluphenazine	–	–	–	45	45	49	63	5	23	10	36	–	–	–	69-23-8
Fluphenazine M	–	–	–	26	33	–	–	–	–	–	–	–	–	–	
Flupirtine	–	–	–	–	–	–	–	–	–	–	88	–	–	–	56995-20-1
Flurazepam	3	3	41	71	52	45	62	30	48	40	64	–	–	–	17617-23-1
Flurazepam M Hy/ACFB/2-Amino-5-chloro-2′-fluorobenzophenone	76	68	74	86	89	93	78	11	75	70	–	–	–	31	784-38-3
Flurazepam Hy/DCFB/5-Chloro-2-(2-diethylamino)-ethylamino-2′-fluorobenzophenone	9	8	41	83	45	51	67	53	60	51	–	–	–	0	36105-18-7
Flurazepam Hy/HCFB/5-Chloro-2′-fluoro-2-(hydroxyethyl-amino)benzophenone	48	50	57	82	87	91	77	7	60	66	–	–	–	2	35231-38-0
Flurazepam M	–	–	–	–	–	–	–	–	–	–	45	–	–	–	
Flurazepam M (Dideethyl-)	0	0	14	40	20	60	58	2	18	29	–	–	–	–	17617-59-3
Flurazepam M (Monodeethyl-)	0	0	15	50	28	62	58	8	19	11	–	–	–	–	17656-74-5
Flurazepam M (N(1)-Dealkyl-)	34	45	60	72	83	88	75	2	56	58	–	–	–	–	2886-65-9
Flurazepam M (N(1)-Hydroxy-ethyl-)	19	28	54	61	81	82	74	2	46	55	–	–	–	–	20971-53-3
Flurbiprofen	30	30	45	6	–	–	–	–	–	–	3	–	–	–	5104-49-4
Flurbiprofen M #1	–	22	–	3	84	–	–	–	–	–	–	–	–	–	
Flurbiprofen M #2	–	20	–	10	84	–	–	–	–	–	–	–	–	–	
Fluspirilene	–	–	–	71	63	78	69	4	59	49	87	–	–	–	1841-19-6
Flutazolam	–	–	–	68	84	–	–	–	–	–	–	–	–	–	27060-91-9
Fluvoxamine	–	–	–	46	18	–	–	12	–	–	35	–	–	–	54739-18-3
Fominoben	–	–	–	82	82	–	–	–	–	–	–	–	–	–	18053-31-1
Fonofos	–	–	–	–	–	–	–	–	–	–	–	59	89	–	944-22-9
Formetanate	–	–	–	–	–	–	–	–	–	–	–	1	1	–	22259-30-9
Fosazepam	–	–	–	27	64	–	–	–	–	–	–	–	–	–	35322-07-7
Fosfomycin	–	–	–	–	0	–	–	–	–	–	–	–	–	–	23155-02-4

Table 8.1. (continued)

Substance Name	1	2	3	4	5	6	7	8	9	10	T	P1	P2	B	CAS No.
Fuberidazole	–	–	–	–	–	–	–	–	–	–	–	9	8	–	3878-19-1
Furaltadone	–	–	–	–	–	–	43	0	40	40	–	–	–	–	139-91-3
Furazolidone	–	–	–	–	–	–	44	0	47	59	–	–	–	–	67-45-8
Furethidine	–	–	–	–	–	–	69	37	60	29	–	–	–	–	2385-81-1
Furmecyclox	–	–	–	–	–	–	–	–	–	–	–	49	42	–	60568-05-0
Furosemide	1	7	7	6	86	–	–	0	–	–	–	–	–	–	54-31-9
Fusaric Acid	3	4	8	0	–	–	–	–	–	–	–	–	–	–	536-69-6
Gallamine	–	–	–	0	0	2	0	0	0	0	–	–	–	–	153-76-4
Gallopamil	–	4	–	71	46	–	–	24	–	–	62	–	–	–	16662-47-8
Gallopamil M (D517)	–	7	–	69	56	–	–	21	–	–	–	–	–	–	
Gallopamil M (Nor-)	–	1	–	56	28	–	–	11	–	–	–	–	–	–	
Gentamycin	–	–	–	–	0	–	–	–	–	–	–	–	–	–	1403-66-3
Gentisic Acid	6	23	19	2	88	69	–	–	–	–	–	–	–	–	490-79-9
Gitoformate	–	–	–	–	88	–	–	–	–	–	–	–	–	–	10176-39-3
Glafenine	–	3	–	46	81	78	67	1	38	40	–	–	–	–	3820-67-5
Glibenclamide	30	30	57	11	90	–	–	0	–	–	–	–	–	–	10238-21-8
Glibornuride	40	60	54	5	92	–	–	–	–	–	25	–	–	–	26944-48-9
Gliclazide	–	–	–	9	84	–	–	0	–	–	–	–	–	–	21187-98-4
Glipizide	–	–	–	–	–	–	87	0	41	5	–	–	–	–	29094-61-9
Gliquidone	–	–	–	–	93	–	–	–	–	–	26	–	–	–	33342-05-1
Glisoxepide	–	–	–	5	85	–	–	0	–	–	–	–	–	–	25046-79-1
Glutaral	46	46	53	39	–	–	–	–	–	–	–	–	–	–	111-30-8
Glutethimide	63	62	70	80	86	89	75	31	–	–	90	–	–	–	77-21-4
Glutethimide M (Amino-)	–	–	–	65	–	–	–	–	–	–	–	–	–	–	50275-61-1
Glyburide → Glibenclamide	30	30	57	11	90	–	–	0	–	–	–	–	–	–	10238-21-8
Glycophene → Iprindole	–	–	–	–	–	–	47	49	34	16	–	–	–	–	5560-72-5
Glycopyrronium Bromide	–	–	–	1	3	–	3	–	–	–	–	–	–	–	596-51-0
Glymidine	–	–	–	5	83	–	76	0	65	29	24	–	–	–	339-44-6
Griseofulvin	52	37	68	69	78	–	–	–	–	–	–	–	–	–	126-07-8
Guaifenesin	11	17	40	39	81	–	–	2	–	–	60	–	–	–	93-14-1
Guanethidine	–	–	–	1	3	30	1	0	2	0	–	–	–	–	55-65-2
Guanoclor	–	–	–	–	–	–	3	0	0	0	–	–	–	–	5001-32-1
Guanoxan	–	–	–	0	3	76	1	0	0	0	–	–	–	–	2165-19-7
HCFB → Flurazepam M Hy	48	50	57	82	87	91	77	7	60	66	–	–	–	2	35231-38-0
Halazepam	59	59	70	81	89	91	78	15	6	71	95	–	–	–	23092-17-3
Halazepam Hy/TCB/2-Amino-5-chloro-(2,2,2-trifluoroethyl) benzophenone	81	74	81	89	94	95	82	49	81	75	–	–	–	74	
Halazepam M → Nordazepam	34	45	57	69	82	83	62	3	55	60	81	–	–	–	1088-11-5
Halazepam M → Oxazepam	22	35	42	45	82	82	56	0	40	51	–	–	–	–	604-75-1
Haloperidol	–	–	–	76	51	75	67	11	27	33	64	–	–	–	52-86-8
Haloperidol M	–	–	–	57	52	–	–	6	–	–	50	–	–	–	
Haloxazolam	46	50	65	75	80	90	74	11	66	62	–	–	–	–	59128-97-1
Haloxazolam Hy/ABFB/2-Amino-5-bromo-2′-fluorobenzophenone	70	69	74	83	81	93	83	11	71	69	–	–	–	32	
Harman	–	–	–	57	72	76	70	2	40	33	–	–	–	–	486-84-0

Table 8.1. (continued)

Substance Name	1	2	3	4	5	6	7	8	9	10	T	P1	P2	B	CAS No.
	TLC System														
Harmine	–	–	–	–	65	68	63	0	22	28	–	–	–	–	442-51-3
Heptabarbital	50	64	59	38	88	–	–	–	–	–	–	–	–	–	509-86-4
Heptachlor	–	–	–	–	–	–	–	–	–	–	–	84	97	–	76-44-8
Heptaminol	–	–	–	22	14	–	23	1	2	5	–	–	–	–	543-15-7
Heptenophos	–	–	–	90	89	–	–	–	–	–	–	20	18	–	23560-59-0
Hexachlorophene	32	34	44	30	94	98	–	0	–	–	–	–	–	–	70-30-4
Hexamethonium	–	–	–	0	0	1	0	0	0	0	–	–	–	–	60-26-4
Hexapropymate	57	65	60	78	86	–	–	9	–	–	–	–	–	–	358-52-1
Hexazinone	–	–	–	–	–	–	–	–	–	–	–	4	2	–	51235-04-2
Hexethal	53	67	60	44	–	–	–	–	–	–	–	–	–	–	77-30-5
Hexetidine	–	–	–	79	30	91	70	48	40	20	–	–	–	–	141-94-6
Hexobarbital	65	65	69	53	85	–	–	–	–	–	–	–	–	–	56-29-1
Hexobendine	–	–	–	16	12	–	47	10	44	6	–	–	–	–	54-03-5
Hexocyclium Metilsulfate	–	–	–	1	3	–	2	–	–	–	–	–	–	–	115-63-9
Hexoprenaline	–	–	–	71	12	–	3	1	0	1	–	–	–	–	3215-70-1
Histamine	–	–	–	7	0	3	13	0	0	0	–	–	–	–	51-45-6
Histapyrrodine	–	–	–	75	32	–	–	–	–	–	–	–	–	–	493-80-1
Homatropine	–	–	–	23	7	27	1	5	1	1	–	–	–	–	87-00-3
Homatropine Methylbromide	–	–	–	0	0	12	0	0	0	0	–	–	–	–	80-49-9
Homofenazine	–	–	–	39	13	–	–	6	–	–	–	–	–	–	3833-99-6
Homofenazine M	–	–	–	23	8	–	–	–	–	–	–	–	–	–	
Hordenine	–	–	–	–	–	–	40	5	6	5	–	–	–	–	539-15-1
Hydralazine	–	–	–	–	–	–	51	38	11	64	–	–	–	–	86-54-4
Hydrastine	–	–	–	71	59	57	61	15	64	52	–	–	–	–	118-08-1
Hydrastinine	–	–	–	37	1	15	0	46	0	0	–	–	–	–	6592-85-4
Hydrochlorothiazide	4	34	11	34	78	–	–	–	–	–	–	–	–	–	58-93-5
Hydrocodone	–	–	–	33	11	13	25	4	20	4	18	–	–	–	125-29-1
Hydrocortisone	–	28	–	45	86	–	–	0	–	–	70	–	–	–	50-23-7
Hydroflumethiazide	7	47	13	36	87	–	–	–	–	–	–	–	–	–	135-09-1
Hydromorphone	–	–	–	18	12	14	23	3	9	2	10	–	–	–	466-99-9
Hydroquinidine	–	–	–	43	20	70	45	3	8	5	–	–	–	–	1435-55-8
Hydroquinine	–	–	–	40	21	69	44	2	7	3	–	–	–	–	522-66-7
Hydroxy-2-methyl-propyl) phenylpropionic acid, 2-4′-(2- → Ibuprofen M	–	1	–	0	84	–	–	–	–	–	–	–	–	–	
Hydroxyamphetamine	–	–	–	–	–	–	35	2	2	11	–	–	–	–	1518-86-1
Hydroxybenzoic Acid, p–	–	–	–	23	90	–	–	–	–	–	–	–	–	–	99-96-7
Hydroxycarisodol	–	–	–	–	–	–	–	–	–	–	70	–	–	–	
Hydroxychloroquine	–	–	–	37	7	–	45	1	2	3	–	–	–	–	118-42-3
Hydroxyhippuric Acid,o–	0	0	4	0	78	–	–	–	–	–	–	–	–	–	487-54-7
Hydroxyphenethylmorphinan	–	–	–	–	51	–	–	–	–	–	–	–	–	–	
Hydroxystilbamidine	–	–	–	–	–	–	1	0	0	0	–	–	–	–	495-99-8
Hydroxyzine	–	–	–	54	57	65	68	10	54	19	52	–	–	–	68-88-2
Hydroxyzine M	–	–	–	–	–	–	–	–	–	–	30	–	–	–	
Hyoscyamine → Atropine	–	–	–	24	5	28	18	5	3	1	5	–	–	–	101-31-5
Ibogaine	–	–	–	86	39	72	65	28	50	62	–	–	–	–	83-74-9
Ibomal	50	66	56	32	91	–	–	–	–	–	–	–	–	–	545-93-7

Table 8.1. (continued)

Substance Name	TLC System														CAS No.
	1	2	3	4	5	6	7	8	9	10	T	P1	P2	B	
Ibuprofen	46	57	54	6	75	–	–	–	–	–	9	–	–	–	15687-27-1
Ibuprofen M/2-4′-(2-Carboxy-propyl)phenylpropionic acid	–	13	–	0	78	–	–	–	–	–	–	–	–	–	
Ibuprofen M/2-4′-(2-Hydroxy-2-methyl-propyl)phenylpropionic acid	–	1	–	0	84	–	–	–	–	–	–	–	–	–	
Idobutal	55	69	59	41	–	–	–	–	–	–	–	–	–	–	3146-66-5
Imazalil → Enilconazole	–	–	–	72	78	–	–	–	–	–	–	5	2	–	35554-44-0
Imiclopazine	–	–	–	39	39	–	–	–	–	–	–	–	–	–	7224-08-0
Imipramine	–	–	–	67	21	47	48	48	23	13	50	–	–	–	50-49-7
Imipramine M	–	–	–	–	–	–	–	–	–	–	33	–	–	–	
Imipramine M → Desipramine	–	–	–	40	7	71	26	19	11	3	24	–	–	–	50-47-5
Indapamide	38	61	46	66	89	–	–	–	–	–	–	–	–	–	26807-65-8
Indolyl Acetate,3–	–	–	–	74	80	–	–	–	–	–	–	–	–	–	608-08-2
Indometacin	16	13	38	5	83	63	–	–	–	–	2	–	–	–	53-86-1
Inositol Nicotinate	–	–	–	–	–	–	57	1	43	16	–	–	–	–	6556-11-2
Ioxynil	–	–	–	–	–	–	–	–	–	–	–	4	19	–	1689-83-4
Iprindole	–	–	–	–	–	–	47	49	34	16	–	–	–	–	5560-72-5
Iprodione	–	–	–	–	–	–	–	–	–	–	–	34	54	–	36734-19-7
Iproniazid	–	–	–	42	70	69	69	1	23	17	–	–	–	–	54-92-2
Isoaminile	–	–	–	81	45	–	68	58	54	55	–	–	–	–	77-51-0
Isobutyl 4-Aminobenzoate	–	–	–	–	–	–	67	7	64	69	–	–	–	–	94-14-4
Isocarboxazid	–	–	–	75	84	86	71	20	74	61	–	–	–	–	59-63-2
Isoetarine	–	–	–	–	73	–	59	0	0	0	–	–	–	–	530-08-5
Isofenphos	–	–	–	99	86	–	–	–	–	–	–	41	67	–	25311-71-1
Isomethadone	–	–	–	–	–	–	74	58	45	43	–	–	–	–	466-40-0
Isometheptene	–	–	–	–	–	–	–	–	–	–	15	–	–	–	503-01-5
Isomethiozin	–	–	–	–	–	–	–	–	–	–	–	50	55	–	57052-04-7
Isoniazid	–	–	18	29	55	49	47	1	11	20	–	–	–	–	54-85-3
Isoprenaline	–	–	–	21	14	69	40	0	1	3	4	–	–	–	7683-59-2
Isopropamide Iodide	–	–	–	3	3	41	5	0	5	0	–	–	–	–	71-81-8
Isopropylaminophenazone	–	–	–	63	78	–	–	–	–	–	–	–	–	–	3615-24-5
Isopyrin → Isopropylaminophenazone	–	–	–	63	78	–	–	–	–	–	–	–	–	–	3615-24-5
Isothebaine	–	–	–	61	49	43	55	14	57	26	–	–	–	–	568-21-8
Isothipendyl	–	–	–	64	22	35	52	41	30	14	51	–	–	–	482-15-5
Isothipendyl M	–	–	–	44	21	–	–	–	–	–	–	–	–	–	
Isovaleryl-1,3-indanedione,2–	–	–	–	42	83	–	–	–	–	–	–	–	–	–	83-28-3
Isoxsuprine	–	–	–	62	62	81	78	3	32	53	81	–	–	–	395-28-8
JB-318 → Ethylpiperidyl Benzilate	–	–	–	78	59	54	75	44	63	55	–	–	–	–	3567-12-2
JB-336 → Methylpiperidyl Benzilate	–	–	–	73	58	50	60	30	58	33	–	–	–	–	3321-80-0
Kanamycin	–	–	–	–	0	–	–	–	–	–	–	–	–	–	8063-07-8
Kavain	–	–	–	77	85	–	–	–	–	–	–	–	–	–	500-64-1
Kebuzone	–	–	–	78	87	–	–	–	–	–	–	–	–	–	853-34-9
Ketamine	–	–	–	79	68	72	63	37	63	64	84	–	–	–	6740-88-1

Table 8.1. (continued)

Substance Name	1	2	3	4	5	6	7	8	9	10	T	P1	P2	B	CAS No.
	TLC System														
Ketanserin	–	–	–	65	73	–	–	3	–	–	–	–	–	–	74050-98-9
Ketanserin M/Ketanserinol	–	–	–	49	63	–	–	0	–	–	–	–	–	–	76330-73-9
Ketanserinol → Ketanserin M	–	–	–	49	63	–	–	0	–	–	–	–	–	–	76330-73-9
Ketazolam	45	45	62	74	83	80	66	14	64	66	89	–	–	–	27223-35-4
Ketazolam M → Diazepam	58	49	72	76	82	85	75	27	73	59	90	–	–	–	439-14-5
Ketazolam M → Nordazepam	34	45	57	69	82	83	62	3	55	60	81	–	–	–	1088-11-5
Ketazolam M → Oxazepam	22	35	42	45	82	82	56	0	40	51	–	–	–	–	604-75-1
Ketazolam M → Temazepam	51	47	65	62	82	82	53	8	59	53	81	–	–	–	846-50-4
Ketobemidone	–	–	–	37	26	–	47	2	9	6	25	–	–	–	469-79-4
Ketoconazole	–	–	–	48	72	–	–	–	–	–	–	–	–	–	65277-42-1
Ketoprofen	27	25	41	6	85	–	–	–	–	–	5	–	–	–	22071-15-4
Ketotifen	–	–	–	52	24	–	–	24	–	–	43	–	–	–	34580-13-7
Khellin	–	–	–	70	81	–	–	–	–	–	–	–	–	–	82-02-0
LSD → Lysergide	3	–	–	56	60	59	60	3	39	18	–	–	–	–	50-37-3
Labetalol	–	1	–	29	32	–	–	0	–	–	34	–	–	–	36894-69-6
Lanatoside C	–	–	–	6	89	–	–	–	–	–	–	–	–	–	17575-22-3
Lenacil	–	–	–	–	–	–	–	–	–	–	–	15	8	–	2164-08-1
Leucinocaine	–	–	–	64	78	–	66	6	45	68	–	–	–	–	92-23-9
Levallorphan	–	–	–	74	42	73	67	19	24	45	–	–	–	–	152-02-3
Levamisole	–	–	–	65	53	52	62	18	48	42	–	–	–	–	14769-73-4
Levarterenol/Norepinephrine	–	–	–	3	43	–	0	0	0	0	–	–	–	–	51-41-2
Levodopa	0	0	0	0	11	–	–	–	–	–	–	–	–	–	59-92-7
Levomepromazine	–	–	–	76	32	49	57	47	38	46	75	–	–	–	60-99-1
Levomepromazine M	–	–	–	48	20	–	–	–	–	–	–	–	–	–	
Levomethadone → Methadone	–	–	–	77	16	60	48	59	20	27	66	–	–	–	125-58-6
Levomethorphan → Dextromethorphan	–	–	–	47	10	42	33	42	18	6	26	–	–	–	125-70-2
Levomoramide → Dextromoramide	–	–	–	79	72	78	73	42	71	60	86	–	–	–	5666-11-3
Levophenacylmorphan	–	–	–	–	–	–	70	9	60	67	–	–	–	–	10061-32-2
Levopropoxyphene → Propoxyphene	–	–	–	82	50	63	68	58	55	54	82	–	–	–	2338-37-6
Levorphanol → Dextrorphan	–	–	–	42	10	49	35	14	4	3	19	–	–	–	77-07-6
Levorphanol M (Nor-) → Dextrorphan M	–	–	–	–	–	–	9	2	1	1	–	–	–	–	1531-12-0
Lidocaine	–	–	–	80	72	69	70	35	71	63	88	–	–	–	137-58-6
Lidocaine M/MEGX	–	–	–	–	–	–	–	–	–	–	50	–	–	–	7728-40-7
Lidoflazine	–	–	–	70	70	77	70	11	63	36	–	–	–	–	3416-26-0
Lincomycin	–	–	–	–	75	–	–	–	–	–	–	–	–	–	154-21-2
Lindane	–	–	–	–	–	–	–	–	–	–	–	51	92	–	58-89-9
Linuron	–	–	–	–	–	–	–	–	–	–	–	22	31	–	330-55-2
Lisinopril	–	0	–	0	27	–	–	0	–	–	–	–	–	–	83915-83-7
Lobeline	–	–	–	75	23	–	61	17	35	29	–	–	–	–	90-69-7
Lofepramine	–	–	–	90	82	–	–	–	–	–	85	–	–	–	23047-25-8
Lofexidine	–	–	–	53	17	–	–	–	–	–	–	–	–	–	31036-80-3
Lonazolac	–	–	–	6	88	–	–	–	–	–	7	–	–	–	53008-88-1
Loperamide	–	–	–	74	52	81	70	9	32	22	67	–	–	–	53179-11-6

Table 8.1. (continued)

Substance Name	1	2	3	4	5	6	7	8	9	10	T	P1	P2	B	CAS No.
	TLC System														
Loprazolam	3	1	36	40	26	15	40	1	48	5	26	–	–	–	61197-73-7
Lorazepam/Lormetazepam M	23	41	42	43	82	82	52	1	36	28	75	–	–	–	846-49-1
Lorazepam Hy/ADB/2-Amino-2′,5-dichlorobenzophenone	77	70	76	87	89	93	78	11	76	70	–	–	–	33	2958-30-3
Lorcainide	–	–	–	80	41	–	–	48	–	–	65	–	–	–	59729-31-6
Lorcainide M	–	–	–	18	99	–	–	3	–	–	–	–	–	–	
Lormetazepam	46	45	60	59	82	82	52	6	61	50	79	–	–	–	848-75-9
Lormetazepam Hy/MDB/2′,5-Dichloro-2-methylaminobenzophenone	81	69	81	87	82	93	75	49	81	7	–	–	–	57	
Lormetazepam M → Lorazepam	23	41	42	43	82	82	52	1	36	28	75	–	–	–	846-49-1
Loxapine	–	–	–	54	49	–	–	36	–	–	66	–	–	–	1977-10-2
Lysergamide	–	–	–	36	57	51	60	0	19	7	–	–	–	–	478-94-4
Lysergic Acid	0	0	0	0	70	16	58	0	0	0	–	–	–	–	82-58-6
Lysergide/LSD	3	–	–	56	60	59	60	3	39	18	–	–	–	–	50-37-3
MACB → Diazepam Hy	82	72	82	88	89	93	79	53	81	71	–	–	–	–	1022-13-5
MCPA/(4-Chloro-o-tolyloxy) acetic Acid	9	11	15	5	–	–	–	–	–	–	–	0	0	–	94-74-6
MCPB	–	–	–	–	–	–	–	–	–	–	–	8	6	–	94-81-5
MCPP → Mecoprop	–	–	–	–	–	–	–	–	–	–	–	5	2	–	7085-19-0
MDA → 3,4-Methylenedioxyamphetamine	–	–	–	42	10	76	39	18	12	17	30	–	–	–	4764-17-4
MDB → Lormetazepam Hy	81	69	81	87	82	93	75	49	81	7	–	–	–	57	
MDMA → 3,4-Methylenedioxymethamphetamine	–	20	–	39	89	–	–	24	–	–	22	–	–	–	42542-10-9
MEGX → Lidocaine M	–	–	–	–	–	–	–	–	–	–	50	–	–	–	7728-40-7
MMDA → 3-Methoxy-4,5-methylenedioxyamphetamine	–	–	–	–	–	–	–	–	–	–	28	–	–	–	13674-05-0
MMDA-2 → 2-Methoxy-4,5-methylenedioxyamphetamine	–	–	–	43	9	74	35	23	10	18	–	–	–	–	64638-05-7
MNB → Nimetazepam Hy	66	66	80	83	76	86	73	31	77	68	–	–	–	29	4958-56-9
MNFB → Flunitrazepam Hy	7	63	80	86	89	90	77	20	79	71	–	–	–	29	735-06-8
Mafenide	1	1	2	27	22	–	49	–	–	–	–	–	–	–	138-39-6
Malathion	–	–	–	99	84	–	–	–	–	–	–	31	53	–	121-75-5
Maprotiline	–	–	–	36	6	71	15	18	5	2	15	–	–	–	10262-69-8
Maprotiline M (Nor-)	–	–	–	43	54	–	–	17	–	–	–	–	–	–	
Mazindol	–	–	–	53	46	65	63	7	13	13	53	–	–	–	22232-71-9
Mebenazine	–	–	–	–	–	–	70	48	69	63	–	–	–	–	65-64-5
Mebendazole	–	–	–	60	80	84	65	0	59	49	–	–	–	–	31431-39-7
Mebeverine	–	–	–	86	32	–	63	40	53	49	–	–	–	–	3625-06-7
Mebhydrolin	–	–	–	65	36	46	57	27	45	20	54	–	–	–	524-81-2
Mebutamate	10	35	35	60	82	85	–	0	33	56	–	–	–	–	64-55-1
Mecamylamine	–	–	–	–	–	–	16	51	2	4	–	–	–	–	60-40-2
Meclofenamic Acid	–	43	–	12	–	–	–	–	–	–	15	–	–	–	644-62-2
Meclofenoxate	–	–	–	67	46	–	77	26	42	22	2	–	–	–	51-68-3
Mecloxamine	–	–	–	62	20	–	–	–	–	–	–	–	–	–	5668-06-4
Meclozine	–	–	–	87	80	88	76	61	79	70	93	–	–	–	569-65-3

Table 8.1. (continued)

Substance Name	TLC System														CAS No.
	1	2	3	4	5	6	7	8	9	10	T	P1	P2	B	
Meconin	–	–	–	71	77	78	–	11	73	63	–	–	–	–	569-31-3
Mecoprop/MCPP	–	–	–	–	–	–	–	–	–	–	–	5	2	–	7085-19-0
Medazepam	56	40	73	78	79	83	67	41	74	62	88	–	–	–	2898-12-6
Medazepam M (Nor-)	9	22	30	67	66	63	73	4	51	47	–	–	–	–	1694-78-6
Medazepam M → Diazepam	58	49	72	76	82	85	75	27	73	59	90	–	–	–	439-14-5
Medazepam M → Nordazepam	34	45	57	69	82	83	62	3	55	60	81	–	–	–	1088-11-5
Medazepam M → Oxazepam	22	35	42	45	82	82	56	0	40	51	–	–	–	–	604-75-1
Medazepam M → Temazepam	51	47	65	62	82	82	53	8	59	53	81	–	–	–	846-50-4
Medrylamine	–	–	–	55	24	–	–	–	–	–	–	–	–	–	524-99-2
Mefenamic Acid	41	48	54	11	87	–	–	–	–	–	15	–	–	–	61-68-7
Mefenorex	–	–	–	75	38	–	–	43	–	–	75	–	–	–	17243-57-1
Mefruside	45	58	55	67	–	–	–	–	–	–	–	–	–	–	7195-27-9
Melitracene	–	–	–	65	27	–	–	52	–	–	61	–	–	–	5118-29-6
Melperone	–	–	–	73	25	–	–	50	–	–	62	–	–	–	3575-80-2
Melperone M (FG 5155)	–	2	–	61	29	–	–	30	–	–	–	–	–	–	68883-08-9
Menthol M	–	–	–	–	–	–	–	–	–	–	70	–	–	–	
Mepacrine	–	–	–	–	–	–	43	15	5	9	–	–	–	–	83-89-6
Mepazine → Pecazine	–	1	–	65	27	–	53	47	44	16	–	–	–	–	60-89-9
Meperidine → Pethidine	–	–	–	60	34	40	52	37	34	11	50	–	–	–	57-42-1
Mephenesin	–	–	–	55	87	89	64	2	43	57	–	–	–	–	59-47-2
Mephenesin Carbamate	14	36	35	55	–	–	–	–	–	–	–	–	–	–	533-06-2
Mephentermine	–	–	–	–	–	–	25	34	8	2	–	–	–	–	100-92-5
Mephenytoin	62	58	66	74	–	–	–	–	–	–	–	–	–	–	50-12-4
Mepindolol	–	–	–	43	20	–	–	3	–	–	35	–	–	–	23694-81-7
Mepivacaine	–	–	–	66	63	60	65	31	62	48	70	–	–	–	96-88-8
Mepivacaine M → Bupivacaine M	–	–	–	49	34	–	–	12	–	–	–	–	–	–	
Meprobamate	9	36	32	56	63	87	75	0	32	58	81	–	–	–	57-53-4
Meproscillarin	–	–	–	78	87	–	–	–	–	–	–	–	–	–	33396-37-1
Mepyramine	–	–	–	58	22	33	51	39	25	14	49	–	–	–	91-84-9
Mequitazine	–	–	–	27	3	–	10	6	6	0	7	–	–	–	29216-28-2
Mequitazine M	–	–	–	8	–	–	–	–	–	–	–	–	–	–	
Mercaptopurine,6–	–	–	–	–	76	–	–	–	–	–	–	–	–	–	50-44-2
Merphalan → Sarcolysis	–	–	–	–	52	–	–	–	–	–	–	–	–	–	531-76-0
Mescaline	–	–	–	24	6	63	20	3	10	12	–	–	–	–	54-04-6
Mesoridazine	–	–	–	30	11	–	38	3	6	1	–	–	–	–	5588-33-0
Mesuximide	–	–	–	87	90	–	76	–	–	–	–	–	–	–	77-41-8
Metaclazepam	47	35	71	79	82	84	77	40	73	62	–	–	–	–	84031-17-4
Metaclazepam M (Dinor-)	–	9	–	46	81	–	–	0	–	–	–	–	–	–	86298-28-4
Metaclazepam M (Nor-)	33	25	65	73	81	84	76	15	68	58	–	–	–	–	86298-26-2
Metamfepramone	–	–	–	68	55	–	–	–	–	–	82	–	–	–	15351-09-4
Metamitron	–	–	–	–	–	–	–	–	–	–	–	5	4	–	41394-05-2
Metamizol	0	0	2	2	85	59	84	0	1	2	53	–	–	–	50567-35-6
Metaraminol	–	–	–	18	13	76	42	1	1	24	–	–	–	–	54-49-9
Metazocine	–	–	–	–	–	–	31	10	5	4	–	–	–	–	3734-52-9
Metformin	–	–	–	0	3	–	–	–	–	–	–	–	–	–	657-24-9
Metformin	–	–	0	80	93	48	1	0	0	0	–	–	–	–	657-24-9

Table 8.1. (continued)

Substance Name	TLC System														CAS No.
	1	2	3	4	5	6	7	8	9	10	T	P1	P2	B	
Methabenzthiazuron	–	–	–	–	–	–	–	–	–	–	–	18	22	–	18691-97-9
Methadone/Levomethadone	–	–	–	77	16	60	48	59	20	27	66	–	–	–	76-99-3
Methadone Intermediate/Pre-Methadone	–	–	–	–	–	–	63	47	55	43	–	–	–	–	125-79-1
Methadone M	–	–	–	–	–	–	–	–	–	–	56	–	–	–	
Methadone M (Nor-)	–	–	–	63	33	–	56	40	34	18	–	–	–	–	467-85-6
Methamfetamine	–	–	–	42	9	63	31	28	13	5	22	–	–	–	537-46-2
Methamidophos	–	–	–	–	–	–	–	–	–	–	–	1	0	–	10265-92-6
Methanthelinium Bromide	–	–	–	76	3	–	2	0	–	–	–	–	–	–	53-46-3
Methaphenilene	–	–	–	–	–	–	54	46	42	21	–	–	–	–	493-78-7
Methapyrilene	–	–	–	66	21	24	52	41	26	13	52	–	–	–	91-80-5
Methaqualone	–	–	–	78	79	84	70	36	80	56	90	–	–	–	72-44-6
Methaqualone M	–	–	–	–	–	–	–	–	–	–	80	–	–	–	
Metharbital	66	65	69	54	87	–	–	–	–	–	–	–	–	–	50-11-3
Methdilazine	–	–	–	–	–	–	29	32	15	6	–	–	–	–	1982-37-2
Methenamine	–	–	–	–	12	12	30	4	13	3	–	–	–	–	100-97-0
Methenolone	–	–	–	89	92	–	–	–	–	–	–	–	–	–	153-00-4
Methidathion	91	92	95	99	80	88	86	55	99	99	–	29	56	–	950-37-8
Methocarbamol	7	23	38	49	84	–	70	–	–	–	68	–	–	–	532-03-6
Methohexital	73	72	71	58	85	–	–	–	–	–	–	–	–	–	151-83-7
Methomyl	–	–	–	–	–	–	–	–	–	–	–	6	6	–	16752-77-5
Methopromazine	–	–	–	–	–	–	43	30	22	12	–	–	–	–	61-01-8
Methoprotryne	–	–	–	–	–	–	–	–	–	–	–	19	11	–	841-06-5
Methoserpidine	–	–	–	–	–	–	72	4	77	64	–	–	–	–	865-04-3
Methotrimeprazine → Levo-mepromazine	–	–	–	76	32	49	57	47	38	46	75	–	–	–	60-99-1
Methoxamine	–	–	–	11	12	73	55	24	4	38	–	–	–	–	390-28-3
Methoxy-4,5-methylenedioxy-amphetamine,2-/MMDA-2	–	–	–	43	9	74	35	23	10	18	–	–	–	–	64638-05-7
Methoxy-4,5-methylenedioxy-amphetamine,3-/MMDA	–	–	–	–	–	–	–	–	–	–	28	–	–	–	13674-05-0
Methoxyamphetamine,4–	–			43	9	74	73	23	77	69	–	–	–	–	23239-32-9
Methoxychlor	–	–	–	–	–	–	–	–	–	–	–	43	84	–	72-43-5
Methoxyphenamine	–	–	–	32	7	–	23	26	4	2	19	–	–	–	93-30-1
Methoxyphenamine M	–	–	–	–	–	–	–	–	–	–	42	–	–	–	
Methscopolamine Bromide	–	–	–	1	3	18	2	0	0	0	–	–	–	–	155-41-9
Methyclothiazide	19	50	27	53	–	–	–	–	–	–	–	–	–	–	135-07-9
Methyl Nicotinate	–	–	–	72	74	77	61	37	66	59	–	–	–	–	93-60-7
Methyl Paraben → Methyl p-Hydroxybenzoate	62	65	55	44	81	–	–	–	–	–	–	–	–	–	99-76-3
Methyl Salicylate	–	–	–	–	–	90	–	–	–	–	–	–	–	–	119-36-8
Methyl p-Hydroxybenzoate	62	65	55	44	81	–	–	–	–	–	–	–	–	–	99-76-3
Methyl-1-phenylcyclohexyl) piperidine,1-(4–	–	–	–	87	31	78	–	72	47	68	–	–	–	–	19420-52-1
Methyl-5-phenylbarbituric Acid,5–	29	61	46	23	–	–	–	–	–	–	–	–	–	–	76-94-8

Table 8.1. (continued)

Substance Name	1	2	3	4	5	6	7	8	9	10	T	P1	P2	B	CAS No.
	TLC System														
Methylamino-2′fluoro-5-nitrobenzophenone,2- → Flunitrazepam M Hy	7	63	80	86	89	90	77	20	79	71	–	–	–	29	735-06-8
Methylamino-5-chlorobenzophenone,2- → Diazepam Hy	82	72	82	88	89	93	79	53	81	71	–	–	–	52	1022-13-5
Methylamino-5-nitrobenzophenone,2- → Nimetazepam Hy	66	66	80	83	76	86	73	31	77	68	–	–	–	29	4958-56-9
Methylbromazepam,1–	–	–	–	56	–	–	–	–	–	–	–	–	–	–	
Methyldesorphine	–	–	–	–	–	–	44	12	15	6	–	–	–	–	16008-36-9
Methyldopa	–	–	–	2	60	75	49	1	1	1	–	–	–	–	555-30-6
Methylenedioxyamphetamine,3,4-/MDA	–	–	–	42	10	76	39	18	12	17	30	–	–	–	4764-17-4
Methylenedioxymethamphetamine,3,4-/MDMA/XTC	–	20	–	39	89	–	–	24	–	–	22	–	–	–	42542-10-9
Methylenedioxyphentermin,3,4–	–	–	–	–	11	75	–	–	–	–	–	–	–	–	39235-63-7
Methylephedrine,N–	–	–	–	35	12	–	32	–	–	–	–	–	–	–	552-79-4
Methylergometrine	–	–	–	41	69	–	62	0	14	12	–	–	–	–	113-42-8
Methylpentynol Carbamate	49	62	57	74	–	–	–	–	–	–	–	–	–	–	302-66-9
Methylphenazone,4–	–	–	–	58	74	–	62	–	–	–	–	–	–	–	5677-84-9
Methylphenidate	–	–	–	66	40	70	57	35	41	23	–	–	–	–	113-45-1
Methylphenobarbital	70	67	70	41	86	–	–	–	–	–	–	–	–	–	115-38-8
Methylpiperidyl Benzilate/JB-336	–	–	–	73	58	50	60	30	58	33	–	–	–	–	3321-80-0
Methylprednisolon	–	27	–	41	87	–	–	0	–	–	70	–	–	–	83-43-2
Methylprimidone,4–	–	–	–	38	82	–	–	–	–	–	–	–	–	–	59026-32-3
Methyltestosterone,17–	–	–	–	–	–	86	–	–	–	–	–	–	–	–	58-18-4
Methyprylon	31	25	55	63	78	–	58	–	–	–	72	–	–	–	125-64-4
Methysergide	–	–	–	–	–	–	65	1	21	15	–	–	–	–	361-37-5
Metipranolol	–	–	–	44	20	–	–	–	–	–	35	–	–	–	22664-55-7
Metisazone	–	–	–	–	–	–	65	3	68	69	–	–	–	–	1910-68-5
Metixene	–	–	–	61	21	–	50	45	25	12	44	–	–	–	4969-02-2
Metixene M	–	–	–	42	11	–	–	–	–	–	–	–	–	–	
Metobromuron	–	–	–	–	–	–	–	–	–	–	–	21	27	–	3060-89-7
Metoclopramide	–	–	–	51	17	–	47	1	7	13	30	–	–	–	364-62-5
Metoclopramide M	–	–	–	–	–	–	–	–	–	–	15	–	–	–	
Metofenazate	–	–	–	76	66	–	–	–	–	–	–	–	–	–	388-51-2
Metolazone	23	51	33	57	–	–	–	–	–	–	–	–	–	–	17560-51-9
Metomidate	–	–	–	74	77	81	71	23	65	48	–	–	–	–	5377-20-8
Metopimazine	–	–	–	–	–	–	56	0	11	12	–	–	–	–	14008-44-7
Metopon	–	–	–	27	11	14	–	5	13	3	–	–	–	–	143-52-2
Metoprolol	–	–	–	44	20	74	49	10	8	9	25	–	–	–	37350-58-6
Metoxuron	–	–	–	–	–	–	–	–	–	–	–	7	4	–	19937-59-8
Metribuzin	–	–	–	–	–	–	–	–	–	–	–	24	31	–	21087-64-9
Metronidazole	–	–	32	46	75	70	58	2	36	40	–	–	–	–	443-48-1
Metyrapone	–	–	–	64	66	–	58	16	58	41	–	–	–	–	54-36-4
Mevinphos	–	–	–	75	86	–	–	–	–	–	–	12	10	–	7786-34-7
Mexazolam	–	55	–	78	86	–	–	12	–	–	–	–	–	–	31868-18-5

Table 8.1. (continued)

Substance Name	TLC System														CAS No.
	1	2	3	4	5	6	7	8	9	10	T	P1	P2	B	
Mexiletine	–	–	–	55	25	78	40	17	4	9	48	–	–	–	31828-71-4
Mianserin	–	–	–	68	48	50	58	39	58	23	66	–	–	–	24219-97-4
Mianserin M (8-Hydroxy-)	–	–	–	54	46	52	60	4	37	18	–	–	–	–	
Mianserin M (N-oxide)	–	–	–	18	23	16	37	0	9	0	–	–	–	–	
Mianserin M (Nor-)	–	–	–	51	25	60	46	11	31	6	38	–	–	–	
Miconazole	–	7	56	80	77	80	73	11	67	37	–	–	–	–	22916-47-8
Midazolam	13	5	53	60	69	70	72	6	60	19	60	–	–	–	59467-70-8
Midazolam M (α,4-Dihydroxy-)	2	5	33	38	78	81	75	4	25	14	–	–	–	–	64740-68-7
Midazolam M (α-Hydroxy-)	3	4	41	53	73	72	70	3	52	8	–	–	–	–	59468-90-5
Midazolam M (4-Hydroxy-)	5	5	37	49	76	77	74	1	43	15	–	–	–	–	59468-85-8
Midodrine	–	–	–	26	26	–	–	–	–	–	–	–	–	–	42794-76-3
Minocycline	–	–	–	–	88	–	–	–	–	–	–	–	–	–	10118-90-8
Minoxidil	–	–	–	–	–	–	51	0	3	0	–	–	–	–	38304-91-5
Mofebutazone	–	–	–	89	95	–	–	–	–	–	9	–	–	–	2210-63-1
Molindone	–	–	–	–	–	–	–	–	–	–	60	–	–	–	7416-34-4
Molsidomine	–	–	–	62	77	–	–	–	–	–	–	–	–	–	25717-80-0
Monocrotophos	–	–	–	–	–	–	–	–	–	–	–	1	1	–	6923-22-4
Monolinuron	–	–	–	–	–	–	–	–	–	–	–	23	30	–	1746-81-2
Monuron	–	–	–	–	–	–	–	–	–	–	–	13	7	–	150-68-5
Moperone	–	–	–	72	42	–	–	–	–	–	–	–	–	–	1050-79-9
Morazone	–	–	–	58	61	–	58	8	46	31	–	–	–	–	6536-18-1
Morinamide	–	–	–	–	–	–	54	8	49	32	–	–	–	–	952-54-5
Moroxydine	–	–	–	0	3	–	–	–	–	–	–	–	–	–	3731-59-7
Morpheridine	–	–	–	–	–	–	62	27	45	11	–	–	–	–	469-81-8
Morphine	–	–	–	20	18	23	37	0	9	1	15	–	–	–	57-27-2
Morphine M (-3-glucuronide)	–	–	–	0	20	0	–	–	–	0	–	–	–	–	20290-09-9
Morphine M (Nor-)	–	–	–	7	5	44	17	0	0	0	–	–	–	–	466-97-7
Morphine-3-acetate	–	–	–	24	82	21	–	6	28	7	–	–	–	–	5140-28-3
Morphine-6-acetate/Diamorphine M	–	–	–	44	25	27	46	6	26	8	30	–	–	–	2784-73-8
Moxaverine	–	–	–	85	80	–	–	–	–	–	–	–	–	–	10539-19-2
Moxisylyte	–	–	–	–	–	–	52	31	44	19	–	–	–	–	54-32-0
Myrophine	–	–	–	–	–	–	50	30	53	16	–	–	–	–	467-18-5
Nadolol	–	–	–	20	14	–	42	1	1	1	1	–	–	–	42200-33-9
Nafcillin	–	–	–	4	86	–	–	–	–	–	1	–	–	–	985-16-0
Nafcillin M	–	–	–	–	–	–	–	–	–	–	85	–	–	–	
Naftazone	–	–	–	–	–	–	66	0	53	23	–	–	–	–	15687-37-3
Naftidrofuryl	–	–	–	78	43	–	64	52	41	35	–	–	–	–	31329-57-4
Nalidixic Acid	39	31	62	2	63	–	–	–	–	–	–	–	–	–	389-08-2
Nalorphine	–	–	–	32	57	59	59	1	23	29	–	–	–	–	62-67-9
Naloxone	–	–	–	47	74	–	65	9	66	63	79	–	–	–	465-65-6
Naphazoline	–	–	–	27	3	52	14	3	6	3	–	–	–	–	835-31-4
Naproxen	33	38	44	6	82	–	–	–	–	–	4	–	–	–	22204-53-1
Naptalam	–	–	–	–	–	–	–	–	–	–	–	21	52	–	132-66-1
Narceine	–	–	–	0	34	14	52	0	3	0	–	–	–	–	131-28-2
Narcobarbital	72	67	69	53	–	–	–	–	–	–	–	–	–	–	125-55-3
Narcophin	–	–	–	–	16	22	–	–	–	–	–	–	–	–	6055-90-9

Table 8.1. (continued)

Substance Name	TLC System														CAS No.
	1	2	3	4	5	6	7	8	9	10	T	P1	P2	B	
Nealbarbital	58	68	60	44	92	–	–	–	–	–	–	–	–	–	561-83-1
Neburon	–	–	–	–	–	–	–	–	–	–	–	26	24	–	555-37-3
Nefopam	–	–	–	59	30	–	50	33	32	17	46	–	–	–	13669-70-0
Neomycin	–	–	–	–	0	–	–	–	–	–	–	–	–	–	1404-04-2
Neopine	–	–	–	–	–	–	35	5	12	4	–	–	–	–	467-14-1
Neostigmine	–	–	–	1	1	13	2	0	–	–	–	–	–	–	59-99-4
Nialamide	–	–	–	–	68	64	70	2	25	4	–	–	–	–	51-12-7
Nicametate	–	–	–	68	35	–	56	41	35	20	–	–	–	–	3099-52-3
Nicardipine	–	–	–	80	82	–	–	1	–	–	–	–	–	–	55985-32-5
Nicergoline	–	–	–	73	43	–	–	–	–	–	–	–	–	–	27848-84-6
Niclosamide	–	–	–	23	91	–	91	–	–	–	–	–	–	–	50-65-7
Nicocodine	–	–	–	47	19	–	43	12	42	10	–	–	–	–	3688-66-2
Nicofuranose	–	–	–	–	–	–	61	42	70	61	–	–	–	–	15351-13-0
Nicomorphine	–	–	–	40	18	–	42	3	34	6	–	–	–	–	639-48-5
Nicotinamide	–	–	–	40	68	66	54	0	21	27	–	–	–	–	98-92-0
Nicotine	–	–	–	61	39	22	54	39	35	13	40	4	1	–	54-11-5
Nicotinic Acid	1	0	4	0	72	–	58	–	–	–	–	–	–	–	59-67-6
Nicotinyl Alcohol	–	–	–	48	74	69	56	4	17	22	–	–	–	–	100-55-0
Nifedipine	–	–	–	71	79	–	68	1	65	68	88	–	–	–	21829-25-4
Nifedipine M	–	–	–	82	83	–	–	26	–	–	–	–	–	–	
Nifenalol	–	–	–	43	22	–	–	–	–	–	36	–	–	–	7413-36-7
Nifenazone	–	–	–	36	58	–	–	0	–	–	25	–	–	–	2139-47-1
Niflumic Acid	3	3	15	11	88	–	–	–	–	–	16	–	–	–	4394-00-7
Nikethamide	–	–	–	59	71	67	59	15	56	29	–	–	–	–	59-26-7
Nimetazepam	53	46	71	77	81	81	74	12	70	55	–	–	–	–	2011-67-8
Nimetazepam Hy/MNB/2-Methylamino-5-nitrobenzophenone	66	66	80	83	76	86	73	31	77	68	–	–	–	29	4958-56-9
Nimodipine	–	–	–	78	87	–	–	1	–	–	92	–	–	–	66085-59-4
Nimodipine M	–	–	–	82	86	–	–	–	–	–	–	–	–	–	
Nimorazole	–	–	–	58	60	–	57	3	44	33	–	–	–	–	6506-37-2
Niridazole	–	–	–	–	–	–	54	1	44	53	–	–	–	–	61-57-4
Nitrazepam	35	46	53	64	84	86	68	0	36	55	85	–	–	–	146-22-5
Nitrazepam Hy/ANB/2-Amino-5-nitrobenzophenone	71	65	68	85	90	91	77	5	69	69	–	–	–	15	1775-95-7
Nitrazepam M (7-Acetamido-)	7	12	24	44	80	83	71	0	20	49	–	–	–	–	4928-03-4
Nitrazepam M (7-Amino-)	10	20	30	54	77	79	71	0	30	46	–	–	–	–	4928-02-3
Nitrazepam M Hy/DAB/2,5-Diaminobenzophenone	37	41	54	81	80	79	72	5	57	57	–	–	–	0	18330-94-4
Nitrendipine	–	–	–	80	87	–	–	2	–	–	–	–	–	–	39562-70-4
Nitrendipine M	–	–	–	82	86	–	–	34	–	–	–	–	–	–	
Nitrofural	–	–	–	–	77	–	–	–	–	–	–	–	–	–	59-87-0
Nitrofurantoin	2	30	33	6	84	–	–	–	–	–	–	–	–	–	67-20-9
Nitrofurazone → Nitrofural	–	–	–	–	77	–	–	–	–	–	–	–	–	–	59-87-0
Nitroglycerin	71	72	69	86	–	–	–	–	–	–	–	–	–	–	55-63-0
Nitrophenol,p–	–	–	–	–	–	–	–	–	–	–	–	14	20	–	100-02-7
Nomifensine	–	–	–	64	53	52	56	9	29	31	64	–	–	–	24526-64-5

Table 8.1. (continued)

Substance Name	TLC System														CAS No.
	1	2	3	4	5	6	7	8	9	10	T	P1	P2	B	
Noracymethadol	–	–	–	65	29	73	55	36	38	13	–	–	–	–	1477-39-0
Norbormide	–	–	–	–	–	–	70	0	62	64	–	–	–	–	991-42-4
Nordazepam	34	45	57	69	82	83	62	3	55	60	81	–	–	–	1088-11-5
Nordefrin Hydrochloride	–	–	–	–	–	–	25	1	0	10	–	–	–	–	138-61-4
Norepinephrine → Levarterenol	–	–	–	3	43	–	0	0	0	0	–	–	–	–	51-41-2
Norfenefrine	–	–		13	12	–	–	–	–	–	–	–	–	–	536-21-0
Norharman	–	–	–	–	–	–	64	0	30	33	–	–	–	–	244-63-3
Norpipanone	–	–	–	–	–	–	68	59	50	38	–	–	–	–	561-48-8
Nortriptyline → Amitriptyline M (Nor-)	–	–	–	46	87	71	34	28	16	4	30	–	–	–	72-69-5
Nortriptyline M #1 → Amitriptyline M	–	–	–	–	–	–	–	–	–	–	17	–	–	–	
Nortriptyline M #2 → Amitriptyline M	–	–	–	–	–	–	–	–	–	–	14	–	–	–	
Noscapine	–	–	–	78	72	75	64	21	74	64	90	–	–	–	128-62-1
Noxiptiline	–	–	–	66	29	–	53	43	35	18	–	–	–	–	3362-45-6
Obidoxime Chloride	–	–	–	0	53	8	1	0	0	0	–	–	–	–	114-90-9
Octacaine	–	–	–	75	78	–	57	32	27	29	–	–	–	–	13912-77-1
Octamylamine	–	–	–	–	–	–	22	28	11	25	–	–	–	–	502-59-0
Octaverine	–	–	–	–	–	–	67	18	77	68	–	–	–	–	549-68-8
Octopamine	–	–	–	17	11	–	–	0	–	–	10	–	–	–	104-14-3
Omethoate	–	–	–	–	–	–	–	–	–	–	–	0	0	–	1113-02-6
Opipramol	–	–	–	38	35	39	54	6	22	7	26	–	–	–	315-72-0
Opipramol M #1	–	–	–	–	–	–	–	–	–	–	95	–	–	–	
Opipramol M #2	–	–	–	–	–	–	–	–	–	–	83	–	–	–	
Opipramol M #3	–	–	–	–	–	–	–	–	–	–	55	–	–	–	
Opipramol M #4	–	–	–	–	–	–	–	–	–	–	18	–	–	–	
Opipramol M #5	–	–	–	–	–	–	–	–	–	–	6	–	–	–	
Orazamide	–	–	–	25	58	–	–	–	–	–	–	–	–	–	2574-78-9
Orciprenaline	–	–	–	18	21	77	48	1	3	6	11	–	–	–	586-06-1
Ordinol-cis → Clopenthixol	–	–	–	44	45	–	56	7	32	11	88	–	–	–	982-24-1
Orientalidine	–	–	–	61	49	43	–	13	59	25	–	–	–	–	23943-90-0
Ornidazole	–	–	–	62	81	–	–	–	–	–	–	–	–	–	16773-42-5
Orphenadrine	–	–	–	68	25	49	55	48	33	16	52	–	–	–	83-98-7
Orthocaine	–	–	–	–	–	–	79	1	37	50	–	–	–	–	536-25-4
Oxatomide	–	–	–	62	74	–	–	4	–	–	75	–	–	–	60607-34-3
Oxazepam	22	35	42	45	82	82	56	0	40	51	–	–	–	–	604-75-1
Oxazepam Hy/ACB/2-Amino-5-chlorobenzophenone	76	69	76	86	90	91	78	12	76	71	–	–	–	27	719-59-5
Oxazolam	53	54	65	71	78	95	77	15	68	66	92	–	–	–	24143-17-7
Oxazolam M → Nordazepam	34	45	57	69	82	83	62	3	55	60	81	–	–	–	1088-11-5
Oxazolam M → Oxazepam	22	35	42	45	82	82	56	0	40	51	–	–	–	–	604-75-1
Oxedrine → Synephrine	–	–	–	11	8	–	25	3	1	18	7	–	–	–	94-07-5
Oxedrine → Synephrine	–	–	–	11	8	–	25	3	1	18	7	–	–	–	94-07-5
Oxeladine	–	–	–	67	19	–	50	51	22	19	48	–	–	–	468-61-1
Oxeladine M	–	–	–	–	–	–	–	–	–	–	28	–	–	–	
Oxetacaine	–	–	–	38	61	–	52	10	7	15	–	–	–	–	126-27-2

Table 8.1. (continued)

Substance Name	TLC System 1	2	3	4	5	6	7	8	9	10	T	P1	P2	B	CAS No.
Oxomemazine	–	–	–	59	24	33	48	18	36	25	45	–	–	–	3689-50-7
Oxprenolol	–	–	–	45	20	78	48	11	11	13	31	–	–	–	6452-71-7
Oxybuprocaine	–	–	–	83	54	–	62	23	41	36	–	–	–	–	99-43-4
Oxycodone	–	–	–	62	30	33	50	25	51	39	62	–	–	–	76-42-6
Oxydemeton	–	–	–	78	74	–	–	–	–	–	–	–	–	–	
Oxydemeton-methyl	–	–	–	–	–	–	–	–	–	–	–	18	0	–	301-12-2
Oxyfedrine	–	–	–	72	37	–	–	–	–	–	–	–	–	–	15687-41-9
Oxymetazoline	–	–	–	34	80	–	9	1	1	1	–	–	–	–	1491-59-4
Oxymorphone	–	–	–	33	27	36	48	10	37	30	–	–	–	–	76-41-5
Oxypendyl	–	–	–	34	31	–	–	–	–	–	–	–	–	–	5585-93-3
Oxypertine	–	–	–	78	74	–	68	4	65	58	84	–	–	–	153-87-7
Oxyphenbutazone	52	62	57	9	90	–	77	0	–	–	16	–	–	–	129-20-4
Oxyphencyclimine	–	–	–	6	2	18	2	1	3	0	–	–	–	–	125-53-1
Oxyphenonium Bromide	–	–	–	1	2	36	3	0	1	0	–	–	–	–	50-10-2
Oxytetracycline	–	–	0	0	8	–	5	–	–	–	–	–	–	–	79-57-2
PCC → Phencyclidine intermediate (1-Piperidino-1-cyclohexanecarbonitrile)	–	–	–	10	80	88	5	62	0	0	–	–	–	–	3867-15-0
PCE → Eticyclidine	–	–	–	79	21	78	–	65	27	37	–	–	–	–	2201-15-2
PCM → 4-(1-Phenylcyclohexyl)morpholine	–	–	–	87	72	73	–	62	74	64	–	–	–	–	2201-40-3
PCNB → Quintozene	–	–	–	–	–	–	–	–	–	–	–	86	95	–	82-68-8
PEMA → Primidone M	–	–	–	51	82	–	–	–	–	–	–	–	–	–	80866-90-6
PHP → Rolicyclidine	–	–	–	79	15	65	–	66	25	26	–	–	–	–	2201-39-0
PMA → 4-Methoxyamphetamine	–	–	–	43	9	74	73	23	77	69	–	–	–	–	23239-32-9
Pancuronium Bromide	–	–	–	0	0	15	1	0	0	0	–	–	–	–	15500-66-0
Papaverine	–	–	–	69	74	74	61	8	65	47	72	–	–	–	58-74-2
Paracetamol	15	32	26	45	77	–	–	0	–	–	70	–	–	–	103-90-2
Paracetamol M/p-Aminophenol	21	40	30	59	75	–	–	3	–	–	–	–	–	–	123-30-8
Paraflutizide	–	–	–	72	94	–	–	–	–	–	–	–	–	–	1580-83-2
Paramethadione	0	60	56	7	–	–	–	–	–	–	–	–	–	–	115-67-3
Paraquat	–	–	–	0	0	0	0	0	0	0	–	–	–	–	4685-14-7
Parathion-ethyl	–	–	–	–	–	–	–	–	–	–	–	41	84	–	56-38-2
Parathion-methyl	–	–	–	–	–	–	–	–	–	–	–	30	73	–	298-00-0
Pargyline	–	–	–	60	77	–	70	–	–	–	–	–	–	–	555-57-7
Pecazine	–	1	–	65	27	–	53	47	44	16	–	–	–	–	60-89-9
Pecazine M	–	–	–	36	14	–	–	–	–	–	–	–	–	–	
Pemoline	–	11	–	36	81	81	60	0	23	40	–	–	–	–	2152-34-3
Pempidine	–	–	–	–	–	–	24	68	3	10	–	–	–	–	79-55-0
Penbutolol	–	0	–	50	22	80	50	13	15	8	40	–	–	–	38363-40-5
Pendimethalin	–	–	–	–	–	–	–	–	–	–	–	60	92	–	40487-42-1
Penfluridol	–	–	–	84	72	89	76	17	60	60	–	–	–	–	26864-56-2
Penicillamine	–	–	–	3	44	–	36	1	3	3	–	–	–	–	52-67-5
Penoxalin → Pendimethalin	–	–	–	–	–	–	–	–	–	–	–	60	92	–	40487-42-1
Pentaerithritol Tetranitrate	–	–	–	72	92	–	–	–	–	–	–	–	–	–	78-11-5
Pentamidine	–	–	0	–	0	–	1	1	0	0	–	–	–	–	100-33-4

Table 8.1. (continued)

Substance Name	TLC System														CAS No.
	1	2	3	4	5	6	7	8	9	10	T	P1	P2	B	
Pentapiperide	–	–	–	–	–	–	1	0	1	0	–	–	–	–	7009-54-3
Pentaquine	–	–	–	–	–	–	33	34	5	3	–	–	–	–	86-78-2
Pentazocine	–	–	–	70	34	72	61	16	12	28	61	–	–	–	359-83-1
Pentazocine M	–	–	–	–	–	–	–	–	–	–	48	–	–	–	
Pentetrazol	–	–	–	60	72	74	72	7	64	63	–	–	–	–	54-95-5
Penthienate Bromide	–	–	–	3	9	–	2	–	–	–	–	–	–	–	60-44-6
Pentifylline	–	–	–	66	72	–	55	6	66	46	72	–	–	–	1028-33-7
Pentobarbital	55	66	59	45	90	–	–	–	–	–	–	–	–	–	76-74-4
Pentolinium	–	–	–	0	0	1	0	0	0	0	–	–	–	–	144-44-5
Pentorex	–	–	–	49	11	–	–	–	–	–	–	–	–	–	434-43-5
Pentoxifylline	–	–	–	55	64	–	–	–	–	–	–	–	–	–	6493-05-6
Pentoxyverine	–	–	–	68	16	49	48	47	22	14	48	–	–	–	77-23-6
Perazine	–	–	–	47	21	23	48	25	37	3	31	–	–	–	84-97-9
Perazine M #1	–	–	–	–	–	–	–	–	–	–	21	–	–	–	
Perazine M #2	–	–	–	–	–	–	–	–	–	–	8	–	–	–	
Perazine M #3	–	–	–	–	–	–	–	–	–	–	3	–	–	–	
Perazine M #4	–	–	–	25	11	–	–	–	–	–	–	–	–	–	
Perhexiline	–	–	–	59	8	–	41	57	8	6	36	–	–	–	6621-47-2
Periciazine	–	–	–	51	46	61	58	4	16	18	43	–	–	–	2622-26-6
Periciazine M	–	–	–	33	35	–	–	–	–	–	–	–	–	–	
Permethrin	–	–	–	–	–	–	–	–	–	–	–	61	94	–	52645-53-1
Perphenazine	–	–	–	42	40	40	55	7	29	9	30	–	–	–	58-39-9
Perphenazine M	–	–	–	23	28	–	–	–	–	–	–	–	–	–	
Pethidine	–	–	–	60	34	40	52	37	34	11	50	–	–	–	57-42-1
Pethidine Intermediate A	15	10	48	65	–	–	–	–	–	–	–	–	–	–	3627-62-1
Pethidine Intermediate C	0	0	0	0	–	–	–	–	–	–	–	–	–	–	3627-48-3
Pethidine M (Nor-)	–	–	–	32	8	63	–	7	8	2	15	–	–	–	77-17-8
Phanquinone	–	–	–	54	86	–	49	3	45	17	–	–	–	–	84-12-8
Phenacemide	22	40	50	65	–	–	–	–	–	–	–	–	–	–	63-98-9
Phenacetin	38	37	52	68	83	–	–	–	–	–	82	–	–	–	62-44-2
Phenadoxone	–	–	–	–	80	83	80	58	77	69	–	–	–	–	467-84-5
Phenaglycodol	–	–	–	71	84	–	–	–	–	–	–	–	–	–	79-93-6
Phenampromide	–	–	–	–	–	–	81	56	67	65	–	–	–	–	129-83-9
Phenanthrene	–	–	–	83	88	92	79	64	84	74	–	–	–	–	85-01-8
Phenazocine	–	–	–	74	50	81	68	16	39	49	–	–	–	–	127-35-5
Phenazone	18	14	50	45	66	66	65	4	–	–	49	–	–	–	60-80-0
Phenazone M	–	–	–	–	–	–	–	–	–	–	33	–	–	–	
Phenazopyridine	–	–	–	70	80	–	59	1	50	53	88	–	–	–	94-78-0
Phencyclidine	–	–	–	84	23	69	59	73	35	66	85	–	–	–	77-10-1
Phencyclidine intermediate/PCC	–	–	–	10	80	88	5	62	0	0	–	–	–	–	3867-15-0
Phendimetrazine	–	–	–	62	49	41	57	36	51	24	55	–	–	–	634-03-7
Phendimetrazine M (Nor-) → Phenmetrazine	–	–	–	46	34	45	50	14	27	14	39	–	–	–	134-49-6
Phenelzine	–	–	–	83	29	82	77	37	12	63	–	–	–	–	51-71-8
Phenethylamine	–	–	–	54	44	–	49	28	28	39	22	–	–	–	64-04-0
Pheneturide	38	53	59	71	–	–	76	–	–	–	–	–	–	–	90-49-3
Phenformin	–	–	–	–	–	–	3	0	0	0	–	–	–	–	114-86-3

Table 8.1. (continued)

Substance Name	TLC System														CAS No.
	1	2	3	4	5	6	7	8	9	10	T	P1	P2	B	
Phenglutarimide	–	–	–	–	–	–	40	17	8	8	–	–	–	–	1156-05-4
Phenindamine	–	–	–	68	41	49	63	45	57	21	–	–	–	–	82-88-2
Phenindione	65	56	70	21	–	–	–	–	–	–	–	–	–	–	83-12-5
Pheniramine	–	–	–	46	14	26	45	35	13	3	–	–	–	–	86-21-5
Phenmedipham	–	–	–	–	–	–	–	–	–	–	–	11	17	–	13684-63-4
Phenmetrazine	–	–	–	46	34	45	50	14	27	14	39	–	–	–	134-49-6
Phenobarbital	47	65	53	28	85	–	–	–	–	–	90	–	–	–	50-06-6
Phenolphthalein	38	59	39	53	86	–	–	–	–	–	90	–	–	–	77-09-8
Phenomorphan	–	–	–	78	51	80	69	18	39	47	–	–	–	–	468-07-5
Phenoperidine	–	–	–	76	70	82	71	26	64	58	–	–	–	–	562-26-5
Phenothiazine	–	–	–	82	82	89	79	11	79	71	–	–	–	–	92-84-2
Phenothiazine M #1	–	–	–	–	–	–	–	–	–	–	5	–	–	–	
Phenothiazine M #2	–	–	–	–	–	–	–	–	–	–	10	–	–	–	
Phenoxybenzamine	–	–	–	87	84	97	73	63	76	68	–	–	–	–	59-96-1
Phenprobamate	47	55	60	73	82	–	75	–	–	–	–	–	–	–	673-31-4
Phenprocoumon	62	58	61	19	93	–	–	–	–	–	29	–	–	–	435-97-2
Phensuximide	71	59	72	77	–	–	75	–	–	–	–	–	–	–	86-34-0
Phentermine	–	–	–	48	11	78	46	26	24	12	34	–	–	–	122-09-8
Phentolamine	–	–	–	33	6	–	32	1	3	2	–	–	–	–	50-60-2
Phenyl Salicylate	–	–	–	–	–	90	–	–	–	–	–	–	–	–	118-55-8
Phenylbutazone	78	68	76	65	87	–	79	–	–	–	–	–	–	–	50-33-9
Phenylbutazone M	–	–	–	–	–	–	–	–	–	–	25	–	–	–	
Phenylbutazone M → Oxyphenbutazone	52	62	57	9	90	–	77	0	–	–	16	–	–	–	129-20-4
Phenylcyclohexyl)morpholine,4- (1-/PCM	–	–	–	87	72	73	–	62	74	64	–	–	–	–	2201-40-3
Phenylcyclohexyl)pyrrolidine,1- (1- → Rolicyclidine	–	–	–	79	15	65	–	66	25	26	–	–	–	–	2201-39-0
Phenylcyclohexylamine,1–	–	–	–	70	23	79	–	41	33	29	–	–	–	–	2201-24-3
Phenyldimethylpyrazolone → Phenazone	18	14	50	45	66	66	65	4	–	–	49	–	–	–	60-80-0
Phenylephrine	–	–	–	12	8	67	33	1	1	0	5	–	–	–	59-42-7
Phenylpropanolamine → Ephedrine M	–	0	–	30	12	75	44	4	4	33	23	–	–	–	492-41-1
Phenyltoloxamine	–	–	–	67	32	–	53	38	48	15	55	–	–	–	92-12-6
Phenytoin	33	55	53	41	86	–	–	–	–	–	90	–	–	–	57-41-0
Phenytoin M (4-Hydroxyphenyl)	–	–	–	26	88	–	–	–	–	–	–	–	–	–	
Pholcodine	–	–	–	25	15	–	36	3	18	2	–	–	–	–	509-67-1
Pholedrine	–	–	–	27	9	–	29	3	3	3	–	–	–	–	370-14-9
Phosalone	–	–	–	99	90	–	–	–	–	–	–	31	67	–	2310-17-0
Phosdrin → Mevinphos	–	–	–	75	86	–	–	–	–	–	–	12	10	–	7786-34-7
Phosphamidon	–	–	–	–	–	–	–	–	–	–	–	7	2	–	13171-21-6
Phoxim	–	–	–	–	–	–	–	–	–	–	–	42	86	–	14816-18-3
Phthalylsulfacetamide	0	0	0	0	–	–	–	–	–	–	–	–	–	–	131-69-1
Phthalylsulfathiazole	0	0	0	0	–	–	–	–	–	–	–	–	–	–	85-73-4
Physostigmine	–	–	–	55	41	38	55	12	36	18	50	–	–	–	57-47-6
Picloram	–	–	–	–	–	–	–	–	–	–	–	0	0	–	1918-02-1

Table 8.1. (continued)

Substance Name	TLC System 1	2	3	4	5	6	7	8	9	10	T	P1	P2	B	CAS No.
Picloxydine	–	–	–	–	–	–	1	1	1	1	–	–	–	–	5636-92-0
Pilocarpine	–	–	–	44	52	45	53	0	32	12	–	–	–	–	92-13-7
Piminodine	–	–	–	88	63	77	67	36	64	59	–	–	–	–	13495-09-5
Pimozide	–	–	–	71	73	82	71	3	60	40	75	–	–	–	2062-78-4
Pinazepam	65	61	73	81	85	92	75	29	72	70	–	–	–	–	52463-83-9
Pinazepam Hy/CPB/5-Chloro-2-(2-propinyl)benzophenone	75	73	80	86	86	91	83	48	77	72	–	–	–	57	
Pindolol	–	–	–	43	18	78	49	2	5	8	32	–	–	–	13523-86-9
Pindone	–	–	–	–	–	–	–	–	–	–	–	17	45	–	83-26-1
Pipamazine	–	–	–	43	41	52	65	0	17	14	–	–	–	–	84-04-8
Pipamperone	–	–	–	43	33	61	56	1	12	8	20	–	–	–	1893-33-0
Pipazetate	–	–	–	48	12	–	47	17	13	6	–	–	–	–	2167-85-3
Pipemidic Acid	0	0	2	0	–	–	–	–	–	–	–	–	–	–	51940-44-4
Piperacetazine	–	–	–	–	–	–	56	6	19	17	–	–	–	–	3819-00-9
Piperazine	–	–	–	–	1	4	5	1	1	0	–	–	–	–	110-85-0
Piperidine	–	–	–	–	1	–	–	–	–	–	–	–	–	–	110-89-4
Piperidino-1-cyclohexanecarbonitrile,1- → Phencyclidine intermediate	–	–	–	10	80	88	5	62	0	0	–	–	–	–	3867-15-0
Piperidolate	–	–	–	82	54	52	69	55	81	55	–	–	–	–	82-98-4
Piperocaine	–	–	–	76	21	56	55	53	37	27	–	–	–	–	136-82-3
Piperonal	–	–	–	–	–	87	–	–	–	–	–	–	–	–	120-57-0
Piperoxan	–	–	–	–	–	–	63	59	67	61	–	–	–	–	59-39-2
Pipobroman	–	–	–	–	–	–	66	2	58	41	–	–	–	–	54-91-1
Pipotiazine	–	–	–	53	40	59	66	3	32	21	–	–	–	–	39860-99-6
Pipoxolan	–	–	–	–	–	–	77	53	68	56	–	–	–	–	23744-24-3
Pipradrol	–	–	–	81	19	79	54	59	38	39	–	–	–	–	467-60-7
Piprinhydrinate	–	–	–	56	19	–	–	–	–	–	39	–	–	–	606-90-6
Piprinhydrinate M	–	–	–	–	–	–	–	–	–	–	28	–	–	–	
Pirenzepine	–	–	–	16	18	–	–	–	–	–	5	–	–	–	28797-61-7
Pirenzepine M (Desamide)	–	–	–	70	81	–	–	4	–	–	–				
Pirenzepine M (N-Desmethyl-)	–	–	–	3	44	–	–	0	–	–	–	–	–	–	
Piribedil	–	–	–	83	72	–	–	–	–	–	–	–	–	–	3605-01-4
Piridoxilate	–	–	–	20	–	–	–	–	–	–	–	–	–	–	
Pirimicarb	–	–	–	–	–	–	–	–	–	–	–	26	17	–	23103-98-2
Pirimiphos-methyl	–	–	–	–	–	–	–	–	–	–	–	50	75	–	29232-93-7
Piritramide	–	–	–	61	73	74	70	1	45	42	–	–	–	–	302-41-0
Piroxicam	51	38	71	17	88	–	–	–	–	–	21	–	–	–	36322-90-4
Piroxicam M (5-Hydroxy-)	8	18	37	4	–	–	–	–	–	–	–	–	–	–	
Pirprofen	–	28	–	5	85	–	–	–	–	–	9	–	–	–	31793-07-4
Pitofenone	–	–	–	73	35	46	66	26	55	38	–	–	–	–	54063-52-4
Pizotifen	–	–	–	64	28	–	48	45	–	–	56	–	–	–	15574-96-6
Polyethyleneglycol	–	–	–	–	–	–	–	–	–	–	10	–	–	–	25322-68-3
Polythiazide	22	60	32	63	–	–	–	–	–	–	–	–	–	–	346-18-9
Practolol	–	–	–	–	–	–	45	0	1	4	–	–	–	–	6673-35-4
Prajmaline	–	–	–	–	–	–	–	–	–	–	48	–	–	–	35080-11-6
Prajmaline tautomeric form	–	–	–	–	–	–	–	–	–	–	39	–	–	–	

Table 8.1. (continued)

Substance Name	TLC System														CAS No.
	1	2	3	4	5	6	7	8	9	10	T	P1	P2	B	
Prajmalium Bitartrate	–	–	–	–	8	–	59	–	–	–	–	–	–	–	2589-47-1
Pramiverine	–	–	–	–	9	–	–	–	–	–	–	–	–	–	14334-40-8
Pramoxine	–	–	–	–	62	60	70	43	55	41	–	–	–	–	140-65-8
Prazepam	64	55	72	81	84	89	65	36	74	63	95	–	–	–	2955-38-6
Prazepam Hy/CCB/5-Chloro-2-cyclopropyl-methylamino-benzophenone	83	74	82	88	92	95	80	58	82	74	–	–	–	68	2897-00-9
Prazepam M (3-Hydroxy-)	55	55	70	72	85	88	71	12	–	52	–	–	–	–	18818-61-6
Prazepam M → Nordazepam	34	45	57	69	82	83	62	3	55	60	81	–	–	–	1088-11-5
Prazepam M → Oxazepam	22	35	42	45	82	82	56	0	40	51	–	–	–	–	604-75-1
Prazosin	–	–	–	59	68	74	60	1	47	49	70	–	–	–	19216-56-9
Pre-Methadone → Methadone Intermediate	–	–	–	–	–	–	63	47	55	43	–	–	–	–	125-79-1
Prenoxdiazine	–	–	–	82	61	–	–	49	–	–	86	–	–	–	982-43-4
Prenylamine	–	–	–	84	43	85	68	55	68	56	89	–	–	–	390-64-7
Pridinol	–	–	–	87	39	–	–	–	–	–	88	–	–	–	511-45-5
Prilocaine	–	–	–	75	62	79	77	29	64	60	–	–	–	–	721-50-6
Primaquine	–	–	–	–	–	–	19	13	5	15	–	–	–	–	90-34-6
Primidone	8	23	28	41	76	–	–	–	–	–	–	–	–	–	125-33-7
Primidone M/PEMA	–	–	–	51	82	–	–	–	–	–	–	–	–	–	80866-90-6
Proadifen	–	–	–	–	–	–	69	60	66	56	–	–	–	–	302-33-0
Probarbital	46	65	56	41	–	–	–	–	–	–	–	–	–	–	76-76-6
Probenecid	13	3	24	5	87	–	–	–	–	–	–	–	–	–	57-66-9
Procainamide	–	–	–	39	17	33	49	1	5	9	20	–	–	–	51-06-9
Procaine	–	–	–	71	36	42	54	5	31	30	64	–	–	–	59-46-1
Procarbazine	–	–	68	80	88	–	49	2	10	4	–	–	–	–	366-70-1
Prochlorperazine	–	–	–	55	26	26	49	34	37	7	–	–	–	–	58-38-8
Proclonol	–	–	–	86	93	97	83	44	78	75	–	–	–	–	14088-71-2
Procyclidine	–	–	–	74	20	68	48	62	31	23	53	–	–	–	77-37-2
Procyclidine M	–	–	–	–	–	–	–	–	–	–	16	–	–	–	
Profadol	–	–	–	–	–	–	42	8	6	8	–	–	–	–	428-37-5
Profenamine	–	–	–	83	31	55	67	64	47	66	–	–	–	–	522-00-9
Progesterone	–	–	–	–	–	89	–	–	–	–	–	–	–	–	57-83-0
Proglumetacin	–	–	–	–	–	–	–	–	–	–	53	–	–	–	57132-53-3
Proguanil	–	–	–	18	7	79	3	0	1	1	–	–	–	–	500-92-5
Proheptazine	–	–	–	–	–	–	46	45	30	10	–	–	–	–	77-14-5
Prolintane	–	–	–	79	22	–	50	67	32	25	64	–	–	–	493-92-5
Promazine	–	–	–	62	18	35	44	38	30	11	41	–	–	–	58-40-2
Promazine Artifact	–	–	–	–	–	–	–	–	–	–	4	–	–	–	
Promazine M	–	–	–	28	9	–	–	–	–	–	–	–	–	–	
Promethazine	–	–	–	65	30	44	50	36	35	17	54	–	–	–	60-87-7
Promethazine M #1	–	–	–	40	18	–	–	–	–	–	18	–	–	–	
Promethazine M #2	–	–	–	40	18	–	–	–	–	–	11	–	–	–	
Prometryn	–	–	–	–	–	–	–	–	–	–	–	32	31	–	7287-19-6
Propafenone	–	–	–	48	17	–	–	3	–	–	36	–	–	–	54063-53-5
Propafenone M (5-Hydroxy-)	–	–	–	39	18	–	–	0	–	–	–	–	–	–	
Propafenone M (N-Desmethyl)	–	–	–	34	12	–	–	1	–	–	–	–	–	–	

Table 8.1. (continued)

Substance Name	1	2	3	4	5	6	7	8	9	10	T	P1	P2	B	CAS No.
Propallylonal → Ibomal	50	66	56	32	91	–	–	–	–	–	–	–	–	–	545-93-7
Propamidine	–	–	–	–	–	–	1	1	1	1	–	–	–	–	104-32-5
Propanidid	–	–	–	72	80	–	66	20	70	55	–	–	–	–	1421-14-3
Propantheline Bromide	–	–	–	4	3	31	4	0	4	0	–	–	–	–	50-34-0
Propantheline M/Xanthanoic Acid	–	8	–	5	87	–	–	–	–	–	–	–	–		
Properidine	–	–	–	–	–	–	56	39	44	16	–	–	–	–	561-76-2
Propham	–	–	–	–	–	–	–	–	–	–	–	39	57	–	122-42-9
Propiconazole	–	–	–	–	–	–	–	–	–	–	–	18	11	–	60207-90-1
Propiomazine	–	–	–	68	30	52	55	34	42	26	–	–	–	–	362-29-8
Propiomazine M	–	–	–	40	16	–	–	–	–	–	–	–	–	–	
Propoxur	–	–	–	–	–	–	–	–	–	–	–	20	21	–	114-26-1
Propoxycaine	–	–	–	–	–	–	58	3	33	28	–	–	–	–	86-43-1
Propoxyphene/Dextropropoxyphene/Levopropoxyphene	–	–	–	82	50	63	68	58	55	54	82	–	–	–	77-50-9
Propoxyphene M (Nor-)	–	–	–	–	–	–	–	–	–	–	20	–	–	–	66796-40-5
Propoxyphene M (Nor-)	–	–	–	–	–	–	–	–	–	–	36	–	–	–	66796-40-5
Propranolol	–	–	–	49	21	79	50	6	10	7	36	–	–	–	525-66-6
Propranolol M	–	–	–	–	–	–	–	–	–	–	75	–	–	–	
Propyl p-Hydroxybenzoate	65	67	56	50	90	–	–	–	–	–	–	–	–	–	94-13-3
Propylhexedrine	–	–	–	–	–	–	26	34	–	–	14	–	–	–	101-40-6
Propyphenazone	61	49	65	74	81	–	71	32	–	–	86	–	–	–	479-92-5
Propyphenazone M (Hydroxy-propyl-)	–	–	–	–	–	–	–	–	–	–	40	–	–	–	
Propyphenazone M (Nor-)	–	–	–	–	–	–	–	–	–	–	16	–	–	–	50993-68-5
Propyphylline → Diprophylline	–	–	–	25	70	59	48	0	12	12	–	–	–	–	479-18-5
Propyzamide	–	–	–	–	–	–	–	–	–	–	–	33	52	–	23950-58-5
Proscillaridine	–	–	–	1	80	–	–	–	–	–	–	–	–	–	466-06-8
Protheobromine	–	–	–	0	60	–	–	–	–	–	–	–	–	–	50-39-5
Prothipendyl	–	–	–	59	15	29	47	43	23	9	43	–	–	–	303-69-5
Prothipendyl M	–	–	–	35	10	–	–	–	–	–	–	–	–	–	
Protionamide	–	–	–	–	77	–	66	1	38	57	–	–	–	–	14222-60-7
Protokylol	–	–	–	–	–	–	65	1	3	6	–	–	–	–	136-70-9
Protriptyline	–	–	–	38	6	69	19	18	7	2	–	–	–	–	438-60-8
Proxymetacaine	–	–	–	–	–	–	62	26	41	35	–	–	–	–	499-67-2
Proxyphylline	–	–	–	49	71	–	58	2	33	29	52	–	–	–	603-00-9
Prozapine	–	–	–	82	27	77	55	66	44	36	–	–	–	–	3426-08-2
Pseudoephedrine	–	–	–	17	99	–	33	54	4	63	14	–	–	–	90-82-4
Pseudoephedrine M (Nor-)	–	–	–	29	12	–	42	25	5	46	–	–	–	–	36393-56-3
Pseudomorphine	–	–	–	–	–	–	46	0	0	0	–	–	–	–	125-24-6
Psilocin	–	–	–	47	14	48	39	5	9	9	33	–	–	–	520-53-6
Psilocin-(eth)/CZ-74	–	–	–	65	15	67	–	9	15	25	–	–	–	–	22204-89-3
Psilocybine	–	–	–	0	80	1	5	0	0	0	–	–	–	–	520-52-5
Pyrazinamide	–	–	44	54	71	70	63	3	42	46	–	–	–	–	98-96-4
Pyrazobutamine	–	–	–	–	–	–	–	–	–	–	61	–	–	–	
Pyrazophos	–	–	–	–	–	–	–	–	–	–	–	32	47	–	13457-18-6
Pyricarbate	–	–	–	72	77	–	–	–	–	–	–	–	–	–	1882-26-4

Table 8.1. (continued)

Substance Name	TLC System														CAS No.
	1	2	3	4	5	6	7	8	9	10	T	P1	P2	B	
Pyridate	–	–	–	–	–	–	–	–	–	–	–	47	64	–	55512-33-9
Pyridoxine Hydrochloride	–	–	–	15	75	67	59	0	8	5	–	–	–	–	58-56-0
Pyrilamine → Mepyramine	–	–	–	58	22	33	51	39	25	14	49	–	–	–	91-84-9
Pyrimethamine	–	–	–	58	66	–	61	2	31	21	–	–	–	–	58-14-0
Pyrithyldione	–	–	–	71	80	–	–	–	–	–	–	–	–	–	77-04-3
Pyritinol	–	–	–	–	76	–	–	–	–	–	–	–	–	–	10049-83-9
Pyrrobutamine	–	–	–	71	25	66	54	54	37	18	61	–	–	–	91-82-7
Quazepam	78	71	78	83	87	96	74	27	75	76	–	–	–	–	36735-22-5
Quazepam Hy/ CFTB/ 5-Chloro-2′-fluoro-2-(2,2,2-trifluoro-ethyl)aminobenzophenone	74	73	80	85	77	93	79	44	77	73	–	–	–	74	
Quazepam M (2-Oxo-)	59	57	70	80	87	90	78	16	89	71	–	–	–	–	
Quazepam M (3-Hydroxy-2-oxo-)	42	52	58	69	88	90	71	2	55	59	–	–	–	–	
Quazepam M (3-Hydroxy-N-de-alkyl-2-oxo-)	15	28	35	49	88	89	67	0	30	38	–	–	–	–	
Quazepam M (N-Dealkyl-2-oxo-)	34	45	60	72	83	88	75	2	56	58	–	–	–	–	2886-65-9
Quinethazone	4	21	15	40	–	–	75	–	–	–	–	–	–	–	73-49-4
Quinidine	–	–	–	49	30	63	51	4	12	6	25	–	–	–	56-54-2
Quinine	–	–	–	45	26	65	51	2	11	4	25	–	–	–	130-95-0
Quinisocaine	–	–	–	–	–	–	61	55	46	28	–	–	–	–	86-80-6
Quinomethionate	–	–	–	–	–	–	–	–	–	–	–	70	82	–	2439-01-2
Quintozene	–	–	–	–	–	–	–	–	–	–	–	86	95	–	82-68-8
RO 15-1788 → Flumazenil	30	14	61	61	76	72	71	3	63	44	–	–	–	–	78755-81-4
RO 15-3890 → Flumazenil M	–	0	–	0	44	–	–	0	–	–	–	–	–	–	84378-44-9
Racemethorphan → Dextro-methorphan	–	–	–	47	10	42	33	42	18	6	26	–	–	–	510-53-2
Racemoramide → Dextromor-amide	–	–	–	79	72	78	73	42	71	60	86	–	–	–	545-59-5
Racemorphan	–	–	–	–	–	–	34	14	9	2	–	–	–	–	297-90-5
Ranitidine	–	–	–	–	–	–	–	–	–	–	5	–	–	–	66357-35-5
Raubasine → Ajmalicine	–	–	–	81	81	–	–	–	–	–	–	–	–	–	483-04-5
Reposal	49	64	56	36	86	–	–	1	–	–	–	–	–	–	3625-25-0
Rescinnamine	–	–	–	81	77	79	73	1	75	64	–	–	–	–	24815-24-5
Reserpine	–	–	–	77	76	80	69	2	74	63	–	–	–	–	50-55-5
Resorcinol	–	–	–	–	–	91	–	–	–	–	–	–	–	–	108-46-3
Rhein	–	–	–	0	–	–	–	–	–	–	–	–	–	–	478-43-3
Rifampin	–	–	–	–	83	–	–	–	–	–	–	–	–	–	13292-46-1
Rimantadine	–	–	–	–	–	–	37	50	11	35	–	–	–	–	13392-28-4
Rolicyclidine	–	–	–	79	15	65	–	66	25	26	–	–	–	–	2201-39-0
Rolicyprine	–	–	–	–	–	–	59	0	21	17	–	–	–	–	2829-19-8
Rolitetracycline	–	–	–	–	0	–	–	–	–	–	–	–	–	–	751-97-3
Rotenone → Ibogaine	–	–	–	86	39	72	65	28	50	62	–	–	–	–	83-74-9
S 421	–	–	–	–	–	–	–	–	–	–	–	78	97	–	127-90-2
STP → 2,5-Dimethoxy-4-methylamphetamine	–	–	–	41	9	76	51	15	17	16	–	–	–	–	15588-95-1
Saccharin	1	1	4	7	87	–	–	–	–	–	–	–	–	–	81-07-2

Table 8.1. (continued)

Substance Name	TLC System														CAS No.
	1	2	3	4	5	6	7	8	9	10	T	P1	P2	B	
Salbutamol	–	0	–	20	16	74	46	1	1	4	11	–	–	–	18559-94-9
Salicylamide	38	55	43	50	83	84	–	0	–	–	85	–	–	–	65-45-2
Salicylic Acid	7	1	24	10	86	–	–	–	–	–	–	–	–	–	69-72-7
Salinazid	–	–	–	–	–	–	84	1	20	25	–	–	–	–	495-84-1
Salverine	–	–	–	79	33	–	–	–	–	–	–	–	–	–	6376-26-7
Santonin	64	50	69	75	–	79	–	–	–	–	–	–	–	–	481-06-1
Sarcolysis	–	–	–	–	52	–	–	–	–	–	–	–	–	–	531-76-0
Scopolamine	–	–	–	48	49	47	55	6	37	18	45	–	–	–	51-34-3
Secbutabarbital	50	64	57	48	88	–	–	–	–	–	–	–	–	–	125-40-6
Secobarbital	55	68	62	45	88	–	–	–	–	–	–	–	–	–	76-73-3
Selegiline	–	–	–	–	–	–	–	–	–	–	92	–	–	–	14611-51-9
Sethoxydim	–	–	–	–	–	–	–	–	–	–	–	50	48	–	74051-80-2
Sigmodal	59	71	63	50	86	–	–	–	–	–	–	–	–	–	1216-40-6
Simazine	–	–	–	–	–	–	–	–	–	–	–	22	18	–	122-34-9
Sisomicin	–	–	–	–	0	–	–	–	–	–	–	–	–	–	32385-11-8
Sotalol	–	–	–	30	19	–	53	1	3	5	18	–	–	–	3930-20-9
Sparteine	–	–	–	40	4	10	5	67	3	5	8	–	–	–	90-39-1
Spiperone	–	–	–	61	51	72	74	4	49	39	–	–	–	–	749-02-0
Spiramycin	–	–	–	–	46	–	–	–	–	–	–	–	–	–	8025-81-8
Spironolactone	66	51	75	78	84	–	–	–	–	–	88	–	–	–	52-01-7
Spironolactone M #1	–	–	–	–	–	–	–	–	–	–	30	–	–	–	
Spironolactone M #2	–	–	–	–	–	–	–	–	–	–	18	–	–	–	
Stilbamidine	–	–	–	–	–	–	1	0	0	1	–	–	–	–	122-06-5
Streptomycin	–	–	–	–	0	–	–	–	–	–	–	–	–	–	57-92-1
Strophanthin	–	–	–	0	81	–	–	0	–	–	–	–	–	–	11005-63-3
Strychnine	–	–	–	32	8	11	26	8	19	2	12	–	–	–	57-24-9
Strychnine M (N(6)-Oxide)	–	–	–	–	7	–	–	–	–	–	–	–	–	–	7248-28-4
Styramate	13	39	30	53	–	–	62	1	20	53	–	–	–	–	94-35-9
Succinylsulfathiazole	0	0	2	0	–	–	–	–	–	–	–	–	–	–	116-43-8
Sulazepam	–	–	–	84	89	–	–	–	–	–	–	–	–	–	2898-13-7
Sulfacarbamid			14		81		76	0	1	8	–	–	–	–	547-44-4
Sulfacetamid	17	42	28	4	87	–	70	–	–	–	–	–	–	–	144-80-9
Sulfadiazine	22	39	38	4	81	–	64	–	–	–	–	–	–	–	68-35-9
Sulfadicramide	–	–	–	–	85	–	–	–	–	–	–	–	–	–	115-68-4
Sulfadimethoxine	31	51	48	10	86	–	65	–	–	–	–	–	–	–	122-11-2
Sulfadimidine	23	45	44	13	79	–	62	–	–	–	–	–	–	–	57-68-1
Sulfadoxine	37	51	55	79	78	–	67	63	–	–	–	–	–	–	2447-57-6
Sulfaethidole	14	35	34	8	–	–	67	–	–	–	–	–	–	–	94-19-9
Sulfafurazole	25	52	33	6	81	–	65	–	–	–	–	–	–	–	127-69-5
Sulfaguanidine	1	6	7	25	75	–	65	–	–	–	–	–	–	–	57-67-0
Sulfaguanole	–	–	40	36	81	–	–	–	–	–	–	–	–	–	27031-08-9
Sulfalene	40	50	54	8	–	–	67	–	–	–	–	–	–	–	152-47-6
Sulfaloxic Acid	–	–	–	–	85	–	–	–	–	–	–	–	–	–	14376-16-0
Sulfamerazine	23	41	42	8	80	–	65	–	–	–	–	–	–	–	127-79-7
Sulfamethizole	12	23	27	4	83	–	65	–	–	–	–	–	–	–	144-82-1
Sulfamethoxazole	26	54	41	5	79	–	65	–	–	–	–	–	–	–	723-46-6
Sulfamethoxypyridazine	19	39	43	9	83	–	65	–	–	–	–	–	–	–	80-35-3

Table 8.1. (continued)

Substance Name	TLC System 1	2	3	4	5	6	7	8	9	10	T	P1	P2	B	CAS No.
Sulfametoxydiazine	24	43	47	12	82	–	60	–	–	–	–	–	–	–	651-06-9
Sulfanilamide	13	46	22	51	83	–	67	–	–	–	–	–	–	–	63-74-1
Sulfaperin	–	–	–	–	80	–	–	–	–	–	–	–	–	–	599-88-2
Sulfaphenazole	29	51	43	9	89	–	69	–	–	–	–	–	–	–	526-08-9
Sulfaproxyline	–	–	–	–	89	–	–	–	–	–	–	–	–	–	116-42-7
Sulfapyridine	16	42	34	24	–	–	67	–	–	–	–	–	–	–	144-83-2
Sulfaquinoxaline	–	–	–	–	–	–	–	–	–	–	–	2	2	–	59-40-5
Sulfasalazine	0	0	2	0	85	–	66	–	–	–	–	–	–	–	599-79-1
Sulfasomizole	23	49	33	6	87	–	–	2	–	–	–	–	–	–	632-00-8
Sulfathiazole	9	20	27	8	79	–	66	–	–	–	–	–	–	–	72-14-0
Sulfathiourea	–	–	–	–	87	–	–	–	–	–	–	–	–	–	515-49-1
Sulfinpyrazone	4	4	30	16	–	–	–	–	–	–	–	–	–	–	57-96-5
Sulfisomidine	5	16	27	5	80	–	64	–	–	–	–	–	–	–	515-64-0
Sulforidazine	–	0	–	54	19	–	–	–	–	–	36	–	–	–	14759-06-9
Sulfotep	–	–	–	–	–	–	–	–	–	–	–	50	84	–	3689-24-5
Sulindac	14	10	34	4	87	–	–	–	–	–	–	–	–	–	38194-50-2
Suloctidil	–	–	–	78	31	–	–	–	–	–	–	–	–	–	54063-56-8
Sulpiride	–	–	–	34	17	–	38	0	–	–	18	–	–	–	15676-16-1
Sultiame	23	43	42	57	81	–	–	–	–	–	–	–	–	–	61-56-3
Suxamethonium Chloride	–	–	–	0	0	0	1	0	0	0	–	–	–	–	71-27-2
Synephrine	–	–	–	11	8	–	25	3	1	18	7	–	–	–	94-07-5
TCB → Halazepam Hy	81	74	81	89	94	95	82	49	81	75	–	–	–	74	
TCP → Tenocyclidine	–	–	–	86	38	70	–	73	54	68	–	–	–	–	21500-98-1
Tacrine	–	–	–	–	–	–	43	5	4	10	–	–	–	–	321-64-2
Talbutal	53	67	60	46	92	–	–	–	–	–	–	–	–	–	115-44-6
Temazepam	51	47	65	62	82	82	53	8	59	53	81	–	–	–	846-50-4
Temephos	–	–	–	–	–	–	–	–	–	–	–	23	75	–	3383-96-8
Tenamfetamin → 3,4-Methylenedioxyamphetamine	–	–	–	42	10	76	39	18	12	17	30	–	–	–	4764-17-4
Tenocyclidine	–	–	–	86	38	70	–	73	54	68	–	–	–	–	21500-98-1
Terbufos	–	–	–	–	–	–	–	–	–	–	–	63	90	–	13071-79-9
Terbutaline	–	–	–	21	18	77	47	1	1	5	14	–	–	–	23031-25-6
Terbutryn	–	–	–	–	–	–	–	–	–	–	–	32	32	–	886-50-0
Terbutylazine	–	–	–	–	–	–	–	–	–	–	–	32	27	–	5915-41-3
Terfenadine	–	–	–	74	45	–	–	13	–	–	75	–	–	–	50679-08-8
Terodiline	–	–	–	80	18	–	52	62	19	26	–	–	–	–	15793-40-5
Terpin hydrate	–	–	–	–	–	–	–	–	–	–	77	–	–	–	2541-01-6
Terpin hydrate M	–	–	–	–	–	–	–	–	–	–	55	–	–	–	
Tertatolol	–	1	–	43	17	–	–	13	–	–	–	–	–	–	34784-64-0
Testosterone	–	–	–	–	–	88	–	–	–	–	–	–	–	–	58-22-0
Tetrabarbital	52	66	57	48	–	–	–	–	–	–	–	–	–	–	76-23-3
Tetrabenazine	–	–	–	79	80	–	69	41	78	67	–	–	–	–	58-46-8
Tetracaine	–	–	–	64	43	39	57	15	32	16	–	–	–	–	94-24-6
Tetrachlorvinphos	–	–	–	–	–	–	–	–	–	–	–	25	29	–	22248-79-9
Tetracycline	–	–	–	–	88	–	5	–	–	–	–	–	–	–	60-54-8
Tetramisole	–	–	–	66	51	53	69	18	54	46	–	–	–	–	5036-02-2
Tetrazepam	57	49	67	79	84	89	78	32	69	66	91	–	–	–	10379-14-3

Table 8.1. (continued)

Substance Name	TLC System 1	2	3	4	5	6	7	8	9	10	T	P1	P2	B	CAS No.
Tetrazepam M (Nor-)	35	45	56	73	85	92	77	5	58	60	–	–	–	–	10379-11-0
Tetryzoline	–	–	–	26	5	60	13	7	2	2	–	–	–	–	84-22-0
Thebacon	–	–	–	49	24	25	45	20	34	11	–	–	–	–	466-90-0
Thebaine	–	–	–	45	23	32	45	24	37	5	–	–	–	–	115-37-7
Thenalidine	–	–	–	52	20	–	50	38	44	12	–	–	–	–	86-12-4
Thenyldiamine	–	–	–	–	19	36	53	42	25	12	–	–	–	–	91-79-2
Theobromine	–	4	–	34	59	54	53	1	31	21	–	–	–	–	83-67-0
Theodrenaline	–	–	–	–	38	–	–	–	–	–	–	–	–	–	13460-98-5
Theophylline	–	9	–	11	74	66	75	1	30	11	40	–	–	–	58-55-9
Thialbarbital	77	72	69	43	–	–	–	–	–	–	–	–	–	–	467-36-7
Thiamazole	–	–	–	41	80	–	62	1	52	59	–	–	–	–	60-56-0
Thiambutosine	–	–	–	–	–	–	76	6	77	69	–	–	–	–	500-89-0
Thiamine Hydrochloride	–	–	–	1	2	18	1	0	0	0	–	–	–	–	67-03-8
Thiazafluron	–	–	–	–	–	–	–	–	–	–	–	10	4	–	25366-23-8
Thiazinamium Metilsulfate	–	–	–	0	1	25	2	0	0	0	–	–	–	–	58-34-4
Thibenzazoline	–	–	–	65	87	–	–	–	–	–	–	–	–	–	6028-35-9
Thienyl)cyclohexyl]piperidine,1-[1-(2- → Tenocyclidine	–	–	–	86	38	70	–	73	54	68	–	–	–	–	21500-98-1
Thiethylperazine	–	–	–	53	27	–	51	30	41	8	32	–	–	–	1420-55-9
Thiethylperazine M	–	–	–	30	13	–	–	–	–	–	–	–	–	–	
Thiobarbital	0	72	0	34	90	–	–	–	–	–	–	–	–	–	77-32-7
Thiobarbituric Acid	–	–	–	0	–	–	–	–	–	–	–	–	–	–	504-17-6
Thioctic Acid	–	–	–	3	83	–	–	–	–	–	–	–	–	–	62-46-4
Thiocyclam	–	–	–	–	–	–	–	–	–	–	–	26	11	–	31895-21-3
Thiofanox	–	–	–	–	–	–	–	–	–	–	–	20	17	–	39196-18-4
Thiopental	77	74	68	49	–	–	–	–	–	–	–	–	–	–	76-75-5
Thiopropazate	–	–	–	74	62	59	61	35	53	42	–	–	–	–	84-06-0
Thiopropazate M	–	–	–	46	48	–	–	–	–	–	–	–	–	–	
Thioproperazine	–	–	–	43	22	22	46	7	34	6	–	–	–	–	316-81-4
Thioproperazine M	–	–	–	26	13	–	–	–	–	–	–	–	–	–	
Thioridazine	–	–	–	67	20	55	48	42	30	13	45	–	–	–	50-52-2
Thioridazine M	–	–	–	38	10	–	–	–	–	–	22	–	–	–	
Thiotepa	–	–	–	–	75	–	–	–	–	–	–	–	–	–	52-24-4
Thiram	–	–	–	–	–	–	–	–	–	–	–	19	51	–	137-26-8
Thonzylamine	–	–	–	65	22	31	55	38	28	14	–	–	–	–	91-85-0
Thurfyl Nicotinate	–	–	–	–	–	–	60	31	63	50	–	–	–	–	70-19-9
Thymol	73	68	65	81	–	92	–	–	–	–	–	–	–	–	89-83-8
Thymol M	–	–	–	–	–	–	–	–	–	–	15	–	–	–	
Tiabendazole	–	–	–	–	75	–	67	7	54	53	–	7	3	–	148-79-8
Tiapride	–	–	–	40	16	–	–	1	–	–	21	–	–	–	51012-32-9
Tiaprofenic Acid	–	5	–	4	86	–	–	–	–	–	–	–	–	–	33005-95-7
Tibexin → Prenoxdiazine	–	–	–	82	61	–	–	49	–	–	86	–	–	–	982-43-4
Tiemonium Iodide	–	–	–	3	2	–	–	–	–	–	–	–	–	–	144-12-7
Tigloidine	–	–	–	–	–	–	42	39	21	7	–	–	–	–	495-83-0
Tiletamine	–	–	–	–	–	–	67	52	68	62	–	–	–	–	14176-49-9
Tilidine	–	–	–	84	61	–	–	–	–	–	93	–	–	–	20380-58-9
Timolol	–	–	–	50	20	75	52	6	11	9	30	–	–	–	26839-75-8

Table 8.1. (continued)

Substance Name	TLC System 1	2	3	4	5	6	7	8	9	10	T	P1	P2	B	CAS No.
Tiocarlide	–	–	–	–	–	–	80	7	78	72	–	–	–	–	910-86-1
Tiotixene	–	–	–	44	26	24	49	10	40	7	27	–	–	–	5591-45-7
Tiotixene M #1	–	–	–	36	12	–	–	–	–	–	25	–	–	–	
Tiotixene M #2	–	–	–	36	12	–	–	–	–	–	10	–	–	–	
Tizanidine	–	–	–	56	51	–	–	–	–	–	–	–	–	–	51322-75-9
Tobramycin	–	–	–	–	0	–	–	–	–	–	–	–	–	–	32986-56-4
Tocainide	–	–	–	44	42	74	60	2	23	–	39	–	–	–	41708-72-9
Tocainide M	–	–	–	–	–	–	–	–	–	–	81	–	–	–	
Tofenacin	–	–	–	48	14	–	45	26	21	7	–	–	–	–	15301-93-6
Tofisopam	–	–	–	72	77	–	–	–	–	–	–	–	–	–	22345-47-7
Tolazamide	52	50	66	7	86	–	–	0	–	–	–	–	–	–	1156-19-0
Tolazoline	–	–	–	25	3	55	13	2	2	2	–	–	–	–	59-98-3
Tolbutamide	51	55	62	12	88	88	76	–	–	–	19	–	–	–	64-77-7
Tolclofos-methyl	–	–	–	–	–	–	–	–	–	–	–	52	89	–	57018-04-9
Toliprolol	–	0	–	44	21	–	–	12	–	–	33	–	–	–	2933-94-0
Tolmetin	13	20	30	5	85	–	–	–	–	–	–	–	–	–	26171-23-3
Tolpropamine	–	–	–	68	26	–	51	52	32	15	50	–	–	–	5632-44-0
Tolycaine	–	–	–	–	–	–	70	39	73	66	–	–	–	–	3686-58-6
Tolylfluanid	–	–	–	82	41	–	–	–	–	–	–	33	67	–	731-27-1
Tozalinone	–	–	–	–	–	–	57	2	54	34	–	–	–	–	655-05-0
Tramadol	–	–	–	78	30	–	–	–	–	–	60	–	–	–	27203-92-5
Tramadol M #1	–	–	–	–	–	–	–	–	–	–	44	–	–	–	
Tramadol M #2	–	–	–	–	–	–	–	–	–	–	30	–	–	–	
Tramadol M #3	–	–	–	–	–	–	–	–	–	–	18	–	–	–	
Tramazoline	–	–	–	–	4	–	6	4	2	2	–	–	–	–	1082-57-1
Tranylcypromine	–	–	–	58	41	67	54	39	33	48	66	–	–	–	155-09-9
Trazodone	–	–	–	66	64	61	63	10	58	37	60	–	–	–	19794-93-5
Trazodone M	–	–	–	–	–	–	–	–	–	–	27	–	–	–	
Triadimefon	–	–	–	–	–	–	–	–	–	–	–	22	23	–	43121-43-3
Triadimenol	–	–	–	–	–	–	–	–	–	–	–	11	3	–	55219-65-3
Triallate	–	–	–	–	–	–	–	–	–	–	–	82	90	–	2303-17-5
Triamterene	–	–	13	30	50	65	51	1	8	4	31	–	–	–	396-01-0
Triazolam	5	2	41	44	68	65	60	1	40	16	39	–	–	–	28911-01-5
Triazolam M (α-Hydroxy-)	4	3	39	42	75	74	71	4	49	21	–	–	–	–	37115-45-0
Triazolam M (4-Hydroxy-)	3	1	29	24	75	71	63	0	29	8	–	–	–	–	65686-11-5
Triazophos	–	–	–	–	–	–	–	–	–	–	–	21	38	–	24017-47-8
Tribenoside	–	–	–	86	94	–	–	–	–	–	–	–	–	–	10310-32-4
Tribromoethanol	64	67	59	75	–	–	–	–	–	–	–	–	–	–	75-80-9
Trichlorfon	–	–	–	61	90	–	–	–	–	–	–	4	2	–	52-68-6
Trichlormethiazide	15	60	23	14	88	–	80	–	–	–	–	–	–	–	133-67-5
Trichlorphos → Trichlorfon	–	–	–	61	90	–	–	–	–	–	–	4	2	–	52-68-6
Triclocarban	60	59	63	75	–	–	–	–	–	–	–	–	–	–	101-20-2
Trietazine	–	–	–	–	–	–	–	–	–	–	–	43	42	–	1912-26-1
Trifenmorph	–	–	–	–	–	–	–	–	–	–	–	58	73	–	1420-06-0
Triflubazam	–	–	–	77	87	–	–	–	–	–	–	–	–	–	22365-40-8
Trifluomeprazine	–	–	–	–	–	–	65	60	58	52	–	–	–	–	2622-37-9
Trifluoperazine	–	–	–	55	30	29	53	33	30	8	39	–	–	–	117-89-5

Table 8.1. (continued)

Substance Name	TLC System														CAS No.
	1	2	3	4	5	6	7	8	9	10	T	P1	P2	B	
Trifluoperazine M	–	–	–	36	21	–	–	–	–	–	–	–	–	–	
Trifluperidol	2	5	26	77	55	76	73	13	–	–	70	–	–	–	749-13-3
Triflupromazine	–	–	–	75	32	49	54	47	35	22	63	–	–	–	146-54-3
Triflupromazine M	–	–	–	–	19	–	–	–	–	–	–	–	–	–	
Trifluralin	–	–	–	–	–	–	–	–	–	–	–	72	97	–	1582-09-8
Triforine	–	–	–	–	–	–	–	–	–	–	–	19	2	–	26644-46-2
Trihexyphenidyl	–	–	–	83	43	75	68	66	61	59	–	–	–	–	144-11-6
Trihexyphenidyl M #1	–	–	–	–	–	–	–	–	–	–	39	–	–	–	
Trihexyphenidyl M #2	–	–	–	–	–	–	–	–	–	–	18	–	–	–	
Trimeperidine	–	–	–	–	–	–	58	41	41	17	–	–	–	–	64-39-1
Trimeprazine	–	–	–	77	32	46	58	54	39	31	73	–	–	–	84-96-8
Trimeprazine M	–	–	–	50	19	–	–	–	–	–	–	–	–	–	
Trimetazidine	–	–	–	–	–	–	22	5	4	1	–	–	–	–	5011-34-7
Trimethobenzamide	–	–	–	47	24	–	–	2	–	–	32	–	–	–	138-56-7
Trimethoprim	–	–	20	45	45	59	55	0	22	12	40	–	–	–	738-70-5
Trimethoxyamphetamine,2,4,6–	–	–	–	29	7	70	33	6	11	12	–	–	–	–	1082-88-8
Trimetozine	–	–	–	68	80	–	61	11	72	52	–	–	–	–	635-41-6
Trimipramine	–	–	–	80	36	56	59	62	54	37	78	–	–	–	739-71-9
Tripelennamine	–	–	–	68	22	34	55	44	27	15	51	–	–	–	91-81-6
Triprolidine	–	–	–	55	19	30	51	11	20	6	–	–	–	–	486-12-4
Tritoqualine	–	–	–	86	87	–	–	–	–	–	–	–	–	–	14504-73-5
Trofosfamide	–	–	–	–	80	–	–	–	–	–	–	–	–	–	22089-22-1
Tropacocaine	–	–	–	50	13	28	42	35	22	6	–	–	–	–	537-26-8
Tropicamide	–	–	–	50	0	–	–	–	–	–	–	–	–	–	1508-75-4
Tropine	–	–	–	8	1	33	–	4	–	–	–	–	–	–	120-29-6
Trospium Chloride	–	–	–	11	85	–	–	–	–	–	–	–	–	–	10405-02-4
Tryptamine	–	–	–	26	6	–	23	1	2	11	–	–	–	–	61-54-1
Tuaminoheptane	–	–	–	–	–	–	33	1	7	24	–	–	–	–	123-82-0
Tubocurarine	–	–	–	0	0	11	0	0	0	0	–	–	–	–	57-95-4
Tybamate	35	68	–	65	–	–	77	–	–	–	–	–	–	–	4268-36-4
Undecylenic Acid	–	–	–	–	67	–	–	–	–	–	–	–	–	–	112-38-9
Valnoctamine	–	–	–	62	85	–	–	1	–	–	–	–	–	–	4171-13-5
Valproic Acid	0	52	0	0	–	–	–	–	–	–	–	–	–	–	99-66-1
Verapamil	–	–	–	73	43	61	59	23	70	42	62	–	–	–	52-53-9
Verapamil M #1	–	–	–	–	–	–	–	–	–	–	42	–	–	–	
Verapamil M #2	–	–	–	–	–	–	–	–	–	–	15	–	–	–	
Veratrine	–	–	–	71	17	–	59	4	35	17	–	–	–	–	8051-02-3
Viloxazine	–	–	–	36	25	–	42	6	23	6	29	–	–	–	46817-91-8
Viloxazine M	–	–	–	–	–	–	–	–	–	–	10	–	–	–	
Viminol	–	–	–	–	79	–	–	–	–	–	–	–	–	–	21363-18-8
Vinbarbital	50	65	57	34	89	–	–	–	–	–	–	–	–	–	125-42-8
Vinblastine	–	–	–	57	46	–	60	1	60	29	–	–	–	–	865-21-4
Vincamine	–	–	–	74	56	–	–	–	–	–	–	–	–	–	1617-90-9
Vinclozolin	–	–	–	31	77	–	–	–	–	–	–	42	76	–	50471-44-8
Vinylbital	38	64	66	39	89	–	–	–	–	–	–	–	–	–	2430-49-1
Viquidil	–	–	–	12	3	–	–	–	–	–	–	–	–	–	84-55-9
Visnadin	–	–	–	84	86	–	–	–	–	–	–	–	–	–	477-32-7

Table 8.1. (continued)

Substance Name	TLC System														CAS No.
	1	2	3	4	5	6	7	8	9	10	T	P1	P2	B	
Volazocine	–	–	–	78	30	73	56	63	37	26	–	–	–	–	15686-68-7
Warfarin	64	62	64	18	–	–	–	–	–	–	–	12	11	–	81-81-2
XTC → 3,4-Methylenedioxy-methamphetamine	–	20	–	39	89	–	–	24	–	–	22	–	–	–	42542-10-9
Xanthanoic Acid → Propan-theline M	–	8	–	5	87	–	–	–	–	–	–	–	–	–	
Xanthinol Nicotinate	–	–	–	23	25	–	–	–	–	–	–	–	–	–	437-74-1
Xipamide	38	64	36	13	93	–	–	0	–	–	–	–	–	–	14293-44-8
Xylometazoline	–	–	–	30	5	64	13	7	5	3	–	–	–	–	526-36-3
Yohimbine	–	–	–	64	66	70	63	6	38	52	–	–	–	–	146-48-5
Zimeldine	–	–	–	48	20	–	47	27	25	–	32	–	–	–	56775-88-3
Zomepirac	12	12	–	4	88	–	–	–	–	–	1	–	–	–	33369-31-2
Zopiclone	–	0	–	47	42	–	–	4	–	–	–	–	–	–	43200-80-2
Zuclopenthixol → Clopenthixol	–	–	–	44	45	–	56	7	32	11	88	–	–	–	982-24-1
Zuclopenthixol Artifact → Clopenthixol Artifact	–	–	–	–	–	–	–	–	–	–	36	–	–	–	

8.2 hR$_f$ Data in TLC System 1 in Ascending Numerical Order vs. Substance Names

TLC System 1	2	3	4	5	6	7	8	9	10	[System 1: Chl–Ac (80+20) / Silica]
0	0	0	0	–	–	–	–	–	–	Cromoglicic Acid
0	0	0	0	–	–	–	–	–	–	Enalapril
0	0	0	0	–	–	–	–	–	–	Pethidine Intermediate C
0	0	0	0	–	–	–	–	–	–	Phthalylsulfacetamide
0	0	0	0	–	–	–	–	–	–	Phthalylsulfathiazole
0	0	0	0	11	–	–	–	–	–	Levodopa
0	0	0	0	70	16	58	0	0	0	Lysergic Acid
0	0	0	0	84	–	–	–	–	–	Barbituric Acid
0	0	2	0	–	–	–	–	–	–	Pipemidic Acid
0	0	2	0	–	–	–	–	–	–	Succinylsulfathiazole
0	0	2	0	85	–	66	–	–	–	Sulfasalazine
0	0	2	2	85	59	84	0	1	2	Metamizol/Dipyrone
0	0	4	0	78	–	–	–	–	–	Hydroxyhippuric Acid,o-
0	0	7	8	82	72	75	0	2	0	Cinchophen
0	0	10	16	–	–	–	–	–	–	Carbazochrome
0	0	14	40	20	60	58	2	18	29	Flurazepam M (Dideethyl-)
0	0	15	50	28	62	58	8	19	11	Flurazepam M (Monodeethyl-)
0	2	4	0	–	–	–	–	–	–	Carbidopa
0	6	26	39	73	61	61	0	24	28	Bromazepam M (3-Hydroxy-)
0	52	0	0	–	–	–	–	–	–	Valproic Acid
0	60	56	7	–	–	–	–	–	–	Paramethadione
0	72	0	34	90	–	–	–	–	–	Thiobarbital
1	0	0	0	–	–	–	–	–	–	Baclofen
1	0	4	0	72	–	58	–	–	–	Nicotinic Acid
1	0	28	40	49	39	63	1	36	2	Adinazolam M (Nor-)
1	1	2	27	22	–	49	–	–	–	Mafenide
1	1	4	7	87	–	–	–	–	–	Saccharin
1	1	6	0	–	–	–	–	–	–	Captopril
1	1	25	31	62	50	64	0	–	–	Bromazepam M (N(py)-Oxide)
1	5	28	33	85	–	–	–	–	–	Digoxin
1	6	7	25	75	–	65	–	–	–	Sulfaguanidine
1	7	7	6	86	–	–	0	–	–	Furosemide
1	10	6	4	87	–	–	–	–	–	Bumetanide
2	5	26	77	55	76	73	13	–	–	Trifluperidol
2	5	33	38	78	81	75	4	25	14	Midazolam M (α,4-Dihydroxy-)
2	16	11	2	–	–	–	–	–	–	Chlorothiazide
2	30	33	6	84	–	–	–	–	–	Nitrofurantoin
3	–	–	56	60	59	60	3	39	18	Lysergide/LSD
3	1	29	24	75	71	63	0	29	8	Triazolam M (4-Hydroxy-)
3	1	36	40	26	15	40	1	48	5	Loprazolam
3	2	5	5	71	–	–	–	–	–	Etacrynic Acid
3	2	26	26	73	69	67	1	25	7	Alprazolam M (4-Hydroxy-)
3	3	15	11	88	–	–	–	–	–	Niflumic Acid
3	3	33	40	68	60	69	0	32	42	Chlordiazepoxide M (Nor-)

Table 8.2. (continued)

TLC System 1	2	3	4	5	6	7	8	9	10	[System 1: Chl–Ac (80+20) / Silica]
3	3	41	71	52	45	62	30	48	40	Flurazepam
3	4	8	0	–	–	–	–	–	–	Fusaric Acid
3	4	41	53	73	72	70	3	52	8	Midazolam M (α-Hydroxy-)
3	10	42	36	88	–	–	–	–	–	Digitoxin
4	3	39	42	75	74	71	4	49	21	Triazolam M (α-Hydroxy-)
4	4	30	16	–	–	–	–	–	–	Sulfinpyrazone
4	4	37	37	76	76	72	1	44	18	Alprazolam M (α-Hydroxy-)
4	5	29	55	42	–	57	4	38	17	Clozapine
4	6	11	4	–	–	–	–	–	–	Dichlorophenoxy)acetic Acid,(2,4-/2,4-D
4	10	22	9	–	–	–	–	–	–	Aloxiprin
4	21	15	40	–	–	75	–	–	–	Quinethazone
4	31	18	5	84	–	85	–	–	–	Acetazolamide
4	32	21	24	–	–	–	–	–	–	Ethyl Biscoumacetate
4	34	11	34	78	–	–	–	–	–	Hydrochlorothiazide
4	40	23	42	88	83	–	–	–	–	Chlortalidone
5	2	41	44	68	65	60	1	40	16	Triazolam
5	4	37	28	76	76	68	1	35	13	Brotizolam M (6-Hydroxy-)
5	5	37	49	76	77	74	1	43	15	Midazolam M (4-Hydroxy-)
5	10	10	0	–	–	–	–	–	–	Cyclamic Acid
5	16	27	5	80	–	64	–	–	–	Sulfisomidine
5	24	15	5	–	–	70	–	–	–	Aminosalicylic Acid,p-
6	23	19	2	88	69	–	–	–	–	Gentisic Acid
7	1	24	10	86	–	–	–	–	–	Salicylic Acid
7	2	40	47	67	66	67	1	57	14	Alprazolam
7	5	44	50	71	60	71	2	53	25	Estazolam
7	6	41	45	78	78	72	2	46	31	Brotizolam M (α-Hydroxy-)
7	11	30	42	79	82	73	0	22	45	Clonazepam M (7-Acetamido-)
7	12	24	44	80	83	71	0	20	49	Nitrazepam M (7-Acetamido-)
7	17	28	0	–	–	–	–	–	–	Carbenoxolone
7	23	38	49	84	–	70	–	–	–	Methocarbamol
7	47	13	36	87	–	–	–	–	–	Hydroflumethiazide
7	63	80	86	89	90	77	20	79	71	Flunitrazepam M Hy/MNFB
8	5	18	16	89	–	–	–	–	–	Diflunisal
8	18	37	4	–	–	–	–	–	–	Piroxicam M (5-Hydroxy-)
8	23	28	41	76	–	–	–	–	–	Primidone
9	2	51	57	68	66	69	6	55	22	Adinazolam
9	8	41	83	45	51	67	53	60	51	Flurazepam Hy/DCFB
9	11	15	5	–	–	–	–	–	–	MCPA/(4-Chloro-o-tolyloxy)acetic Acid
9	20	27	8	79	–	66	–	–	–	Sulfathiazole
9	22	30	67	66	63	73	4	51	47	Medazepam M (Nor-)
9	36	32	56	63	87	75	0	32	58	Meprobamate
10	10	53	52	76	77	62	2	50	22	Chlordiazepoxide
10	20	30	54	77	79	71	0	30	46	Nitrazepam M (7-Amino-)
10	35	35	60	82	85	–	0	33	56	Mebutamate
11	17	40	39	81	–	–	2	–	–	Guaifenesin
11	19	31	18	–	–	–	–	–	–	Bufexamac

Table 8.2. (continued)

TLC System 1	2	3	4	5	6	7	8	9	10	[System 1: Chl–Ac (80+20) / Silica]
11	19	35	54	78	75	72	0	34	47	Flunitrazepam M (7-Aminonor-)
11	20	40	52	76	77	73	0	30	47	Clonazepam M (7-Amino-)
11	29	46	86	–	–	68	–	–	–	Butalamine
11	31	34	36	86	90	69	0	31	48	Chloramphenicol
11	33	53	46	–	–	–	–	–	–	Diphenadione
11	34	32	52	87	–	82	0	–	–	Chlorphenesin Carbamate
11	50	22	50	–	–	–	–	–	–	Ethiazide
12	12	–	4	88	–	–	–	–	–	Zomepirac
12	23	27	4	83	–	65	–	–	–	Sulfamethizole
13	3	24	5	87	–	–	–	–	–	Probenecid
13	5	53	60	69	70	72	6	60	19	Midazolam
13	18	47	63	73	69	61	6	41	53	Bromazepam
13	20	30	5	85	–	–	–	–	–	Tolmetin
13	39	30	53	–	–	62	1	20	53	Styramate
13	46	22	51	83	–	67	–	–	–	Sulfanilamide
13	62	25	44	–	–	–	–	–	–	Epithiazide
14	10	34	4	87	–	–	–	–	–	Sulindac
14	35	34	8	–	–	67	–	–	–	Sulfaethidole
14	36	35	55	–	–	–	–	–	–	Mephenesin Carbamate
14	51	30	9	–	–	–	–	–	–	Benzthiazide
14	64	23	33	–	–	82	–	–	–	Diclofenamide
15	5	53	52	72	71	72	5	52	27	Brotizolam
15	10	48	65	–	–	–	–	–	–	Pethidine Intermediate A
15	10	55	52	59	55	52	3	58	25	Caffeine
15	24	42	41	81	83	63	0	35	51	Demoxepam/Chlordiazepoxide M
15	28	35	49	88	89	67	0	30	38	Quazepam M (3-Hydroxy-N-dealkyl-2-oxo-)
15	32	26	45	77	–	–	0	–	–	Paracetamol
15	60	23	14	88	–	80	–	–	–	Trichlormethiazide
16	13	38	5	83	63	–	–	–	–	Indometacin
16	42	34	24	–	–	67	–	–	–	Sulfapyridine
17	19	72	28	–	–	–	–	–	–	Diiodohydroxyquinoline
17	42	28	4	87	–	70		–	–	Sulfacetamid
18	14	50	45	66	66	65	4	–	–	Phenazone/Antipyrine
18	28	33	4	–	–	–	–	–	–	Alclofenac
18	30	31	9	78	–	–	–	–	–	Acetylsalicylic Acid
18	30	39	4	–	–	–	–	–	–	Fenbufen
18	32	33	30	–	–	–	–	–	–	Dicoumarol
18	60	26	59	–	–	77	–	–	–	Cyclothiazide
19	11	41	51	78	76	73	0	40	49	Flunitrazepam M (7-Acetamido-)
19	28	54	61	81	82	74	2	46	55	Flurazepam M (N(1)-Hydroxyethyl-)
19	36	50	9	–	–	–	–	–	–	Dantrolene
19	38	39	55	–	–	79	–	–	–	Clopamide
19	39	43	9	83	–	65	–	–	–	Sulfamethoxypyridazine
19	43	31	1	–	–	58	–	–	–	Aminobenzoic Acid,p-
19	50	27	53	–	–	–	–	–	–	Methyclothiazide
21	21	52	63	77	74	74	1	55	52	Flunitrazepam M (7-Amino-)

Table 8.2. (continued)

TLC System 1	2	3	4	5	6	7	8	9	10	[System 1: Chl–Ac (80+20) / Silica]
21	40	30	59	75	–	–	3	–	–	Paracetamol M/p-Aminophenol
21	46	47	7	–	–	–	–	–	–	Enoxolone
21	62	27	66	–	–	–	–	–	–	Cyclopenthiazide
22	35	42	45	82	82	56	0	40	51	Oxazepam/M of Chlordiazepoxide, Clorazepic Acid, Camazepam, Diazepam, Halazepam, Ketazolam, Medazepam, Oxazolam, Prazepam
22	39	38	4	81	–	64	–	–	–	Sulfadiazine
22	40	50	65	–	–	–	–	–	–	Phenacemide
22	60	32	63	–	–	–	–	–	–	Polythiazide
23	26	32	5	–	–	–	–	–	–	Chlorobenzoic Acid,p-
23	41	42	8	80	–	65	–	–	–	Sulfamerazine
23	41	42	43	82	82	52	1	36	28	Lorazepam/Lormetazepam M
23	43	42	57	81	–	–	–	–	–	Sultiame
23	45	44	13	79	–	62	–	–	–	Sulfadimidine
23	49	33	6	87	–	–	2	–	–	Sulfasomizole
23	51	33	57	–	–	–	–	–	–	Metolazone
24	43	47	12	82	–	60	–	–	–	Sulfametoxydiazine
25	10	58	62	70	68	66	21	–	–	Aminophenazone/Amidopyrine
25	27	47	12	90	–	–	–	–	–	Diclofenac
25	52	33	6	81	–	65	–	–	–	Sulfafurazole
25	71	30	57	–	–	–	–	–	–	Bendroflumethiazide/Bendrofluthiazide
26	28	35	4	–	–	–	–	–	–	Benzoic Acid
26	42	52	63	89	93	77	0	52	56	Budesonide
26	54	41	5	79	–	65	–	–	–	Sulfamethoxazole
27	25	41	6	85	–	–	–	–	–	Ketoprofen
29	51	43	9	89	–	69	–	–	–	Sulfaphenazole
29	61	46	23	–	–	–	–	–	–	Methyl-5-phenylbarbituric Acid,5-
30	14	61	61	76	72	71	3	63	44	Flumazenil/RO 15-1788
30	30	45	6	–	–	–	–	–	–	Flurbiprofen
30	30	57	11	90	–	–	0	–	–	Glibenclamide/Glyburide
31	25	55	63	78	–	58	–	–	–	Methyprylon
31	51	47	60	79	–	76	–	–	–	Clorexolone
31	51	48	10	86	–	65	–	–	–	Sulfadimethoxine
32	34	44	30	94	98	–	0	–	–	Hexachlorophene
32	43	54	64	84	90	75	0	52	61	Clobazam M (Nor-)
32	47	53	65	–	–	–	–	–	–	Aminoglutethimide
33	25	65	73	81	84	76	15	68	58	Metaclazepam M (N-Nor-)
33	38	44	6	82	–	–	–	–	–	Naproxen
33	40	50	6	84	–	–	–	–	–	Chlorambucil
33	55	53	41	86	–	–	–	–	–	Phenytoin
34	45	57	69	82	83	62	3	55	60	Nordazepam/M of Chlordiazepoxide, Clorazepic Acid, Diazepam, Halazepam, Ketazolam, Medazepam, Oxazolam, Prazepam
34	45	60	72	83	88	75	2	56	58	Flurazepam M (N(1)-Dealkyl-), Quazepam M (N-Dealkyl-2-oxo-)

Table 8.2. (continued)

TLC System										
1	2	3	4	5	6	7	8	9	10	[System 1: Chl–Ac (80+20) / Silica]
34	46	57	68	83	87	84	3	56	60	Clorazepic Acid
35	40	62	71	–	–	–	–	–	–	Alfadolone Acetate
35	41	57	72	82	86	73	5	58	56	Delorazepam
35	45	56	67	85	87	72	0	53	61	Clonazepam
35	45	56	73	85	92	77	5	58	60	Tetrazepam M (Nor-)
35	46	53	64	84	86	68	0	36	55	Nitrazepam
35	46	56	68	84	85	76	0	58	59	Flunitrazepam M (Nor-)
35	51	52	68	84	–	–	6	–	–	Bromisoval
35	68	–	65	–	–	77	–	–	–	Tybamate
36	53	59	75	85	79	–	4	–	–	Carisoprodol
37	41	54	81	80	79	72	5	57	57	Nitrazepam M Hy/DAB
37	51	55	79	78	–	67	63	–	–	Sulfadoxine
38	35	59	41	85	–	–	–	–	–	Etamivan
38	37	52	68	83	–	–	–	–	–	Phenacetin
38	37	64	57	81	80	71	1	59	52	Flunitrazepam M (3-Hydroxy-)
38	43	49	10	87	88	72	0	–	–	Chlorpropamide
38	53	59	71	–	–	76	–	–	–	Pheneturide
38	55	43	50	83	84	–	0	–	–	Salicylamide
38	59	39	53	86	–	–	–	–	–	Phenolphthalein
38	61	46	66	89	–	–	–	–	–	Indapamide
38	64	36	13	93	–	–	0	–	–	Xipamide
38	64	66	39	89	–	–	–	–	–	Vinylbital
39	31	62	2	63	–	–	–	–	–	Nalidixic Acid
39	43	53	12	–	–	–	–	–	–	Acetohexamide
39	45	63	75	78	84	73	0	66	59	Cloxazolam
40	50	54	8	–	–	67	–	–	–	Sulfalene
40	60	54	5	92	–	–	–	–	–	Glibornuride
41	46	55	82	80	80	74	5	59	61	Clonazepam Hy/DCB
41	48	54	11	87	–	–	–	–	–	Mefenamic Acid
41	61	57	32	84	–	–	–	–	–	Barbital
42	38	50	6	–	–	–	–	–	–	Fenoprofen
42	52	58	69	88	90	71	2	55	59	Quazepam M (3-Hydroxy-2-oxo-)
43	65	51	43	–	–	76	–	–	–	Ethoxzolamide
45	45	52	66	80	84	–	–	–	–	Acetanilide
45	45	62	74	83	80	66	14	64	66	Ketazolam
45	58	55	67	–	–	–	–	–	–	Mefruside
46	45	60	59	82	82	52	6	61	50	Lormetazepam
46	46	53	39	–	–	–	–	–	–	Glutaral
46	50	65	75	80	90	74	11	66	62	Haloxazolam
46	57	54	6	75	–	–	–	–	–	Ibuprofen
46	65	56	41	–	–	–	–	–	–	Probarbital
47	35	71	79	82	84	77	40	73	62	Metaclazepam
47	55	60	73	82	–	75	–	–	–	Phenprobamate
47	65	53	28	85	–	–	–	–	–	Phenobarbital
48	50	57	82	87	91	77	7	60	66	Flurazepam Hy/HCFB
48	65	57	40	86	–	–	–	–	–	Aprobarbital

Table 8.2. (continued)

TLC System 1	2	3	4	5	6	7	8	9	10	[System 1: Chl–Ac (80+20) / Silica]
49	48	60	59	84	84	–	–	–	–	Acecarbromal
49	59	58	76	86	87	76	5	–	–	Ethinamate
49	62	57	74	–	–	–	–	–	–	Methylpentynol Carbamate
49	64	56	36	86	–	–	1	–	–	Reposal
50	53	59	66	84	–	70	5	–	–	Ethosuximide
50	64	57	48	88	–	–	–	–	–	Secbutabarbital
50	64	58	40	88	–	–	–	–	–	Cyclobarbital
50	64	59	38	88	–	–	–	–	–	Heptabarbital
50	65	57	34	89	–	–	–	–	–	Vinbarbital
50	65	58	41	86	–	–	–	–	–	Butobarbital/Butethal
50	65	59	39	90	–	–	–	–	–	Cyclopentobarbital
50	66	56	32	91	–	–	–	–	–	Ibomal/Propallylonal
50	66	56	34	87	–	–	–	–	–	Allobarbital
51	38	71	17	88	–	–	–	–	–	Piroxicam
51	47	65	62	82	82	53	8	59	53	Temazepam/M of Camazepam, Diazepam, Ketazolam, Medazepam
51	55	62	12	88	88	76	–	–	–	Tolbutamide
52	37	68	69	78	–	–	–	–	–	Griseofulvin
52	48	60	16	92	92	–	0	–	–	Acenocoumarol
52	50	66	7	86	–	–	0	–	–	Tolazamide
52	53	64	68	–	88	–	–	–	–	Bemegride
52	62	57	9	90	–	77	0	–	–	Oxyphenbutazone/Phenylbutazone M (Hydroxy-)
52	66	57	48	–	–	–	–	–	–	Tetrabarbital
52	66	58	44	88	–	–	–	–	–	Amobarbital
52	68	57	29	87	–	–	–	–	–	Brallobarbital
53	46	71	77	81	81	74	12	70	55	Nimetazepam
53	47	70	75	84	85	62	8	70	62	Clobazam
53	54	60	71	–	–	–	–	–	–	Ethotoin
53	54	65	71	78	95	77	15	68	66	Oxazolam
53	55	64	75	85	87	–	12	–	–	Carbromal
53	58	62	74	85	93	76	0	62	67	Ethyl Loflazepate
53	67	60	44	–	–	–	–	–	–	Hexethal
53	67	60	46	92	–	–	–	–	–	Talbutal
53	68	57	45	88	–	–	–	–	–	Butallylonal
54	47	72	74	80	82	63	10	72	63	Flunitrazepam
54	61	56	33	88	90	85	0	51	47	Chlorzoxazone
54	67	57	44	87	–	–	1	–	–	Butalbital
54	67	59	35	–	–	–	–	–	–	Allylbarbituric Acid,5-
55	32	69	75	82	83	76	12	73	65	Camazepam
55	48	66	80	84	87	78	31	69	66	Clotiazepam
55	55	70	72	85	88	71	12	–	52	Prazepam M (3-Hydroxy-)
55	66	59	45	90	–	–	–	–	–	Pentobarbital
55	68	62	45	88	–	–	–	–	–	Secobarbital
55	69	59	41	–	–	–	–	–	–	Idobutal (5-Allyl-5-butylbarbituric Acid)
56	40	73	78	79	83	67	41	74	62	Medazepam

Table 8.2. (continued)

TLC System 1	2	3	4	5	6	7	8	9	10	[System 1: Chl–Ac (80+20) / Silica]
56	50	67	75	78	86	76	24	69	63	Fludiazepam
56	62	63	76	84	87	67	5	57	66	Benzocaine
57	49	67	79	84	89	78	32	69	66	Tetrazepam
57	65	60	78	86	–	–	9	–	–	Hexapropymate
58	49	72	76	82	85	75	27	73	59	Diazepam/Ketazolam M/Medazepam M
58	67	57	34	91	–	–	–	–	–	Dichlorophen
58	68	60	44	92	–	–	–	–	–	Nealbarbital
59	57	70	80	87	90	78	16	89	71	Quazepam M (2-Oxo-)
59	59	70	81	89	91	78	15	6	71	Halazepam
59	71	63	50	86	–	–	–	–	–	Sigmodal (5-(2-Bromoallyl)-5-(1-methylbutyl) barbituric Acid)
60	58	63	79	–	–	–	–	–	–	Ergosterol
60	59	63	75	–	–	–	–	–	–	Triclocarban
61	49	65	74	81	–	71	32	–	–	Propyphenazone
61	58	64	78	–	–	–	–	–	–	Estradiol
61	60	64	76	84	96	–	22	64	69	Cholesterol
62	58	61	19	93	–	–	–	–	–	Phenprocoumon
62	58	65	23	–	–	–	–	–	–	Benziodarone
62	58	66	74	–	–	–	–	–	–	Mephenytoin
62	65	55	44	81	–	–	–	–	–	Methyl p-Hydroxybenzoate/Methyl Paraben
62	65	65	58	–	–	–	–	–	–	Fenimide
63	47	68	42	75	75	72	1	–	–	Carbimazole
63	62	70	80	86	89	75	31	–	–	Glutethimide
64	50	69	75	–	79	–	–	–	–	Santonin
64	55	72	81	84	89	65	36	74	63	Prazepam
64	61	66	85	86	88	77	9	68	69	Bromazepam Hy/ABP
64	62	64	18	–	–	–	–	–	–	Warfarin
64	66	57	46	88	–	–	–	–	–	Ethyl p-Hydroxybenzoate
64	67	59	75	–	–	–	–	–	–	Tribromoethanol
65	56	70	21	–	–	–	–	–	–	Phenindione
65	61	73	81	85	92	75	29	72	70	Pinazepam
65	65	69	53	85	–	–	–	–	–	Hexobarbital
65	67	56	50	90	–	–	–	–	–	Propyl p-Hydroxybenzoate
66	51	75	78	84	–	–	–	–	–	Spironolactone
66	65	69	54	87	–	–	–	–	–	Metharbital
66	66	80	83	76	86	73	31	77	68	Nimetazepam Hy/MNB
70	67	70	41	86	–	–	–	–	–	Methylphenobarbital
70	69	74	83	81	93	83	11	71	69	Haloxazolam Hy/ABFB
71	59	72	77	–	–	75	–	–	–	Phensuximide
71	65	68	85	90	91	77	5	69	69	Nitrazepam Hy/ANB
71	66	68	86	87	91	77	4	69	70	Flunitrazepam M Hy/ANFB
71	69	81	83	77	90	75	48	77	69	Fludiazepam Hy/CFMB
71	71	70	58	–	–	–	–	–	–	Enallylpropymal
71	72	69	86	–	–	–	–	–	–	Nitroglycerin
72	66	70	86	87	93	77	4	69	70	Clonazepam Hy/ANCB
72	67	69	53	–	–	–	–	–	–	Narcobarbital

Table 8.2. (continued)

TLC System 1	2	3	4	5	6	7	8	9	10	[System 1: Chl–Ac (80+20) / Silica]
73	68	65	81	–	92	–	–	–	–	Thymol
73	72	71	58	85	–	–	–	–	–	Methohexital
74	73	80	85	77	93	79	44	77	73	Quazepam Hy/CFTB
74	77	73	80	87	95	–	37	–	–	Cyclandelate
75	66	71	84	81	–	–	–	–	–	Clofibrate
75	73	80	86	86	91	83	48	77	72	Pinazepam Hy/CPB
76	68	74	86	89	93	78	11	75	70	Flurazepam Hy/ACFB
76	69	76	86	90	91	78	12	76	71	Oxazepam Hy/ACB
77	70	76	87	89	93	78	11	76	70	Lorazepam Hy/ADB
77	72	69	43	–	–	–	–	–	–	Thialbarbital
77	74	68	49	–	–	–	–	–	–	Thiopental
78	68	76	65	87	–	79	–	–	–	Phenylbutazone
78	71	78	83	87	96	74	27	75	76	Quazepam
80	69	78	43	–	–	–	–	–	–	Danthron
81	69	81	87	82	93	75	49	81	7	Lormetazepam Hy/MDB
81	74	81	89	94	95	82	49	81	75	Halazepam Hy/TCB
81	74	82	86	87	–	–	0	–	–	Ethchlorvynol
82	72	82	88	89	93	79	53	81	71	Diazepam Hy/MACB
82	74	80	87	–	–	–	–	–	–	Clofenotane (p,p′-Isomer)/p,p′-DDT
83	74	82	88	92	95	80	58	82	74	Prazepam Hy/CCB
91	92	95	99	80	88	86	55	99	99	Methidathion
91	93	94	99	81	87	84	40	99	99	Azinphos-methyl

8.3 hR$_f$ Data in TLC System 2 in Ascending Numerical Order vs. Substance Names

TLC System 2	1	3	4	5	6	7	8	9	10	[System 2: Etac / Silica]
0	–	–	0	27	–	–	0	–	–	Lisinopril
0	–	–	0	44	–	–	0	–	–	Flumazenil M/RO 15-3890
0	–	–	2	87	–	–	0	–	–	Clenbuterol M #2
0	–	–	20	16	74	46	1	1	4	Salbutamol/Albuterol
0	–	–	25	38	81	76	0	1	4	Fenoterol
0	–	–	30	12	75	44	4	4	33	Ephedrine M/Phenylpropanolamine
0	–	–	44	21	–	–	12	–	–	Toliprolol
0	–	–	47	21	–	–	1	–	–	Carazolol
0	–	–	47	42	–	–	4	–	–	Zopiclone
0	–	–	48	22	–	–	8	–	–	Bisoprolol
0	–	–	49	20	–	–	12	–	–	Bupranolol
0	–	–	50	22	80	50	13	15	8	Penbutolol
0	–	–	51	22	–	–	11	–	–	Betaxolol
0	–	–	54	19	–	–	–	–	–	Sulforidazine
0	0	0	0	–	–	–	–	–	–	Cromoglicic Acid
0	0	0	0	–	–	–	–	–	–	Enalapril
0	0	0	0	–	–	–	–	–	–	Pethidine Intermediate C
0	0	0	0	–	–	–	–	–	–	Phthalylsulfacetamide
0	0	0	0	–	–	–	–	–	–	Phthalylsulfathiazole
0	0	0	0	11	–	–	–	–	–	Levodopa
0	0	0	0	70	16	58	0	0	0	Lysergic Acid
0	0	0	0	84	–	–	–	–	–	Barbituric Acid
0	0	2	0	–	–	–	–	–	–	Pipemidic Acid
0	0	2	0	–	–	–	–	–	–	Succinylsulfathiazole
0	0	2	0	85	–	66	–	–	–	Sulfasalazine
0	0	2	2	85	59	84	0	1	2	Metamizol/Dipyrone
0	0	4	0	78	–	–	–	–	–	Hydroxyhippuric Acid,o-
0	0	7	8	82	72	75	0	2	0	Cinchophen
0	0	10	16	–	–	–	–	–	–	Carbazochrome
0	0	14	40	20	60	58	2	18	29	Flurazepam M (Dideethyl-)
0	0	15	50	28	62	58	8	19	11	Flurazepam M (Monodeethyl-)
0	1	0	0	–	–	–	–	–	–	Baclofen
0	1	4	0	72	–	58	–	–	–	Nicotinic Acid
0	1	28	40	49	39	63	1	36	2	Adinazolam M (Nor-)
1	–	–	0	84	–	–	–	–	–	Ibuprofen M (2-4′-(2-Hydroxy-2-methyl-propyl)phenylpropionic Acid)
1	–	–	29	32	–	–	0	–	–	Labetalol
1	–	–	43	17	–	–	13	–	–	Tertatolol
1	–	–	56	28	–	–	11	–	–	Gallopamil M (Nor-)
1	–	–	65	27	–	53	47	44	16	Pecazine/Mepazine
1	1	2	27	22	–	49	–	–	–	Mafenide
1	1	4	7	87	–	–	–	–	–	Saccharin
1	1	6	0	–	–	–	–	–	–	Captopril
1	1	25	31	62	50	64	0	–	–	Bromazepam M (N(py)-Oxide)

Table 8.3. (continued)

TLC System 2	1	3	4	5	6	7	8	9	10	[System 2: Etac / Silica]
1	3	29	24	75	71	63	0	29	8	Triazolam M (4-Hydroxy-)
1	3	36	40	26	15	40	1	48	5	Loprazolam
1	7	24	10	86	–	–	–	–	–	Salicylic Acid
2	–	–	61	29	–	–	30	–	–	Melperone M (FG 5155)
2	0	4	0	–	–	–	–	–	–	Carbidopa
2	3	5	5	71	–	–	–	–	–	Etacrynic Acid
2	3	26	26	73	69	67	1	25	7	Alprazolam M (4-Hydroxy-)
2	5	41	44	68	65	60	1	40	16	Triazolam
2	7	40	47	67	66	67	1	57	14	Alprazolam
2	9	51	57	68	66	69	6	55	22	Adinazolam
3	–	–	46	81	78	67	1	38	40	Glafenine
3	–	–	60	62	69	78	3	51	32	Benperidol
3	–	–	73	50	–	–	13	–	–	Bromperidol
3	3	15	11	88	–	–	–	–	–	Niflumic Acid
3	3	33	40	68	60	69	0	32	42	Chlordiazepoxide M (Nor-)
3	3	41	71	52	45	62	30	48	40	Flurazepam
3	4	39	42	75	74	71	4	49	21	Triazolam M (α-Hydroxy-)
3	13	24	5	87	–	–	–	–	–	Probenecid
4	–	–	34	59	54	53	1	31	21	Theobromine
4	–	–	71	46	–	–	24	–	–	Gallopamil
4	3	8	0	–	–	–	–	–	–	Fusaric Acid
4	3	41	53	73	72	70	3	52	8	Midazolam M (α-Hydroxy-)
4	4	30	16	–	–	–	–	–	–	Sulfinpyrazone
4	4	37	37	76	76	72	1	44	18	Alprazolam M (α-Hydroxy-)
4	5	37	28	76	76	68	1	35	13	Brotizolam M (6-Hydroxy-)
5	–	–	4	86	–	–	–	–	–	Tiaprofenic Acid
5	–	–	71	64	–	–	30	–	–	Buspirone
5	1	28	33	85	–	–	–	–	–	Digoxin
5	2	26	77	55	76	73	13	–	–	Trifluperidol
5	2	33	38	78	81	75	4	25	14	Midazolam M (α,4-Dihydroxy-)
5	4	29	55	42	–	57	4	38	17	Clozapine
5	5	37	49	76	77	74	1	43	15	Midazolam M (4-Hydroxy-)
5	7	44	50	71	60	71	2	53	25	Estazolam
5	8	18	16	89	–	–	–	–	–	Diflunisal
5	13	53	60	69	70	72	6	60	19	Midazolam
5	15	53	52	72	71	72	5	52	27	Brotizolam
6	–	–	38	66	–	–	0	–	–	Etofylline
6	0	26	39	73	61	61	0	24	28	Bromazepam M (3-Hydroxy-)
6	1	7	25	75	–	65	–	–	–	Sulfaguanidine
6	4	11	4	–	–	–	–	–	–	Dichlorophenoxy)acetic Acid,(2,4-/2,4-D
6	7	41	45	78	78	72	2	46	31	Brotizolam M (α-Hydroxy-)
7	–	–	69	56	–	–	21	–	–	Gallopamil M (D517)
7	–	56	80	77	80	73	11	67	37	Miconazole
7	1	7	6	86	–	–	0	–	–	Furosemide
8	–	–	5	87	–	–	–	–	–	Propantheline M/Xanthanoic Acid
8	9	41	83	45	51	67	53	60	51	Flurazepam Hy/DCFB

Table 8.3. (continued)

TLC System 2	1	3	4	5	6	7	8	9	10	[System 2: Etac / Silica]
9	–	–	11	74	66	75	1	30	11	Theophylline
9	–	–	46	81	–	–	0	–	–	Metaclazepam M (Dinor-)
10	1	6	4	87	–	–	–	–	–	Bumetanide
10	3	42	36	88	–	–	–	–	–	Digitoxin
10	4	22	9	–	–	–	–	–	–	Aloxiprin
10	5	10	0	–	–	–	–	–	–	Cyclamic Acid
10	10	53	52	76	77	62	2	50	22	Chlordiazepoxide
10	14	34	4	87	–	–	–	–	–	Sulindac
10	15	48	65	–	–	–	–	–	–	Pethidine Intermediate A
10	15	55	52	59	55	52	3	58	25	Caffeine
10	25	58	62	70	68	66	21	–	–	Aminophenazone/Amidopyrine
11	–	–	36	81	81	60	0	23	40	Pemoline
11	7	30	42	79	82	73	0	22	45	Clonazepam M (7-Acetamido-)
11	9	15	5	–	–	–	–	–	–	MCPA/(4-Chloro-o-tolyloxy)acetic Acid
11	19	41	51	78	76	73	0	40	49	Flunitrazepam M (7-Acetamido-)
12	7	24	44	80	83	71	0	20	49	Nitrazepam M (7-Acetamido-)
12	12	–	4	88	–	–	–	–	–	Zomepirac
13	–	–	0	78	–	–	–	–	–	Ibuprofen M (2-4′-(2-Carboxypropyl)phenyl-propionic Acid)
13	16	38	5	83	63	–	–	–	–	Indometacin
14	18	50	45	66	66	65	4	–	–	Phenazone/Antipyrine
14	30	61	61	76	72	71	3	63	44	Flumazenil/RO 15-1788
16	2	11	2	–	–	–	–	–	–	Chlorothiazide
16	5	27	5	80	–	64	–	–	–	Sulfisomidine
17	7	28	0	–	–	–	–	–	–	Carbenoxolone
17	11	40	39	81	–	–	2	–	–	Guaifenesin
18	8	37	4	–	–	–	–	–	–	Piroxicam M (5-Hydroxy-)
18	13	47	63	73	69	61	6	41	53	Bromazepam
19	–	–	6	87	–	–	0	–	–	Clenbuterol M #3
19	11	31	18	–	–	–	–	–	–	Bufexamac
19	11	35	54	78	75	72	0	34	47	Flunitrazepam M (7-Aminonor-)
19	17	72	28	–	–	–	–	–	–	Diiodohydroxyquinoline
20	–	–	10	84	–	–	–	–	–	Flurbiprofen M #2
20	–	–	39	89	–	–	24	–	–	Methylenedioxymethamphetamine,3,4-/MDMA/XTC
20	9	27	8	79	–	66	–	–	–	Sulfathiazole
20	10	30	54	77	79	71	0	30	46	Nitrazepam M (7-Amino-)
20	11	40	52	76	77	73	0	30	47	Clonazepam M (7-Amino-)
20	13	30	5	85	–	–	–	–	–	Tolmetin
21	4	15	40	–	–	75	–	–	–	Quinethazone
21	21	52	63	77	74	74	1	55	52	Flunitrazepam M (7-Amino-)
22	–	–	3	84	–	–	–	–	–	Flurbiprofen M #1
22	9	30	67	66	63	73	4	51	47	Medazepam M (Nor-)
23	–	–	82	67	75	73	39	68	60	Fluanisone
23	6	19	2	88	69	–	–	–	–	Gentisic Acid
23	7	38	49	84	–	70	–	–	–	Methocarbamol

Table 8.3. (continued)

TLC System 2	1	3	4	5	6	7	8	9	10	[System 2: Etac / Silica]
23	8	28	41	76	–	–	–	–	–	Primidone
23	12	27	4	83	–	65	–	–	–	Sulfamethizole
24	–	–	75	25	–	–	58	–	–	Bornaprine
24	5	15	5	–	–	70	–	–	–	Aminosalicylic Acid,p-
24	15	42	41	81	83	63	0	35	51	Demoxepam/Chlordiazepoxide M
25	27	41	6	85	–	–	–	–	–	Ketoprofen
25	31	55	63	78	–	58	–	–	–	Methyprylon
25	33	65	73	81	84	76	15	68	58	Metaclazepam M (N-Nor-)
26	23	32	5	–	–	–	–	–	–	Chlorobenzoic Acid,p-
27	–	–	41	87	–	–	0	–	–	Methylprednisolon
27	25	47	12	90	–	–	–	–	–	Diclofenac
28	–	–	5	85	–	–	–	–	–	Pirprofen
28	–	–	45	86	–	–	0	–	–	Hydrocortisone
28	15	35	49	88	89	67	0	30	38	Quazepam M (3-Hydroxy-N-dealkyl-2-oxo-)
28	18	33	4	–	–	–	–	–	–	Alclofenac
28	19	54	61	81	82	74	2	46	55	Flurazepam M (N(1)-Hydroxyethyl-)
28	26	35	4	–	–	–	–	–	–	Benzoic Acid
29	11	46	86	–	–	68	–	–	–	Butalamine
30	2	33	6	84	–	–	–	–	–	Nitrofurantoin
30	18	31	9	78	–	–	–	–	–	Acetylsalicylic Acid
30	18	39	4	–	–	–	–	–	–	Fenbufen
30	30	45	6	–	–	–	–	–	–	Flurbiprofen
30	30	57	11	90	–	–	0	–	–	Glibenclamide/Glyburide
31	4	18	5	84	–	85	–	–	–	Acetazolamide
31	11	34	36	86	90	69	0	31	48	Chloramphenicol
31	39	62	2	63	–	–	–	–	–	Nalidixic Acid
32	4	21	24	–	–	–	–	–	–	Ethyl Biscoumacetate
32	15	26	45	77	–	–	0	–	–	Paracetamol
32	18	33	30	–	–	–	–	–	–	Dicoumarol
32	55	69	75	82	83	76	12	73	65	Camazepam
33	11	53	46	–	–	–	–	–	–	Diphenadione
34	4	11	34	78	–	–	–	–	–	Hydrochlorothiazide
34	11	32	52	87	–	82	0	–	–	Chlorphenesin Carbamate
34	32	44	30	94	98	–	0	–	–	Hexachlorophene
35	10	35	60	82	85	–	0	33	56	Mebutamate
35	14	34	8	–	–	67	–	–	–	Sulfaethidole
35	22	42	45	82	82	56	0	40	51	Oxazepam/M of Chlordiazepoxide, Clorazepic Acid, Camazepam, Diazepam, Halazepam, Ketazolam, Medazepam, Oxazolam, Prazepam
35	38	59	41	85	–	–	–	–	–	Etamivan
35	47	71	79	82	84	77	40	73	62	Metaclazepam
36	9	32	56	63	87	75	0	32	58	Meprobamate
36	14	35	55	–	–	–	–	–	–	Mephenesin Carbamate
36	19	50	9	–	–	–	–	–	–	Dantrolene
37	38	52	68	83	–	–	–	–	–	Phenacetin

Table 8.3. (continued)

TLC System 2	1	3	4	5	6	7	8	9	10	[System 2: Etac / Silica]
37	38	64	57	81	80	71	1	59	52	Flunitrazepam M (3-Hydroxy-)
37	52	68	69	78	–	–	–	–	–	Griseofulvin
38	19	39	55	–	–	79	–	–	–	Clopamide
38	33	44	6	82	–	–	–	–	–	Naproxen
38	42	50	6	–	–	–	–	–	–	Fenoprofen
38	51	71	17	88	–	–	–	–	–	Piroxicam
39	13	30	53	–	–	62	1	20	53	Styramate
39	19	43	9	83	–	65	–	–	–	Sulfamethoxypyridazine
39	22	38	4	81	–	64	–	–	–	Sulfadiazine
40	4	23	42	88	83	–	–	–	–	Chlortalidone
40	21	30	59	75	–	–	3	–	–	Paracetamol M/p-Aminophenol
40	22	50	65	–	–	–	–	–	–	Phenacemide
40	33	50	6	84	–	–	–	–	–	Chlorambucil
40	35	62	71	–	–	–	–	–	–	Alfadolone Acetate
40	56	73	78	79	83	67	41	74	62	Medazepam
41	23	42	8	80	–	65	–	–	–	Sulfamerazine
41	23	42	43	82	82	52	1	36	28	Lorazepam/Lormetazepam M
41	35	57	72	82	86	73	5	58	56	Delorazepam
41	37	54	81	80	79	72	5	57	57	Nitrazepam M Hy/DAB
42	16	34	24	–	–	67	–	–	–	Sulfapyridine
42	17	28	4	87	–	70	–	–	–	Sulfacetamid
42	26	52	63	89	93	77	0	52	56	Budesonide
43	–	–	12	–	–	–	–	–	–	Meclofenamic Acid
43	19	31	1	–	–	58	–	–	–	Aminobenzoic Acid,p-
43	23	42	57	81	–	–	–	–	–	Sultiame
43	24	47	12	82	–	60	–	–	–	Sulfametoxydiazine
43	32	54	64	84	90	75	0	52	61	Clobazam M (Nor-)
43	38	49	10	87	88	72	0	–	–	Chlorpropamide
43	39	53	12	–	–	–	–	–	–	Acetohexamide
45	23	44	13	79	–	62	–	–	–	Sulfadimidine
45	34	57	69	82	83	62	3	55	60	Nordazepam/M of Chlordiazepoxide, Clorazepic Acid, Diazepam, Halazepam, Ketazolam, Medazepam, Oxazolam, Prazepam
45	34	60	72	83	88	75	2	56	58	Flurazepam M (N(1)-Dealkyl-)/Quazepam M (N-Dealkyl-2-oxo-)
45	35	56	67	85	87	72	0	53	61	Clonazepam
45	35	56	73	85	92	77	5	58	60	Tetrazepam M (Nor-)
45	39	63	75	78	84	73	0	66	59	Cloxazolam
45	45	52	66	80	84	–	–	–	–	Acetanilide
45	45	62	74	83	80	66	14	64	66	Ketazolam
45	46	60	59	82	82	52	6	61	50	Lormetazepam
46	13	22	51	83	–	67	–	–	–	Sulfanilamide
46	21	47	7	–	–	–	–	–	–	Enoxolone
46	34	57	68	83	87	84	3	56	60	Clorazepic Acid
46	35	53	64	84	86	68	0	36	55	Nitrazepam
46	35	56	68	84	85	76	0	58	59	Flunitrazepam M (Nor-)

Table 8.3. (continued)

TLC System 2	1	3	4	5	6	7	8	9	10	[System 2: Etac / Silica]
46	41	55	82	80	80	74	5	59	61	Clonazepam Hy/DCB
46	46	53	39	–	–	–	–	–	–	Glutaral
46	53	71	77	81	81	74	12	70	55	Nimetazepam
47	7	13	36	87	–	–	–	–	–	Hydroflumethiazide
47	32	53	65	–	–	–	–	–	–	Aminoglutethimide
47	51	65	62	82	82	53	8	59	53	Temazepam/M of Camazepam, Diazepam, Ketazolam, Medazepam
47	53	70	75	84	85	62	8	70	62	Clobazam
47	54	72	74	80	82	63	10	72	63	Flunitrazepam
47	63	68	42	75	75	72	1	–	–	Carbimazole
48	41	54	11	87	–	–	–	–	–	Mefenamic Acid
48	49	60	59	84	84	–	–	–	–	Acecarbromal
48	52	60	16	92	92	–	0	–	–	Acenocoumarol
48	55	66	80	84	87	78	31	69	66	Clotiazepam
49	23	33	6	87	–	–	2	–	–	Sulfasomizole
49	57	67	79	84	89	78	32	69	66	Tetrazepam
49	58	72	76	82	85	75	27	73	59	Diazepam/Ketazolam M/Medazepam M
49	61	65	74	81	–	71	32	–	–	Propyphenazone
50	–	–	17	93	–	–	0	–	–	Chlorphenprocoumon,p-
50	11	22	50	–	–	–	–	–	–	Ethiazide
50	19	27	53	–	–	–	–	–	–	Methyclothiazide
50	40	54	8	–	–	67	–	–	–	Sulfalene
50	46	65	75	80	90	74	11	66	62	Haloxazolam
50	48	57	82	87	91	77	7	60	66	Flurazepam Hy/HCFB
50	52	66	7	86	–	–	0	–	–	Tolazamide
50	56	67	75	78	86	76	24	69	63	Fludiazepam
50	64	69	75	–	79	–	–	–	–	Santonin
51	14	30	9	–	–	–	–	–	–	Benzthiazide
51	23	33	57	–	–	–	–	–	–	Metolazone
51	29	43	9	89	–	69	–	–	–	Sulfaphenazole
51	31	47	60	79	–	76	–	–	–	Clorexolone
51	31	48	10	86	–	65	–	–	–	Sulfadimethoxine
51	35	52	68	84	–	–	6	–	–	Bromisoval
51	37	55	79	78	–	67	63	–	–	Sulfadoxine
51	66	75	78	84	–	–	–	–	–	Spironolactone
52	0	0	0	–	–	–	–	–	–	Valproic Acid
52	25	33	6	81	–	65	–	–	–	Sulfafurazole
52	42	58	69	88	90	71	2	55	59	Quazepam M (3-Hydroxy-2-oxo-)
53	36	59	75	85	79	–	4	–	–	Carisoprodol
53	38	59	71	–	–	76	–	–	–	Pheneturide
53	50	59	66	84	–	70	5	–	–	Ethosuximide
53	52	64	68	–	88	–	–	–	–	Bemegride
54	26	41	5	79	–	65	–	–	–	Sulfamethoxazole
54	53	60	71	–	–	–	–	–	–	Ethotoin
54	53	65	71	78	95	77	15	68	66	Oxazolam
55	–	–	78	86	–	–	12	–	–	Mexazolam

Table 8.3. (continued)

TLC System 2	1	3	4	5	6	7	8	9	10	[System 2: Etac / Silica]
55	33	53	41	86	–	–	–	–	–	Phenytoin
55	38	43	50	83	84	–	0	–	–	Salicylamide
55	47	60	73	82	–	75	–	–	–	Phenprobamate
55	51	62	12	88	88	76	–	–	–	Tolbutamide
55	53	64	75	85	87	–	12	–	–	Carbromal
55	55	70	72	85	88	71	12	–	52	Prazepam M (3-Hydroxy-)
55	64	72	81	84	89	65	36	74	63	Prazepam
56	65	70	21	–	–	–	–	–	–	Phenindione
57	46	54	6	75	–	–	–	–	–	Ibuprofen
57	59	70	80	87	90	78	16	89	71	Quazepam M (2-Oxo-)
58	45	55	67	–	–	–	–	–	–	Mefruside
58	53	62	74	85	93	76	0	62	67	Ethyl Loflazepate
58	60	63	79	–	–	–	–	–	–	Ergosterol
58	61	64	78	–	–	–	–	–	–	Estradiol
58	62	61	19	93	–	–	–	–	–	Phenprocoumon
58	62	65	23	–	–	–	–	–	–	Benziodarone
58	62	66	74	–	–	–	–	–	–	Mephenytoin
59	38	39	53	86	–	–	–	–	–	Phenolphthalein
59	49	58	76	86	87	76	5	–	–	Ethinamate
59	59	70	81	89	91	78	15	6	71	Halazepam
59	60	63	75	–	–	–	–	–	–	Triclocarban
59	71	72	77	–	–	75	–	–	–	Phensuximide
60	0	56	7	–	–	–	–	–	–	Paramethadione
60	15	23	14	88	–	80	–	–	–	Trichlormethiazide
60	18	26	59	–	–	77	–	–	–	Cyclothiazide
60	22	32	63	–	–	–	–	–	–	Polythiazide
60	40	54	5	92	–	–	–	–	–	Glibornuride
60	61	64	76	84	96	–	22	64	69	Cholesterol
61	29	46	23	–	–	–	–	–	–	Methyl-5-phenylbarbituric Acid,5-
61	38	46	66	89	–	–	–	–	–	Indapamide
61	41	57	32	84	–	–	–	–	–	Barbital
61	54	56	33	88	90	85	0	51	47	Chlorzoxazone
61	64	66	85	86	88	77	9	68	69	Bromazepam Hy/ABP
61	65	73	81	85	92	75	29	72	70	Pinazepam
62	13	25	44	–	–	–	–	–	–	Epithiazide
62	21	27	66	–	–	–	–	–	–	Cyclopenthiazide
62	49	57	74	–	–	–	–	–	–	Methylpentynol Carbamate
62	52	57	9	90	–	77	0	–	–	Oxyphenbutazone/Phenylbutazone M (Hy-droxy-)
62	56	63	76	84	87	67	5	57	66	Benzocaine
62	63	70	80	86	89	75	31	–	–	Glutethimide
62	64	64	18	–	–	–	–	–	–	Warfarin
63	7	80	86	89	90	77	20	79	71	Flunitrazepam M Hy/MNFB
64	14	23	33	–	–	82	–	–	–	Diclofenamide
64	38	36	13	93	–	–	0	–	–	Xipamide
64	38	66	39	89	–	–	–	–	–	Vinylbital

Table 8.3. (continued)

TLC System 2	1	3	4	5	6	7	8	9	10	[System 2: Etac / Silica]
64	49	56	36	86	–	–	1	–	–	Reposal
64	50	57	48	88	–	–	–	–	–	Secbutabarbital
64	50	58	40	88	–	–	–	–	–	Cyclobarbital
64	50	59	38	88	–	–	–	–	–	Heptabarbital
65	–	–	43	81	–	–	–	–	–	Crotylbarbital
65	43	51	43	–	–	76	–	–	–	Ethoxzolamide
65	46	56	41	–	–	–	–	–	–	Probarbital
65	47	53	28	85	–	–	–	–	–	Phenobarbital
65	48	57	40	86	–	–	–	–	–	Aprobarbital
65	50	57	34	89	–	–	–	–	–	Vinbarbital
65	50	58	41	86	–	–	–	–	–	Butobarbital/Butethal
65	50	59	39	90	–	–	–	–	–	Cyclopentobarbital
65	57	60	78	86	–	–	9	–	–	Hexapropymate
65	62	55	44	81	–	–	–	–	–	Methyl p-Hydroxybenzoate/Methyl Paraben
65	62	65	58	–	–	–	–	–	–	Fenimide
65	65	69	53	85	–	–	–	–	–	Hexobarbital
65	66	69	54	87	–	–	–	–	–	Metharbital
65	71	68	85	90	91	77	5	69	69	Nitrazepam Hy/ANB
66	50	56	32	91	–	–	–	–	–	Ibomal/Propallylonal
66	50	56	34	87	–	–	–	–	–	Allobarbital
66	52	57	48	–	–	–	–	–	–	Tetrabarbital
66	52	58	44	88	–	–	–	–	–	Amobarbital
66	55	59	45	90	–	–	–	–	–	Pentobarbital
66	64	57	46	88	–	–	–	–	–	Ethyl p-Hydroxybenzoate
66	66	80	83	76	86	73	31	77	68	Nimetazepam Hy/MNB
66	71	68	86	87	91	77	4	69	70	Flunitrazepam M Hy/ANFB
66	72	70	86	87	93	77	4	69	70	Clonazepam Hy/ANCB
66	75	71	84	81	–	–	–	–	–	Clofibrate
67	53	60	44	–	–	–	–	–	–	Hexethal
67	53	60	46	92	–	–	–	–	–	Talbutal
67	54	57	44	87	–	–	1	–	–	Butalbital
67	54	59	35	–	–	–	–	–	–	Allylbarbituric Acid,5-
67	58	57	34	91	–	–	–	–	–	Dichlorophen
67	64	59	75	–	–	–	–	–	–	Tribromoethanol
67	65	56	50	90	–	–	–	–	–	Propyl p-Hydroxybenzoate
67	70	70	41	86	–	–	–	–	–	Methylphenobarbital
67	72	69	53	–	–	–	–	–	–	Narcobarbital
68	35	–	65	–	–	77	–	–	–	Tybamate
68	52	57	29	87	–	–	–	–	–	Brallobarbital
68	53	57	45	88	–	–	–	–	–	Butallylonal
68	55	62	45	88	–	–	–	–	–	Secobarbital
68	58	60	44	92	–	–	–	–	–	Nealbarbital
68	73	65	81	–	92	–	–	–	–	Thymol
68	76	74	86	89	93	78	11	75	70	Flurazepam Hy/ACFB
68	78	76	65	87	–	79	–	–	–	Phenylbutazone
69	55	59	41	–	–	–	–	–	–	Idobutal (5-Allyl-5-butylbarbituric Acid)

Table 8.3. (continued)

TLC System 2	1	3	4	5	6	7	8	9	10	[System 2: Etac / Silica]
69	70	74	83	81	93	83	11	71	69	Haloxazolam Hy/ABFB
69	71	81	83	77	90	75	48	77	69	Fludiazepam Hy/CFMB
69	76	76	86	90	91	78	12	76	71	Oxazepam Hy/ACB
69	80	78	43	–	–	–	–	–	–	Danthron
69	81	81	87	82	93	75	49	81	7	Lormetazepam Hy/MDB
70	77	76	87	89	93	78	11	76	70	Lorazepam Hy/ADB
71	25	30	57	–	–	–	–	–	–	Bendroflumethiazide/Bendrofluthiazide
71	59	63	50	86	–	–	–	–	–	Sigmodal (5-(2-Bromoallyl)-5-(1-methylbutyl)barbituric Acid)
71	71	70	58	–	–	–	–	–	–	Enallylpropymal
71	78	78	83	87	96	74	27	75	76	Quazepam
72	0	0	34	90	–	–	–	–	–	Thiobarbital
72	71	69	86	–	–	–	–	–	–	Nitroglycerin
72	73	71	58	85	–	–	–	–	–	Methohexital
72	77	69	43	–	–	–	–	–	–	Thialbarbital
72	82	82	88	89	93	79	53	81	71	Diazepam Hy/MACB
73	74	80	85	77	93	79	44	77	73	Quazepam Hy/CFTB
73	75	80	86	86	91	83	48	77	72	Pinazepam Hy/CPB
74	77	68	49	–	–	–	–	–	–	Thiopental
74	81	81	89	94	95	82	49	81	75	Halazepam Hy/TCB
74	81	82	86	87	–	–	0	–	–	Ethchlorvynol
74	82	80	87	–	–	–	–	–	–	Clofenotane (p,p′-Isomer)/p,p′-DDT
74	83	82	88	92	95	80	58	82	74	Prazepam Hy/CCB
77	74	73	80	87	95	–	37	–	–	Cyclandelate
92	91	95	99	80	88	86	55	99	99	Methidathion
93	91	94	99	81	87	84	40	99	99	Azinphos-methyl

8.4 hRf Data in TLC System 3 in Ascending Numerical Order vs. Substance Names

TLC System 3	1	2	4	5	6	7	8	9	10	[System 3: Chl–MeOH (90+10) / Silica]
0	–	–	–	0	–	1	1	0	0	Pentamidine
0	–	–	–	69	–	5	0	1	1	Cytarabine
0	–	–	0	8	–	5	–	–	–	Oxytetracycline
0	–	–	24	6	74	24	0	1	2	Amiloride
0	–	–	38	70	–	55	0	4	8	Ambazone
0	–	–	46	4	14	38	14	4	2	Chloroquine
0	–	–	63	22	–	52	41	28	17	Chloropyramine
0	–	–	76	12	–	30	3	2	2	Ethambutol
0	–	–	80	93	48	1	0	0	0	Metformin
0	0	0	0	–	–	–	–	–	–	Cromoglicic Acid
0	0	0	0	–	–	–	–	–	–	Enalapril
0	0	0	0	–	–	–	–	–	–	Pethidine Intermediate C
0	0	0	0	–	–	–	–	–	–	Phthalylsulfacetamide
0	0	0	0	–	–	–	–	–	–	Phthalylsulfathiazole
0	0	0	0	11	–	–	–	–	–	Levodopa
0	0	0	0	70	16	58	0	0	0	Lysergic Acid
0	0	0	0	84	–	–	–	–	–	Barbituric Acid
0	0	52	0	–	–	–	–	–	–	Valproic Acid
0	0	72	34	90	–	–	–	–	–	Thiobarbital
0	1	0	0	–	–	–	–	–	–	Baclofen
2	0	0	0	–	–	–	–	–	–	Pipemidic Acid
2	0	0	0	–	–	–	–	–	–	Succinylsulfathiazole
2	0	0	0	85	–	66	–	–	–	Sulfasalazine
2	0	0	2	85	59	84	0	1	2	Metamizol/Dipyrone
2	1	1	27	22	–	49	–	–	–	Mafenide
4	0	0	0	78	–	–	–	–	–	Hydroxyhippuric Acid,o-
4	0	2	0	–	–	–	–	–	–	Carbidopa
4	1	0	0	72	–	58	–	–	–	Nicotinic Acid
4	1	1	7	87	–	–	–	–	–	Saccharin
5	3	2	5	71	–	–	–	–	–	Etacrynic Acid
6	1	1	0	–	–	–	–	–	–	Captopril
6	1	10	4	87	–	–	–	–	–	Bumetanide
7	0	0	8	82	72	75	0	2	0	Cinchophen
7	1	6	25	75	–	65	–	–	–	Sulfaguanidine
7	1	7	6	86	–	–	0	–	–	Furosemide
8	3	4	0	–	–	–	–	–	–	Fusaric Acid
10	0	0	16	–	–	–	–	–	–	Carbazochrome
10	5	10	0	–	–	–	–	–	–	Cyclamic Acid
11	2	16	2	–	–	–	–	–	–	Chlorothiazide
11	4	6	4	–	–	–	–	–	–	Dichlorophenoxy)acetic Acid,(2,4-/2,4-D
11	4	34	34	78	–	–	–	–	–	Hydrochlorothiazide
13	–	–	30	50	65	51	1	8	4	Triamterene
13	7	47	36	87	–	–	–	–	–	Hydroflumethiazide
14	–	–	–	81	–	76	0	1	8	Sulfacarbamid

Table 8.4. (Continued)

TLC System 3	1	2	4	5	6	7	8	9	10	[System 3: Chl–Ac (80+20) / Silica]
14	0	0	40	20	60	58	2	18	29	Flurazepam M (Dideethyl-)
15	0	0	50	28	62	58	8	19	11	Flurazepam M (Monodeethyl-)
15	3	3	11	88	–	–	–	–	–	Niflumic Acid
15	4	21	40	–	–	75	–	–	–	Quinethazone
15	5	24	5	–	–	70	–	–	–	Aminosalicylic Acid,p-
15	9	11	5	–	–	–	–	–	–	MCPA/(4-Chloro-o-tolyloxy)acetic Acid
18	–	–	29	55	49	47	1	11	20	Isoniazid
18	4	31	5	84	–	85	–	–	–	Acetazolamide
18	8	5	16	89	–	–	–	–	–	Diflunisal
19	6	23	2	88	69	–	–	–	–	Gentisic Acid
20	–	–	45	45	59	55	0	22	12	Trimethoprim
21	4	32	24	–	–	–	–	–	–	Ethyl Biscoumacetate
22	4	10	9	–	–	–	–	–	–	Aloxiprin
22	11	50	50	–	–	–	–	–	–	Ethiazide
22	13	46	51	83	–	67	–	–	–	Sulfanilamide
23	4	40	42	88	83	–	–	–	–	Chlortalidone
23	14	64	33	–	–	82	–	–	–	Diclofenamide
23	15	60	14	88	–	80	–	–	–	Trichlormethiazide
24	7	1	10	86	–	–	–	–	–	Salicylic Acid
24	7	12	44	80	83	71	0	20	49	Nitrazepam M (7-Acetamido-)
24	13	3	5	87	–	–	–	–	–	Probenecid
25	1	1	31	62	50	64	0	–	–	Bromazepam M (N(py)-Oxide)
25	13	62	44	–	–	–	–	–	–	Epithiazide
26	0	6	39	73	61	61	0	24	28	Bromazepam M (3-Hydroxy-)
26	2	5	77	55	76	73	13	–	–	Trifluperidol
26	3	2	26	73	69	67	1	25	7	Alprazolam M (4-Hydroxy-)
26	15	32	45	77	–	–	0	–	–	Paracetamol
26	18	60	59	–	–	77	–	–	–	Cyclothiazide
27	5	16	5	80	–	64	–	–	–	Sulfisomidine
27	9	20	8	79	–	66	–	–	–	Sulfathiazole
27	12	23	4	83	–	65	–	–	–	Sulfamethizole
27	19	50	53	–	–	–	–	–	–	Methyclothiazide
27	21	62	66	–	–	–	–	–	–	Cyclopenthiazide
28	1	0	40	49	39	63	1	36	2	Adinazolam M (Nor-)
28	1	5	33	85	–	–	–	–	–	Digoxin
28	7	17	0	–	–	–	–	–	–	Carbenoxolone
28	8	23	41	76	–	–	–	–	–	Primidone
28	17	42	4	87	–	70	–	–	–	Sulfacetamid
29	3	1	24	75	71	63	0	29	8	Triazolam M (4-Hydroxy-)
29	4	5	55	42	–	57	4	38	17	Clozapine
30	4	4	16	–	–	–	–	–	–	Sulfinpyrazone
30	7	11	42	79	82	73	0	22	45	Clonazepam M (7-Acetamido-)
30	9	22	67	66	63	73	4	51	47	Medazepam M (Nor-)
30	10	20	54	77	79	71	0	30	46	Nitrazepam M (7-Amino-)
30	13	20	5	85	–	–	–	–	–	Tolmetin
30	13	39	53	–	–	62	1	20	53	Styramate

Table 8.4. (Continued)

TLC System 3	1	2	4	5	6	7	8	9	10	[System 3: Chl–Ac (80+20) / Silica]
30	14	51	9	–	–	–	–	–	–	Benzthiazide
30	21	40	59	75	–	–	3	–	–	Paracetamol M/p-Aminophenol
30	25	71	57	–	–	–	–	–	–	Bendroflumethiazide/Bendrofluthiazide
31	11	19	18	–	–	–	–	–	–	Bufexamac
31	18	30	9	78	–	–	–	–	–	Acetylsalicylic Acid
31	19	43	1	–	–	58	–	–	–	Aminobenzoic Acid,p-
32	–	–	46	75	70	58	2	36	40	Metronidazole
32	9	36	56	63	87	75	0	32	58	Meprobamate
32	11	34	52	87	–	82	0	–	–	Chlorphenesin Carbamate
32	22	60	63	–	–	–	–	–	–	Polythiazide
32	23	26	5	–	–	–	–	–	–	Chlorobenzoic Acid,p-
33	2	5	38	78	81	75	4	25	14	Midazolam M (α,4-Dihydroxy-)
33	2	30	6	84	–	–	–	–	–	Nitrofurantoin
33	3	3	40	68	60	69	0	32	42	Chlordiazepoxide M (Nor-)
33	18	28	4	–	–	–	–	–	–	Alclofenac
33	18	32	30	–	–	–	–	–	–	Dicoumarol
33	23	49	6	87	–	–	2	–	–	Sulfasomizole
33	23	51	57	–	–	–	–	–	–	Metolazone
33	25	52	6	81	–	65	–	–	–	Sulfafurazole
34	11	31	36	86	90	69	0	31	48	Chloramphenicol
34	14	10	4	87	–	–	–	–	–	Sulindac
34	14	35	8	–	–	67	–	–	–	Sulfaethidole
34	16	42	24	–	–	67	–	–	–	Sulfapyridine
35	10	35	60	82	85	–	0	33	56	Mebutamate
35	11	19	54	78	75	72	0	34	47	Flunitrazepam M (7-Aminonor-)
35	14	36	55	–	–	–	–	–	–	Mephenesin Carbamate
35	15	28	49	88	89	67	0	30	38	Quazepam M (3-Hydroxy-N-dealkyl-2-oxo-)
35	26	28	4	–	–	–	–	–	–	Benzoic Acid
36	3	1	40	26	15	40	1	48	5	Loprazolam
36	38	64	13	93	–	–	0	–	–	Xipamide
37	4	4	37	76	76	72	1	44	18	Alprazolam M (α-Hydroxy-)
37	5	4	28	76	76	68	1	35	13	Brotizolam M (6-Hydroxy-)
37	5	5	49	76	77	74	1	43	15	Midazolam M (4-Hydroxy-)
37	8	18	4	–	–	–	–	–	–	Piroxicam M (5-Hydroxy-)
38	7	23	49	84	–	70	–	–	–	Methocarbamol
38	16	13	5	83	63	–	–	–	–	Indometacin
38	22	39	4	81	–	64	–	–	–	Sulfadiazine
39	4	3	42	75	74	71	4	49	21	Triazolam M (α-Hydroxy-)
39	18	30	4	–	–	–	–	–	–	Fenbufen
39	19	38	55	–	–	79	–	–	–	Clopamide
39	38	59	53	86	–	–	–	–	–	Phenolphthalein
40	–	–	36	81	–	–	–	–	–	Sulfaguanole
40	7	2	47	67	66	67	1	57	14	Alprazolam
40	11	17	39	81	–	–	2	–	–	Guaifenesin
40	11	20	52	76	77	73	0	30	47	Clonazepam M (7-Amino-)
41	3	3	71	52	45	62	30	48	40	Flurazepam

Table 8.4. (Continued)

TLC System 3	1	2	4	5	6	7	8	9	10	[System 3: Chl–Ac (80+20) / Silica]
41	3	4	53	73	72	70	3	52	8	Midazolam M (α-Hydroxy-)
41	5	2	44	68	65	60	1	40	16	Triazolam
41	7	6	45	78	78	72	2	46	31	Brotizolam M (α-Hydroxy-)
41	9	8	83	45	51	67	53	60	51	Flurazepam Hy/DCFB
41	19	11	51	78	76	73	0	40	49	Flunitrazepam M (7-Acetamido-)
41	26	54	5	79	–	65	–	–	–	Sulfamethoxazole
41	27	25	6	85	–	–	–	–	–	Ketoprofen
42	3	10	36	88	–	–	–	–	–	Digitoxin
42	15	24	41	81	83	63	0	35	51	Demoxepam/Chlordiazepoxide M
42	22	35	45	82	82	56	0	40	51	Oxazepam/M of Chlordiazepoxide, Clorazepic Acid, Camazepam, Diazepam, Halazepam, Ketazolam, Medazepam, Oxazolam, Prazepam
42	23	41	8	80	–	65	–	–	–	Sulfamerazine
42	23	41	43	82	82	52	1	36	28	Lorazepam/Lormetazepam M
42	23	43	57	81	–	–	–	–	–	Sultiame
43	19	39	9	83	–	65	–	–	–	Sulfamethoxypyridazine
43	29	51	9	89	–	69	–	–	–	Sulfaphenazole
43	38	55	50	83	84	–	0	–	–	Salicylamide
44	–	–	54	71	70	63	3	42	46	Pyrazinamide
44	7	5	50	71	60	71	2	53	25	Estazolam
44	23	45	13	79	–	62	–	–	–	Sulfadimidine
44	32	34	30	94	98	–	0	–	–	Hexachlorophene
44	33	38	6	82	–	–	–	–	–	Naproxen
45	30	30	6	–	–	–	–	–	–	Flurbiprofen
46	11	29	86	–	–	68	–	–	–	Butalamine
46	29	61	23	–	–	–	–	–	–	Methyl-5-phenylbarbituric Acid,5-
46	38	61	66	89	–	–	–	–	–	Indapamide
47	13	18	63	73	69	61	6	41	53	Bromazepam
47	21	46	7	–	–	–	–	–	–	Enoxolone
47	24	43	12	82	–	60	–	–	–	Sulfametoxydiazine
47	25	27	12	90	–	–	–	–	–	Diclofenac
47	31	51	60	79	–	76	–	–	–	Clorexolone
48	15	10	65	–	–	–	–	–	–	Pethidine Intermediate A
48	31	51	10	86	–	65	–	–	–	Sulfadimethoxine
49	38	43	10	87	88	72	0	–	–	Chlorpropamide
50	18	14	45	66	66	65	4	–	–	Phenazone/Antipyrine
50	19	36	9	–	–	–	–	–	–	Dantrolene
50	22	40	65	–	–	–	–	–	–	Phenacemide
50	33	40	6	84	–	–	–	–	–	Chlorambucil
50	42	38	6	–	–	–	–	–	–	Fenoprofen
51	9	2	57	68	66	69	6	55	22	Adinazolam
51	43	65	43	–	–	76	–	–	–	Ethoxzolamide
52	21	21	63	77	74	74	1	55	52	Flunitrazepam M (7-Amino-)
52	26	42	63	89	93	77	0	52	56	Budesonide
52	35	51	68	84	–	–	6	–	–	Bromisoval

Table 8.4. (Continued)

TLC System 3	1	2	4	5	6	7	8	9	10	[System 3: Chl–Ac (80+20) / Silica]
52	38	37	68	83	–	–	–	–	–	Phenacetin
52	45	45	66	80	84	–	–	–	–	Acetanilide
53	10	10	52	76	77	62	2	50	22	Chlordiazepoxide
53	11	33	46	–	–	–	–	–	–	Diphenadione
53	13	5	60	69	70	72	6	60	19	Midazolam
53	15	5	52	72	71	72	5	52	27	Brotizolam
53	32	47	65	–	–	–	–	–	–	Aminoglutethimide
53	33	55	41	86	–	–	–	–	–	Phenytoin
53	35	46	64	84	86	68	0	36	55	Nitrazepam
53	39	43	12	–	–	–	–	–	–	Acetohexamide
53	46	46	39	–	–	–	–	–	–	Glutaral
53	47	65	28	85	–	–	–	–	–	Phenobarbital
54	19	28	61	81	82	74	2	46	55	Flurazepam M (N(1)-Hydroxyethyl-)
54	32	43	64	84	90	75	0	52	61	Clobazam M (Nor-)
54	37	41	81	80	79	72	5	57	57	Nitrazepam M Hy/DAB
54	40	50	8	–	–	67	–	–	–	Sulfalene
54	40	60	5	92	–	–	–	–	–	Glibornuride
54	41	48	11	87	–	–	–	–	–	Mefenamic Acid
54	46	57	6	75	–	–	–	–	–	Ibuprofen
55	15	10	52	59	55	52	3	58	25	Caffeine
55	31	25	63	78	–	58	–	–	–	Methyprylon
55	37	51	79	78	–	67	63	–	–	Sulfadoxine
55	41	46	82	80	80	74	5	59	61	Clonazepam Hy/DCB
55	45	58	67	–	–	–	–	–	–	Mefruside
55	62	65	44	81	–	–	–	–	–	Methyl p-Hydroxybenzoate/Methyl Paraben
56	–	7	80	77	80	73	11	67	37	Miconazole
56	0	60	7	–	–	–	–	–	–	Paramethadione
56	35	45	67	85	87	72	0	53	61	Clonazepam
56	35	45	73	85	92	77	5	58	60	Tetrazepam M (Nor-)
56	35	46	68	84	85	76	0	58	59	Flunitrazepam M (Nor-)
56	46	65	41	–	–	–	–	–	–	Probarbital
56	49	64	36	86	–	–	1	–	–	Reposal
56	50	66	32	91	–	–	–	–	–	Ibomal/Propallylonal
56	50	66	34	87	–	–	–	–	–	Allobarbital
56	54	61	33	88	90	85	0	51	47	Chlorzoxazone
56	65	67	50	90	–	–	–	–	–	Propyl p-Hydroxybenzoate
57	30	30	11	90	–	–	0	–	–	Glibenclamide/Glyburide
57	34	45	69	82	83	62	3	55	60	Nordazepam/M of Chlordiazepoxide, Clorazepic Acid, Diazepam, Halazepam, Ketazolam, Medazepam, Oxazolam, Prazepam
57	34	46	68	83	87	84	3	56	60	Clorazepic Acid
57	35	41	72	82	86	73	5	58	56	Delorazepam
57	41	61	32	84	–	–	–	–	–	Barbital
57	48	50	82	87	91	77	7	60	66	Flurazepam Hy/HCFB
57	48	65	40	86	–	–	–	–	–	Aprobarbital
57	49	62	74	–	–	–	–	–	–	Methylpentynol Carbamate

Table 8.4. (Continued)

TLC System 3	1	2	4	5	6	7	8	9	10	[System 3: Chl–Ac (80+20) / Silica]
57	50	64	48	88	–	–	–	–	–	Secbutabarbital
57	50	65	34	89	–	–	–	–	–	Vinbarbital
57	52	62	9	90	–	77	0	–	–	Oxyphenbutazone/Phenylbutazone M (Hy-droxy-)
57	52	66	48	–	–	–	–	–	–	Tetrabarbital
57	52	68	29	87	–	–	–	–	–	Brallobarbital
57	53	68	45	88	–	–	–	–	–	Butallylonal
57	54	67	44	87	–	–	1	–	–	Butalbital
57	58	67	34	91	–	–	–	–	–	Dichlorophen
57	64	66	46	88	–	–	–	–	–	Ethyl p-Hydroxybenzoate
58	25	10	62	70	68	66	21	–	–	Aminophenazone/Amidopyrine
58	42	52	69	88	90	71	2	55	59	Quazepam M (3-Hydroxy-2-oxo-)
58	49	59	76	86	87	76	5	–	–	Ethinamate
58	50	64	40	88	–	–	–	–	–	Cyclobarbital
58	50	65	41	86	–	–	–	–	–	Butobarbital/Butethal
58	52	66	44	88	–	–	–	–	–	Amobarbital
59	36	53	75	85	79	–	4	–	–	Carisoprodol
59	38	35	41	85	–	–	–	–	–	Etamivan
59	38	53	71	–	–	76	–	–	–	Pheneturide
59	50	53	66	84	–	70	5	–	–	Ethosuximide
59	50	64	38	88	–	–	–	–	–	Heptabarbital
59	50	65	39	90	–	–	–	–	–	Cyclopentobarbital
59	54	67	35	–	–	–	–	–	–	Allylbarbituric Acid,5-
59	55	66	45	90	–	–	–	–	–	Pentobarbital
59	55	69	41	–	–	–	–	–	–	Idobutal (5-Allyl-5-butylbarbituric Acid)
59	64	67	75	–	–	–	–	–	–	Tribromoethanol
60	34	45	72	83	88	75	2	56	58	Flurazepam M (N(1)-Dealkyl-)/Quazepam M (N-Dealkyl-2-oxo-)
60	46	45	59	82	82	52	6	61	50	Lormetazepam
60	47	55	73	82	–	75	–	–	–	Phenprobamate
60	49	48	59	84	84	–	–	–	–	Acecarbromal
60	52	48	16	92	92	–	0	–	–	Acenocoumarol
60	53	54	71	–	–	–	–	–	–	Ethotoin
60	53	67	44	–	–	–	–	–	–	Hexethal
60	53	67	46	92	–	–	–	–	–	Talbutal
60	57	65	78	86	–	–	9	–	–	Hexapropymate
60	58	68	44	92	–	–	–	–	–	Nealbarbital
61	30	14	61	76	72	71	3	63	44	Flumazenil/RO 15-1788
61	62	58	19	93	–	–	–	–	–	Phenprocoumon
62	35	40	71	–	–	–	–	–	–	Alfadolone Acetate
62	39	31	2	63	–	–	–	–	–	Nalidixic Acid
62	45	45	74	83	80	66	14	64	66	Ketazolam
62	51	55	12	88	88	76	–	–	–	Tolbutamide
62	53	58	74	85	93	76	0	62	67	Ethyl Loflazepate
62	55	68	45	88	–	–	–	–	–	Secobarbital
63	39	45	75	78	84	73	0	66	59	Cloxazolam

Table 8.4. (Continued)

TLC System 3	1	2	4	5	6	7	8	9	10	[System 3: Chl–Ac (80+20) / Silica]
63	56	62	76	84	87	67	5	57	66	Benzocaine
63	59	71	50	86	–	–	–	–	–	Sigmodal (5-(2-Bromoallyl)-5-(1-methyl-butyl)barbituric Acid)
63	60	58	79	–	–	–	–	–	–	Ergosterol
63	60	59	75	–	–	–	–	–	–	Triclocarban
64	38	37	57	81	80	71	1	59	52	Flunitrazepam M (3-Hydroxy-)
64	52	53	68	–	88	–	–	–	–	Bemegride
64	53	55	75	85	87	–	12	–	–	Carbromal
64	61	58	78	–	–	–	–	–	–	Estradiol
64	61	60	76	84	96	–	22	64	69	Cholesterol
64	64	62	18	–	–	–	–	–	–	Warfarin
65	33	25	73	81	84	76	15	68	58	Metaclazepam M (N-Nor-)
65	46	50	75	80	90	74	11	66	62	Haloxazolam
65	51	47	62	82	82	53	8	59	53	Temazepam/M of Camazepam, Diazepam, Ketazolam, Medazepam
65	53	54	71	78	95	77	15	68	66	Oxazolam
65	61	49	74	81	–	71	32	–	–	Propyphenazone
65	62	58	23	–	–	–	–	–	–	Benziodarone
65	62	65	58	–	–	–	–	–	–	Fenimide
65	73	68	81	–	92	–	–	–	–	Thymol
66	38	64	39	89	–	–	–	–	–	Vinylbital
66	52	50	7	86	–	–	0	–	–	Tolazamide
66	55	48	80	84	87	78	31	69	66	Clotiazepam
66	62	58	74	–	–	–	–	–	–	Mephenytoin
66	64	61	85	86	88	77	9	68	69	Bromazepam Hy/ABP
67	56	50	75	78	86	76	24	69	63	Fludiazepam
67	57	49	79	84	89	78	32	69	66	Tetrazepam
68	–	–	76	80	–	–	–	–	–	Clotrimazole
68	–	–	80	88	–	49	2	10	4	Procarbazine
68	52	37	69	78	–	–	–	–	–	Griseofulvin
68	63	47	42	75	75	72	1	–	–	Carbimazole
68	71	65	85	90	91	77	5	69	69	Nitrazepam Hy/ANB
68	71	66	86	87	91	77	4	69	70	Flunitrazepam M Hy/ANFB
68	77	74	49	–	–	–	–	–	–	Thiopental
69	55	32	75	82	83	76	12	73	65	Camazepam
69	64	50	75	–	79	–	–	–	–	Santonin
69	65	65	53	85	–	–	–	–	–	Hexobarbital
69	66	65	54	87	–	–	–	–	–	Metharbital
69	71	72	86	–	–	–	–	–	–	Nitroglycerin
69	72	67	53	–	–	–	–	–	–	Narcobarbital
69	77	72	43	–	–	–	–	–	–	Thialbarbital
70	–	–	30	40	88	56	0	5	9	Clioquinol
70	53	47	75	84	85	62	8	70	62	Clobazam
70	55	55	72	85	88	71	12	–	52	Prazepam M (3-Hydroxy-)
70	59	57	80	87	90	78	16	89	71	Quazepam M (2-Oxo-)
70	59	59	81	89	91	78	15	6	71	Halazepam

Table 8.4. (Continued)

TLC System 3	1	2	4	5	6	7	8	9	10	[System 3: Chl–Ac (80+20) / Silica]
70	63	62	80	86	89	75	31	–	–	Glutethimide
70	65	56	21	–	–	–	–	–	–	Phenindione
70	70	67	41	86	–	–	–	–	–	Methylphenobarbital
70	71	71	58	–	–	–	–	–	–	Enallylpropymal
70	72	66	86	87	93	77	4	69	70	Clonazepam Hy/ANCB
71	–	–	6	80	–	53	3	8	4	Azathioprine
71	47	35	79	82	84	77	40	73	62	Metaclazepam
71	51	38	17	88	–	–	–	–	–	Piroxicam
71	53	46	77	81	81	74	12	70	55	Nimetazepam
71	73	72	58	85	–	–	–	–	–	Methohexital
71	75	66	84	81	–	–	–	–	–	Clofibrate
72	17	19	28	–	–	–	–	–	–	Diiodohydroxyquinoline
72	54	47	74	80	82	63	10	72	63	Flunitrazepam
72	58	49	76	82	85	75	27	73	59	Diazepam/Ketazolam M/Medazepam M
72	64	55	81	84	89	65	36	74	63	Prazepam
72	71	59	77	–	–	75	–	–	–	Phensuximide
73	56	40	78	79	83	67	41	74	62	Medazepam
73	65	61	81	85	92	75	29	72	70	Pinazepam
73	74	77	80	87	95	–	37	–	–	Cyclandelate
74	70	69	83	81	93	83	11	71	69	Haloxazolam Hy/ABFB
74	76	68	86	89	93	78	11	75	70	Flurazepam Hy/ACFB
75	66	51	78	84	–	–	–	–	–	Spironolactone
76	76	69	86	90	91	78	12	76	71	Oxazepam Hy/ACB
76	77	70	87	89	93	78	11	76	70	Lorazepam Hy/ADB
76	78	68	65	87	–	79	–	–	–	Phenylbutazone
78	78	71	83	87	96	74	27	75	76	Quazepam
78	80	69	43	–	–	–	–	–	–	Danthron
80	7	63	86	89	90	77	20	79	71	Flunitrazepam M Hy/MNFB
80	66	66	83	76	86	73	31	77	68	Nimetazepam Hy/MNB
80	74	73	85	77	93	79	44	77	73	Quazepam Hy/CFTB
80	75	73	86	86	91	83	48	77	72	Pinazepam Hy/CPB
80	82	74	87	–	–	–	–	–	–	Clofenotane (p,p′-Isomer)/p,p′-DDT
81	71	69	83	77	90	75	48	77	69	Fludiazepam Hy/CFMB
81	81	69	87	82	93	75	49	81	7	Lormetazepam Hy/MDB
81	81	74	89	94	95	82	49	81	75	Halazepam Hy/TCB
82	81	74	86	87	–	–	0	–	–	Ethchlorvynol
82	82	72	88	89	93	79	53	81	71	Diazepam Hy/MACB
82	83	74	88	92	95	80	58	82	74	Prazepam Hy/CCB
94	91	93	99	81	87	84	40	99	99	Azinphos-methyl
95	91	92	99	80	88	86	55	99	99	Methidathion

8.5 hR$_f$ Data in TLC System 4 in Ascending Numerical Order vs. Substance Names

TLC System 4	1	2	3	5	6	7	8	9	10	[System 4: Etac–MeOH–NH4OH (85+10+5) / Silica]
0	–	–	–	–	–	–	–	–	–	Rhein
0	–	–	–	–	–	–	–	–	–	Thiobarbituric Acid
0	–	–	–	0	–	–	0	–	–	Atropine Methonitrate
0	–	–	–	0	29	0	0	1	0	Cetrimide
0	–	–	–	0	29	0	0	1	0	Cetylpyridinium
0	–	–	–	0	2	0	0	0	0	Decamethonium
0	–	–	–	0	0	0	0	0	0	Diquat Dibromide
0	–	–	–	0	2	0	0	0	0	Gallamine
0	–	–	–	0	1	0	0	0	0	Hexamethonium
0	–	–	–	0	12	0	0	0	0	Homatropine Methylbromide
0	–	–	–	0	15	1	0	0	0	Pancuronium Bromide
0	–	–	–	0	0	0	0	0	0	Paraquat
0	–	–	–	0	1	0	0	0	0	Pentolinium
0	–	–	–	0	0	1	0	0	0	Suxamethonium Chloride
0	–	–	–	0	11	0	0	0	0	Tubocurarine
0	–	–	–	1	–	15	0	1	2	Amidefrine
0	–	–	–	1	25	2	0	0	0	Thiazinamium Metilsulfate
0	–	–	–	3	–	–	–	–	–	Metformin
0	–	–	–	3	–	–	–	–	–	Moroxydine
0	–	–	–	3	76	1	0	0	0	Guanoxan
0	–	–	–	4	23	0	0	0	0	Carbachol
0	–	–	–	16	11	–	0	0	–	Betaine
0	–	–	–	16	11	17	0	0	0	Cocaine M/Ecgonine
0	–	–	–	20	0	–	–	–	0	Morphine M (-3-glucuronide)
0	–	–	–	30	2	–	0	0	0	Codeine M (-6-glucuronide)
0	–	–	–	34	14	52	0	3	0	Narceine
0	–	–	–	53	8	1	0	0	0	Obidoxime Chloride
0	–	–	–	60	–	–	–	–	–	Protheobromine
0	–	–	–	80	1	5	0	0	0	Psilocybine
0	–	–	–	81	–	–	0	–	–	Strophanthin
0	–	–	0	8	–	5	–	–	–	Oxytetracycline
0	–	0	–	27	–	–	0	–	–	Lisinopril
0	–	0	–	44	–	–	0	–	–	Flumazenil M/RO 15-3890
0	–	1	–	84	–	–	–	–	–	Ibuprofen M (2-4′-(2-Hydroxy-2-methyl-propyl)phenylpropionic Acid)
0	–	13	–	78	–	–	–	–	–	Ibuprofen M (2-4′-(2-Carboxypropyl)phenyl-propionic Acid)
0	0	0	0	–	–	–	–	–	–	Cromoglicic Acid
0	0	0	0	–	–	–	–	–	–	Enalapril
0	0	0	0	–	–	–	–	–	–	Pethidine Intermediate C
0	0	0	0	–	–	–	–	–	–	Phthalylsulfacetamide
0	0	0	0	–	–	–	–	–	–	Phthalylsulfathiazole
0	0	0	0	11	–	–	–	–	–	Levodopa
0	0	0	0	70	16	58	0	0	0	Lysergic Acid

Table 8.5. (Continued)

TLC System 4	1	2	3	5	6	7	8	9	10	[System 4: Etac–MeOH–NH4OH (85+10+5) / Silica]
0	0	0	0	84	–	–	–	–	–	Barbituric Acid
0	0	0	2	–	–	–	–	–	–	Pipemidic Acid
0	0	0	2	–	–	–	–	–	–	Succinylsulfathiazole
0	0	0	2	85	–	66	–	–	–	Sulfasalazine
0	0	0	4	78	–	–	–	–	–	Hydroxyhippuric Acid,o-
0	0	2	4	–	–	–	–	–	–	Carbidopa
0	0	52	0	–	–	–	–	–	–	Valproic Acid
0	1	0	0	–	–	–	–	–	–	Baclofen
0	1	0	4	72	–	58	–	–	–	Nicotinic Acid
0	1	1	6	–	–	–	–	–	–	Captopril
0	3	4	8	–	–	–	–	–	–	Fusaric Acid
0	5	10	10	–	–	–	–	–	–	Cyclamic Acid
0	7	17	28	–	–	–	–	–	–	Carbenoxolone
1	–	–	–	1	13	2	0	–	–	Neostigmine
1	–	–	–	2	36	3	0	1	0	Oxyphenonium Bromide
1	–	–	–	2	18	1	0	0	0	Thiamine Hydrochloride
1	–	–	–	3	–	2	–	–	–	Clidinium
1	–	–	–	3	–	3	–	–	–	Glycopyrronium Bromide
1	–	–	–	3	–	2	–	–	–	Hexocyclium Metilsulfate
1	–	–	–	3	30	1	0	2	0	Guanethidine
1	–	–	–	3	18	2	0	0	0	Methscopolamine Bromide
1	–	–	–	80	–	–	–	–	–	Proscillaridine
1	19	43	31	–	–	58	–	–	–	Aminobenzoic Acid,p-
2	–	–	–	3	41	8	0	0	0	Butylscopolammonium Bromide,N-
2	–	–	–	3	–	5	0	–	–	Emepronium Bromide
2	–	–	–	4	66	2	0	4	0	Benzalkonium Chloride
2	–	–	–	22	14	21	0	1	3	Cocaine M/N-Benzoylecgonine
2	–	–	–	60	75	49	1	1	1	Methyldopa
2	–	–	–	87	46	88	0	10	10	Dimenhydrinate
2	–	0	–	87	–	–	0	–	–	Clenbuterol M #2
2	0	0	2	85	59	84	0	1	2	Metamizol/Dipyrone
2	2	16	11	–	–	–	–	–	–	Chlorothiazide
2	6	23	19	88	69	–	–	–	–	Gentisic Acid
2	39	31	62	63	–	–	–	–	–	Nalidixic Acid
3	–	–	–	2	–	–	–	–	–	Distigmine Bromide
3	–	–	–	2	–	–	–	–	–	Tiemonium Iodide
3	–	–	–	3	41	5	0	5	0	Isopropamide Iodide
3	–	–	–	9	–	2	–	–	–	Penthienate Bromide
3	–	–	–	43	–	0	0	0	0	Levarterenol/Norepinephrine
3	–	–	–	44	–	–	0	–	–	Pirenzepine M (N-Desmethyl-)
3	–	–	–	44	–	36	1	3	3	Penicillamine
3	–	–	–	83	–	–	–	–	–	Thioctic Acid
3	–	22	–	84	–	–	–	–	–	Flurbiprofen M #1
4	–	–	–	3	31	4	0	4	0	Propantheline Bromide
4	–	–	–	86	–	–	–	–	–	Nafcillin
4	–	5	–	86	–	–	–	–	–	Tiaprofenic Acid

Table 8.5. (Continued)

TLC System 4	1	2	3	5	6	7	8	9	10	[System 4: Etac–MeOH–NH4OH (85+10+5) / Silica]
4	1	10	6	87	–	–	–	–	–	Bumetanide
4	4	6	11	–	–	–	–	–	–	Dichlorophenoxy)acetic Acid,(2,4-/2,4-D
4	8	18	37	–	–	–	–	–	–	Piroxicam M (5-Hydroxy-)
4	12	12	–	88	–	–	–	–	–	Zomepirac
4	12	23	27	83	–	65	–	–	–	Sulfamethizole
4	14	10	34	87	–	–	–	–	–	Sulindac
4	17	42	28	87	–	70	–	–	–	Sulfacetamid
4	18	28	33	–	–	–	–	–	–	Alclofenac
4	18	30	39	–	–	–	–	–	–	Fenbufen
4	22	39	38	81	–	64	–	–	–	Sulfadiazine
4	26	28	35	–	–	–	–	–	–	Benzoic Acid
5	–	–	–	4	39	2	0	4	0	Fenpiverinium
5	–	–	–	83	–	76	0	65	29	Glymidine
5	–	–	–	85	–	–	0	–	–	Glisoxepide
5	–	–	–	89	–	–	–	–	–	Bezafibrate
5	–	8	–	87	–	–	–	–	–	Propantheline M/Xanthanoic Acid
5	–	28	–	85	–	–	–	–	–	Pirprofen
5	3	2	5	71	–	–	–	–	–	Etacrynic Acid
5	4	31	18	84	–	85	–	–	–	Acetazolamide
5	5	16	27	80	–	64	–	–	–	Sulfisomidine
5	5	24	15	–	–	70	–	–	–	Aminosalicylic Acid,p-
5	9	11	15	–	–	–	–	–	–	MCPA/(4-Chloro-o-tolyloxy)acetic Acid
5	13	3	24	87	–	–	–	–	–	Probenecid
5	13	20	30	85	–	–	–	–	–	Tolmetin
5	16	13	38	83	63	–	–	–	–	Indometacin
5	23	26	32	–	–	–	–	–	–	Chlorobenzoic Acid,p-
5	26	54	41	79	–	65	–	–	–	Sulfamethoxazole
5	40	60	54	92	–	–	–	–	–	Glibornuride
6	–	–	–	2	18	2	1	3	0	Oxyphencyclimine
6	–	–	–	3	–	–	–	–	–	Buzepide
6	–	–	–	88	–	–	–	–	–	Lonazolac
6	–	–	–	89	–	–	–	–	–	Lanatoside C
6	–	–	71	80	–	53	3	8	4	Azathioprine
6	–	19	–	87	–	–	0	–	–	Clenbuterol M #3
6	1	7	7	86	–	–	0	–	–	Furosemide
6	2	30	33	84	–	–	–	–	–	Nitrofurantoin
6	23	49	33	87	–	–	2	–	–	Sulfasomizole
6	25	52	33	81	–	65	–	–	–	Sulfafurazole
6	27	25	41	85	–	–	–	–	–	Ketoprofen
6	30	30	45	–	–	–	–	–	–	Flurbiprofen
6	33	38	44	82	–	–	–	–	–	Naproxen
6	33	40	50	84	–	–	–	–	–	Chlorambucil
6	42	38	50	–	–	–	–	–	–	Fenoprofen
6	46	57	54	75	–	–	–	–	–	Ibuprofen
7	–	–	–	0	3	13	0	0	0	Histamine
7	–	–	–	5	44	17	0	0	0	Morphine M (Nor-)

Table 8.5. (Continued)

TLC System 4	1	2	3	5	6	7	8	9	10	[System 4: Etac–MeOH–NH4OH (85+10+5) / Silica]
7	0	60	56	–	–	–	–	–	–	Paramethadione
7	1	1	4	87	–	–	–	–	–	Saccharin
7	21	46	47	–	–	–	–	–	–	Enoxolone
7	52	50	66	86	–	–	0	–	–	Tolazamide
8	–	–	–	–	–	10	–	–	–	Betahistine
8	–	–	–	–	–	–	–	–	–	Mequitazine M
8	–	–	–	1	33	–	4	–	–	Tropine
8	–	–	–	4	56	7	38	0	0	Berberine
8	–	–	–	19	–	–	0	–	–	Clozapine M (N-Oxyde)
8	–	–	–	86	–	–	–	–	–	Bumadizon
8	–	–	–	88	–	68	53	5	67	Azapropazone
8	0	0	7	82	72	75	0	2	0	Cinchophen
8	9	20	27	79	–	66	–	–	–	Sulfathiazole
8	14	35	34	–	–	67	–	–	–	Sulfaethidole
8	23	41	42	80	–	65	–	–	–	Sulfamerazine
8	40	50	54	–	–	67	–	–	–	Sulfalene
9	–	–	–	48	–	–	–	–	–	Cimetidine M (Sulfoxide)
9	–	–	–	84	–	–	0	–	–	Gliclazide
9	4	10	22	–	–	–	–	–	–	Aloxiprin
9	14	51	30	–	–	–	–	–	–	Benzthiazide
9	18	30	31	78	–	–	–	–	–	Acetylsalicylic Acid
9	19	36	50	–	–	–	–	–	–	Dantrolene
9	19	39	43	83	–	65	–	–	–	Sulfamethoxypyridazine
9	29	51	43	89	–	69	–	–	–	Sulfaphenazole
9	52	62	57	90	–	77	0	–	–	Oxyphenbutazone/Phenylbutazone M (Hydroxy-)
10	–	–	–	80	88	5	62	0	0	Phencyclidine interm./1-Piperidino-1-cyclohexanecarbonitrile/PCC
10	–	20	–	84	–	–	–	–	–	Flurbiprofen M #2
10	7	1	24	86	–	–	–	–	–	Salicylic Acid
10	31	51	48	86	–	65	–	–	–	Sulfadimethoxine
10	38	43	49	87	88	72	0	–	–	Chlorpropamide
11	–	–	–	8	–	25	3	1	18	Synephrine/Oxedrine
11	–	–	–	12	73	55	24	4	38	Methoxamine
11	–	–	–	85	–	–	–	–	–	Trospium Chloride
11	–	9	–	74	66	75	1	30	11	Theophylline
11	3	3	15	88	–	–	–	–	–	Niflumic Acid
11	30	30	57	90	–	–	0	–	–	Glibenclamide/Glyburide
11	41	48	54	87	–	–	–	–	–	Mefenamic Acid
12	–	–	–	3	–	–	–	–	–	Viquidil
12	–	–	–	8	67	33	1	1	0	Phenylephrine
12	–	43	–	–	–	–	–	–	–	Meclofenamic Acid
12	24	43	47	82	–	60	–	–	–	Sulfametoxydiazine
12	25	27	47	90	–	–	–	–	–	Diclofenac
12	39	43	53	–	–	–	–	–	–	Acetohexamide
12	51	55	62	88	88	76	–	–	–	Tolbutamide

Table 8.5. (Continued)

TLC System										[System 4:
4	1	2	3	5	6	7	8	9	10	Etac–MeOH–NH4OH (85+10+5) / Silica]
13	–	–	–	1	–	33	–	–	–	Chlorhexidine
13	–	–	–	3	–	0	0	1	0	Epinephrine
13	–	–	–	12	–	–	–	–	–	Norfenefrine
13	23	45	44	79	–	62	–	–	–	Sulfadimidine
13	38	64	36	93	–	–	0	–	–	Xipamide
14	–	–	–	3	–	10	0	3	1	Azacyclonol
14	–	–	–	85	–	–	–	–	–	Chlorotheophylline,8-
14	15	60	23	88	–	80	–	–	–	Trichlormethiazide
15	–	–	–	–	–	55	36	2	1	Dihydralazine
15	–	–	–	75	67	59	0	8	5	Pyridoxine Hydrochloride
16	–	–	–	12	–	47	10	44	6	Hexobendine
16	–	–	–	18	–	–	–	–	–	Pirenzepine
16	–	–	–	27	20	41	0	9	0	Clomipramine M (N-Oxid)
16	0	0	10	–	–	–	–	–	–	Carbazochrome
16	4	4	30	–	–	–	–	–	–	Sulfinpyrazone
16	8	5	18	89	–	–	–	–	–	Diflunisal
16	52	48	60	92	92	–	0	–	–	Acenocoumarol
17	–	–	–	11	–	–	0	–	–	Octopamine
17	–	–	–	99	–	33	54	4	63	Pseudoephedrine
17	–	50	–	93	–	–	0	–	–	Chlorphenprocoumon,p-
17	51	38	71	88	–	–	–	–	–	Piroxicam
18	–	–	–	7	79	3	0	1	1	Proguanil/Chlorguanide
18	–	–	–	12	–	25	2	3	1	Dihydromorphine
18	–	–	–	12	14	23	3	9	2	Hydromorphone
18	–	–	–	13	76	42	1	1	24	Metaraminol
18	–	–	–	21	77	48	1	3	6	Orciprenaline
18	–	–	–	23	16	37	0	9	0	Mianserin M (N-Oxide)
18	–	–	–	84	–	–	–	–	–	Flufenamic Acid
18	–	–	–	99	–	–	3	–	–	Lorcainide M
18	11	19	31	–	–	–	–	–	–	Bufexamac
18	64	62	64	–	–	–	–	–	–	Warfarin
19	–	–	–	92	–	–	–	–	–	Feprazone
19	62	58	61	93	–	–	–	–	–	Phenprocoumon
20	–	–	–	–	–	–	–	–	–	Piridoxilate
20	–	–	–	14	–	42	1	1	1	Nadolol
20	–	–	–	18	23	37	0	9	1	Morphine
20	–	0	–	16	74	46	1	1	4	Salbutamol/Albuterol
21	–	–	–	14	69	40	0	1	3	Isoprenaline
21	–	–	–	18	77	47	1	1	5	Terbutaline
21	–	–	–	75	–	–	–	–	–	Allopurinol
21	65	56	70	–	–	–	–	–	–	Phenindione
22	–	–	–	14	–	45	0	2	2	Atenolol
22	–	–	–	14	–	23	1	2	5	Heptaminol
22	–	–	–	15	74	41	2	2	3	Etilefrine
23	–	–	–	7	27	1	5	1	1	Homatropine
23	–	–	–	8	–	–	–	–	–	Homofenazine M

Table 8.5. (Continued)

TLC System 4	1	2	3	5	6	7	8	9	10	[System 4: Etac–MeOH–NH4OH (85+10+5) / Silica]
23	–	–	–	25	–	–	–	–	–	Xanthinol Nicotinate
23	–	–	–	28	–	–	–	–	–	Perphenazine M
23	–	–	–	86	–	–	–	–	–	Amino-5-bromo-3-hydroxyphenyl)(2-pyridyl) ketone,(2-
23	–	–	–	90	–	–	–	–	–	Hydroxybenzoic Acid, p-
23	–	–	–	91	–	91	–	–	–	Niclosamide
23	29	61	46	–	–	–	–	–	–	Methyl-5-phenylbarbituric Acid,5-
23	62	58	65	–	–	–	–	–	–	Benziodarone
24	–	–	–	5	28	18	5	3	1	Atropine/Hyoscyamine
24	–	–	–	6	63	20	3	10	12	Mescaline
24	–	–	–	82	21	–	6	28	7	Morphine-3-acetate
24	–	–	–	94	–	–	–	–	–	Benzbromarone
24	–	–	0	6	74	24	0	1	2	Amiloride
24	3	1	29	75	71	63	0	29	8	Triazolam M (4-Hydroxy-)
24	4	32	21	–	–	–	–	–	–	Ethyl Biscoumacetate
24	16	42	34	–	–	67	–	–	–	Sulfapyridine
25	–	–	–	–	–	–	–	–	–	Aloe-emodin
25	–	–	–	3	55	13	2	2	2	Tolazoline
25	–	–	–	5	7	16	0	17	1	Brucine
25	–	–	–	10	64	30	5	5	1	Ephedrine
25	–	–	–	11	–	–	–	–	–	Perazine M #4
25	–	–	–	15	–	36	3	18	2	Pholcodine
25	–	–	–	58	–	–	–	–	–	Orazamide
25	–	–	–	70	59	48	0	12	12	Diprophylline/Propyphylline
25	–	0	–	38	81	76	0	1	4	Fenoterol
25	1	6	7	75	–	65	–	–	–	Sulfaguanidine
26	–	–	–	1	22	2	38	1	0	Cotarnine
26	–	–	–	5	60	13	7	2	2	Tetryzoline
26	–	–	–	6	–	23	1	2	11	Tryptamine
26	–	–	–	13	–	–	–	–	–	Thioproperazine M
26	–	–	–	26	–	–	–	–	–	Midodrine
26	–	–	–	33	–	–	–	–	–	Fluphenazine M
26	–	–	–	88	–	–	–	–	–	Phenytoin M (4-Hydroxyphenyl-)
26	–	–	–	96	98	95	0	10	90	Closantel
26	3	2	26	73	69	67	1	25	7	Alprazolam M (4-Hydroxy-)
27	–	–	–	3	52	14	3	6	3	Naphazoline
27	–	–	–	3	–	10	6	6	0	Mequitazine
27	–	–	–	9	–	29	3	3	3	Pholedrine
27	–	–	–	11	14	–	5	13	3	Metopon
27	–	–	–	53	55	54	0	9	12	Cimetidine
27	–	–	–	64	–	–	–	–	–	Fosazepam
27	1	1	2	22	–	49	–	–	–	Mafenide
28	–	–	–	6	37	27	2	8	1	Dothiepin M (Nor-,Sulfoxide)
28	–	–	–	9	–	–	–	–	–	Promazine M
28	–	–	–	81	–	–	0	–	–	Clindamycin
28	5	4	37	76	76	68	1	35	13	Brotizolam M (6-Hydroxy-)

Table 8.5. (Continued)

TLC System 4	1	2	3	5	6	7	8	9	10	[System 4: Etac–MeOH–NH4OH (85+10+5) / Silica]
28	17	19	72	–	–	–	–	–	–	Diiodohydroxyquinoline
28	47	65	53	85	–	–	–	–	–	Phenobarbital
29	–	–	–	7	70	33	6	11	12	Trimethoxyamphetamine,3,4,5-
29	–	–	–	11	19	26	8	13	2	Dihydrocodeine
29	–	–	–	12	–	42	25	5	46	Pseudoephedrine M (Nor-)/Cathine
29	–	–	–	31	–	–	–	–	–	Dixyracine M
29	–	–	18	55	49	47	1	11	20	Isoniazid
29	–	1	–	32	–	–	0	–	–	Labetalol
29	52	68	57	87	–	–	–	–	–	Brallobarbital
30	–	–	–	5	64	13	7	5	3	Xylometazoline
30	–	–	–	11	–	38	3	6	1	Mesoridazine
30	–	–	–	13	–	–	–	–	–	Thiethylperazine M
30	–	–	–	19	–	53	1	3	5	Sotalol
30	–	–	13	50	65	51	1	8	4	Triamterene
30	–	–	70	40	88	56	0	5	9	Clioquinol
30	–	0	–	12	75	44	4	4	33	Ephedrine M/Phenylpropanolamine
30	18	32	33	–	–	–	–	–	–	Dicoumarol
30	32	34	44	94	98	–	0	–	–	Hexachlorophene
31	–	–	–	6	61	26	5	11	14	Dimethoxyphenethylamine,3,4-
31	–	–	–	77	–	–	–	–	–	Vinclozolin
31	1	1	25	62	50	64	0	–	–	Bromazepam M (N(py)-Oxide)
32	–	–	–	–	–	–	–	–	–	Allyl-5-ethylbarbituric Acid,5-
32	–	–	–	7	–	23	26	4	2	Methoxyphenamine
32	–	–	–	8	75	33	1	4	1	Clomipramine M (8-Hydroxy-nor-)
32	–	–	–	8	63	–	7	8	2	Pethidine M (Nor-)
32	–	–	–	8	11	26	8	19	2	Strychnine
32	–	–	–	57	59	59	1	23	29	Nalorphine
32	41	61	57	84	–	–	–	–	–	Barbital
32	50	66	56	91	–	–	–	–	–	Ibomal/Propallylonal
33	–	–	–	6	–	32	1	3	2	Phentolamine
33	–	–	–	10	34	35	0	1	1	Bufotenine
33	–	–	–	11	13	25	4	20	4	Hydrocodone
33	–	–	–	13	71	47	0	3	6	Acebutolol
33	–	–	–	14	–	–	–	–	–	Carteolol
33	–	–	–	15	–	–	–	–	–	Celiprolol
33	–	–	–	27	36	48	10	37	30	Oxymorphone
33	–	–	–	32	–	–	–	–	–	Clopenthixol M
33	–	–	–	35	–	–	–	–	–	Periciazine M
33	–	–	–	62	60	57	0	12	8	Ergometrine
33	–	–	–	69	63	55	0	37	12	Colchicine
33	1	5	28	85	–	–	–	–	–	Digoxin
33	14	64	23	–	–	82	–	–	–	Diclofenamide
33	54	61	56	88	90	85	0	51	47	Chlorzoxazone
34	–	–	–	12	–	–	1	–	–	Propafenone M (N-Desmethyl)
34	–	–	–	17	–	38	0	–	–	Sulpiride
34	–	–	–	31	–	–	–	–	–	Oxypendyl

Table 8.5. (Continued)

TLC System 4	1	2	3	5	6	7	8	9	10	[System 4: Etac–MeOH–NH4OH (85+10+5) / Silica]
34	–	–	–	80	–	9	1	1	1	Oxymetazoline
34	–	–	–	83	–	–	–	–	–	Etiroxate
34	–	4	–	59	54	53	1	31	21	Theobromine
34	0	72	0	90	–	–	–	–	–	Thiobarbital
34	4	34	11	78	–	–	–	–	–	Hydrochlorothiazide
34	50	65	57	89	–	–	–	–	–	Vinbarbital
34	50	66	56	87	–	–	–	–	–	Allobarbital
34	58	67	57	91	–	–	–	–	–	Dichlorophen
35	–	–	–	7	77	23	17	7	4	Amantadine
35	–	–	–	10	–	–	–	–	–	Prothipendyl M
35	–	–	–	12	–	32	–	–	–	Methylephedrine,N-
35	–	–	–	21	22	33	6	18	3	Codeine
35	–	–	–	56	–	53	0	41	11	Demecolcine
35	54	67	59	–	–	–	–	–	–	Allylbarbituric Acid,5-
36	–	–	–	6	71	15	18	5	2	Maprotiline
36	–	–	–	6	–	13	26	6	2	Benzatropine
36	–	–	–	12	–	–	–	–	–	Tiotixene M
36	–	–	–	12	–	–	–	–	–	Tiotixene M #1
36	–	–	–	12	–	–	–	–	–	Tiotixene M #2
36	–	–	–	14	–	–	–	–	–	Pecazine M
36	–	–	–	21	–	–	–	–	–	Trifluoperazine M
36	–	–	–	21	26	40	7	22	6	Ethylmorphine
36	–	–	–	25	–	42	6	23	6	Viloxazine
36	–	–	–	57	51	60	0	19	7	Lysergamide
36	–	–	–	58	–	–	0	–	–	Nifenazone
36	–	–	40	81	–	–	–	–	–	Sulfaguanole
36	–	11	–	81	81	60	0	23	40	Pemoline
36	3	10	42	88	–	–	–	–	–	Digitoxin
36	7	47	13	87	–	–	–	–	–	Hydroflumethiazide
36	11	31	34	86	90	69	0	31	48	Chloramphenicol
36	49	64	56	86	–	–	1	–	–	Reposal
37	–	–	–	1	15	0	46	0	0	Hydrastinine
37	–	–	–	5	70	26	39	13	5	Coniine
37	–	–	–	7	–	45	1	2	3	Hydroxychloroquine
37	–	–	–	26	–	47	2	9	6	Ketobemidone
37	–	–	–	40	–	–	–	–	–	Flupentixol M
37	–	–	–	73	–	–	–	–	–	Clonazepam M (7-Amino-3-hydroxy-)
37	4	4	37	76	76	72	1	44	18	Alprazolam M (α-Hydroxy-)
38	–	–	–	6	69	19	18	7	2	Protriptyline
38	–	–	–	10	–	–	–	–	–	Thioridazine M
38	–	–	–	20	–	–	0	–	–	Dibenzepin M (Dinor-,5-desmethyl-)
38	–	–	–	20	–	–	1	–	–	Dibenzepin M (Nor-,5-desmethyl-)
38	–	–	–	23	–	55	3	6	0	Bamethan
38	–	–	–	35	39	54	6	22	7	Opipramol
38	–	–	–	59	–	65	–	–	–	Dropropizine
38	–	–	–	61	–	52	10	7	15	Oxetacaine

Table 8.5. (Continued)

TLC System 4	1	2	3	5	6	7	8	9	10	[System 4: Etac–MeOH–NH4OH (85+10+5) / Silica]
38	–	–	–	69	–	–	0	–	–	Benzquinamide M (Desacetyl-,N-deethyl)
38	–	–	–	82	–	–	–	–	–	Methylprimidone,4-
38	–	–	0	70	–	55	0	4	8	Ambazone
38	–	6	–	66	–	–	0	–	–	Etofylline
38	2	5	33	78	81	75	4	25	14	Midazolam M (α,4-Dihydroxy-)
38	50	64	59	88	–	–	–	–	–	Heptabarbital
39	–	–	–	13	–	–	6	–	–	Homofenazine
39	–	–	–	17	33	49	1	5	9	Procainamide
39	–	–	–	18	–	–	0	–	–	Propafenone M (5-OH)
39	–	–	–	21	–	–	3	–	–	Dibenzepin M (Dinor-)
39	–	–	–	39	–	–	–	–	–	Imiclopazine/Chlorimiphenine
39	–	–	–	39	–	54	5	27	7	Carfenazine
39	–	20	–	89	–	–	24	–	–	Methylenedioxymethamphetamine,3,4-/MDMA/XTC
39	0	6	26	73	61	61	0	24	28	Bromazepam M (3-Hydroxy-)
39	11	17	40	81	–	–	2	–	–	Guaifenesin
39	38	64	66	89	–	–	–	–	–	Vinylbital
39	46	46	53	–	–	–	–	–	–	Glutaral
39	50	65	59	90	–	–	–	–	–	Cyclopentobarbital
40	–	–	–	4	10	5	67	3	5	Sparteine
40	–	–	–	7	71	26	19	11	3	Desipramine/Imipramine M
40	–	–	–	16	–	–	–	–	–	Bunitrolol
40	–	–	–	16	–	–	–	–	–	Propiomazine M
40	–	–	–	16	–	–	1	–	–	Tiapride
40	–	–	–	18	–	–	–	–	–	Promethazine M #1
40	–	–	–	18	–	–	–	–	–	Promethazine M #2
40	–	–	–	18	–	42	3	34	6	Nicomorphine
40	–	–	–	19	–	–	6	–	–	Dibenzepin M (Nor-)
40	–	–	–	21	69	44	2	7	3	Hydroquinine
40	–	–	–	60	–	–	0	–	–	Dihydroergotamine M (8′-OH)
40	–	–	–	68	66	54	0	21	27	Nicotinamide
40	0	0	14	20	60	58	2	18	29	Flurazepam M (Dideethyl-)
40	1	0	28	49	39	63	1	36	2	Adinazolam M (Nor-)
40	3	1	36	26	15	40	1	48	5	Loprazolam
40	3	3	33	68	60	69	0	32	42	Chlordiazepoxide M (Nor-)
40	4	21	15	–	–	75	–	–	–	Quinethazone
40	48	65	57	86	–	–	–	–	–	Aprobarbital
40	50	64	58	88	–	–	–	–	–	Cyclobarbital
41	–	–	–	9	76	51	15	17	16	Dimethoxy-4-methylamphetamine,2,5-/STP/DOM
41	–	–	–	19	–	–	1	–	–	Clozapine M (Nor-)
41	–	–	–	20	23	41	6	23	8	Benzylmorphine
41	–	–	–	53	–	–	4	–	–	Cotinine
41	–	–	–	69	66	76	0	41	8	Domperidone
41	–	–	–	69	–	62	0	14	12	Methylergometrine
41	–	–	–	80	–	62	1	52	59	Thiamazole

Table 8.5. (Continued)

TLC System 4	1	2	3	5	6	7	8	9	10	[System 4: Etac–MeOH–NH4OH (85+10+5) / Silica]
41	–	27	–	87	–	–	0	–	–	Methylprednisolon
41	8	23	28	76	–	–	–	–	–	Primidone
41	15	24	42	81	83	63	0	35	51	Demoxepam/Chlordiazepoxide M
41	33	55	53	86	–	–	–	–	–	Phenytoin
41	38	35	59	85	–	–	–	–	–	Etamivan
41	46	65	56	–	–	–	–	–	–	Probarbital
41	50	65	58	86	–	–	–	–	–	Butobarbital/Butethal
41	55	69	59	–	–	–	–	–	–	Idobutal (5-Allyl-5-butylbarbituric Acid)
41	70	67	70	86	–	–	–	–	–	Methylphenobarbital
42	–	–	–	9	63	31	28	13	5	Deoxyephedrine,(+/-)-
42	–	–	–	9	63	31	28	13	5	Methamfetamine
42	–	–	–	10	49	35	14	4	3	Dextrorphan/Levorphanol
42	–	–	–	10	76	39	18	12	17	Methylenedioxyamphetamine,3,4-/MDA
42	–	–	–	11	–	–	–	–	–	Metixene M
42	–	–	–	16	–	–	–	–	–	Chlorpromazine M
42	–	–	–	40	40	55	7	29	9	Perphenazine
42	–	–	–	58	–	60	1	28	16	Dihydroergotamine
42	–	–	–	70	69	69	1	23	17	Iproniazid
42	–	–	–	83	–	–	–	–	–	Isovaleryl-1,3-indanedione,2-
42	4	3	39	75	74	71	4	49	21	Triazolam M (α-Hydroxy-)
42	4	40	23	88	83	–	–	–	–	Chlortalidone
42	7	11	30	79	82	73	0	22	45	Clonazepam M (7-Acetamido-)
42	63	47	68	75	75	72	1	–	–	Carbimazole
43	–	–	–	8	75	36	24	10	14	Dimethoxy-4-ethylamphetamine,2,5-/DOET
43	–	–	–	9	74	35	23	10	18	Methoxy-4,5-methylenedioxyamphetamine,2-/MMDA-2
43	–	–	–	9	74	73	23	77	69	Methoxyamphetamine,4-/PMA
43	–	–	–	10	76	36	15	14	17	Bromo-2,5-dimethoxyamphetamine,4-/DOB
43	–	–	–	12	–	–	–	–	–	Etryptamin
43	–	–	–	12	75	43	20	9	18	Amphetamine/Amphetaminil M artifact
43	–	–	–	14	59	18	0	0	0	Dopamine
43	–	–	–	18	78	49	2	5	8	Pindolol
43	–	–	–	20	70	45	3	8	5	Hydroquinidine
43	–	–	–	20	–	–	3	–	–	Mepindolol
43	–	–	–	22	–	–	–	–	–	Nifenalol
43	–	–	–	22	22	46	7	34	6	Thioproperazine
43	–	–	–	30	–	–	1	–	–	Clenbuterol M #1 (Hydroxy-)
43	–	–	–	33	61	56	1	12	8	Pipamperone
43	–	–	–	41	52	65	0	17	14	Pipamazine
43	–	–	–	54	–	–	17	–	–	Maprotiline M (Nor-)
43	–	1	–	17	–	–	13	–	–	Tertatolol
43	–	65	–	81	–	–	–	–	–	Crotylbarbital
43	23	41	42	82	82	52	1	36	28	Lorazepam/Lormetazepam M
43	43	65	51	–	–	76	–	–	–	Ethoxzolamide
43	77	72	69	–	–	–	–	–	–	Thialbarbital
43	80	69	78	–	–	–	–	–	–	Danthron

Table 8.5. (Continued)

TLC System 4	1	2	3	5	6	7	8	9	10	[System 4: Etac–MeOH–NH4OH (85+10+5) / Silica]
44	–	–	–	19	61	49	6	12	5	Cinchonine
44	–	–	–	19	22	45	17	28	5	Dothiepin M (Nor-)
44	–	–	–	20	–	–	–	–	–	Metipranolol
44	–	–	–	20	74	49	10	8	9	Metoprolol
44	–	–	–	21	–	–	–	–	–	Isothipendyl M
44	–	–	–	24	55	49	6	8	6	Cinchonidine
44	–	–	–	25	27	46	6	26	8	Morphine-6-acetate/Diamorphine M
44	–	–	–	26	24	49	10	40	7	Tiotixene
44	–	–	–	42	74	60	2	23	–	Tocainide
44	–	–	–	45	–	56	7	32	11	Clopenthixol/Zuclopenthixol/cis-Ordinol
44	–	–	–	52	45	53	0	32	12	Pilocarpine
44	–	–	–	68	64	63	1	34	22	Ergotamine
44	–	–	–	82	87	68	0	37	42	Dipyridamole
44	–	0	–	21	–	–	12	–	–	Toliprolol
44	5	2	41	68	65	60	1	40	16	Triazolam
44	7	12	24	80	83	71	0	20	49	Nitrazepam M (7-Acetamido-)
44	13	62	25	–	–	–	–	–	–	Epithiazide
44	52	66	58	88	–	–	–	–	–	Amobarbital
44	53	67	60	–	–	–	–	–	–	Hexethal
44	54	67	57	87	–	–	1	–	–	Butalbital
44	58	68	60	92	–	–	–	–	–	Nealbarbital
44	62	65	55	81	–	–	–	–	–	Methyl p-Hydroxybenzoate/Methyl Paraben
45	–	–	–	8	75	36	19	10	17	Dimethoxyamphetamine,2,5-
45	–	–	–	8	75	34	24	13	3	Clomipramine M (Nor-)
45	–	–	–	20	78	48	11	11	13	Oxprenolol
45	–	–	–	23	32	45	24	37	5	Thebaine
45	–	–	–	24	–	–	–	–	–	Aceprometazine M
45	–	–	–	26	51	70	1	24	11	Clomipramine M (2-Hydroxy-)
45	–	–	–	26	65	51	2	11	4	Quinine
45	–	–	–	45	49	63	5	23	10	Fluphenazine
45	–	–	20	45	59	55	0	22	12	Trimethoprim
45	–	28	–	86	–	–	0	–	–	Hydrocortisone
45	7	6	41	78	78	72	2	46	31	Brotizolam M (α-Hydroxy-)
45	15	32	26	77	–	–	0	–	–	Paracetamol
45	18	14	50	66	66	65	4	–	–	Phenazone/Antipyrine
45	22	35	42	82	82	56	0	40	51	Oxazepam/M of Chlordiazepoxide, Clorazepic Acid, Camazepam, Diazepam, Halazepam, Ketazolam, Medazepam, Oxazolam, Prazepam
45	53	68	57	88	–	–	–	–	–	Butallylonal
45	55	66	59	90	–	–	–	–	–	Pentobarbital
45	55	68	62	88	–	–	–	–	–	Secobarbital
46	–	–	–	12	21	45	35	18	2	Chlorphenamine/Chlorpheniramine
46	–	–	–	14	26	45	35	13	3	Pheniramine
46	–	–	–	18	–	–	12	–	–	Fluvoxamine
46	–	–	–	20	22	45	11	28	5	Dothiepin M (Sulfoxide)

Table 8.5. (Continued)

TLC System 4	1	2	3	5	6	7	8	9	10	[System 4: Etac–MeOH–NH4OH (85+10+5) / Silica]
46	–	–	–	24	–	–	11	–	–	Erythromycin
46	–	–	–	34	45	50	14	27	14	Phenmetrazine/Phendimetrazine M (Nor-)
46	–	–	–	48	–	–	–	–	–	Thiopropazate M
46	–	–	–	50	–	62	6	33	–	Flupentixol
46	–	–	–	65	14	52	2	52	21	Endralazine Mesilate
46	–	–	–	87	71	34	28	16	4	Amitriptyline M/Nortriptyline
46	–	–	0	4	14	38	14	4	2	Chloroquine
46	–	–	32	75	70	58	2	36	40	Metronidazole
46	–	3	–	81	78	67	1	38	40	Glafenine
46	–	9	–	81	–	–	0	–	–	Metaclazepam M (Dinor-)
46	11	33	53	–	–	–	–	–	–	Diphenadione
46	53	67	60	92	–	–	–	–	–	Talbutal
46	64	66	57	88	–	–	–	–	–	Ethyl p-Hydroxybenzoate
47	–	–	–	5	66	31	6	7	3	Antazoline
47	–	–	–	10	–	42	35	13	6	Dimetindene
47	–	–	–	10	42	33	42	18	6	Dextromethorphan/Racemethorphan/Levo-methorphan
47	–	–	–	14	48	39	5	9	9	Psilocin
47	–	–	–	19	–	43	12	42	10	Nicocodine
47	–	–	–	21	23	48	25	37	3	Perazine
47	–	–	–	24	–	–	2	–	–	Trimethobenzamide
47	–	–	–	50	–	–	–	–	–	Cafedrine
47	–	–	–	61	–	–	–	–	–	Ampyrone
47	–	–	–	68	–	–	2	–	–	Benzquinamide M (Nor-)
47	–	–	–	74	–	65	9	66	63	Naloxone
47	–	–	–	90	–	90	2	67	70	Buclosamide
47	–	0	–	21	–	–	1	–	–	Carazolol
47	–	0	–	42	–	–	4	–	–	Zopiclone
47	7	2	40	67	66	67	1	57	14	Alprazolam
48	–	–	–	11	78	46	26	24	12	Phentermine
48	–	–	–	12	–	47	17	13	6	Pipazetate
48	–	–	–	14	77	44	18	17	8	Chlorphentermine
48	–	–	–	14	–	45	26	21	7	Tofenacin
48	–	–	–	17	–	–	3	–	–	Propafenone
48	–	–	–	20	–	–	–	–	–	Levomepromazine M
48	–	–	–	20	–	47	27	25	–	Zimeldine
48	–	–	–	25	–	54	3	16	15	Fenpipramide
48	–	–	–	49	47	55	6	37	18	Scopolamine
48	–	–	–	72	–	–	–	–	–	Ketoconazole
48	–	–	–	74	69	56	4	17	22	Nicotinyl Alcohol
48	–	0	–	22	–	–	8	–	–	Bisoprolol
48	50	64	57	88	–	–	–	–	–	Secbutabarbital
48	52	66	57	–	–	–	–	–	–	Tetrabarbital
49	–	–	–	11	–	–	–	–	–	Pentorex
49	–	–	–	12	–	45	33	16	6	Brompheniramine
49	–	–	–	21	79	50	6	10	7	Propranolol

Table 8.5. (Continued)

TLC System 4	1	2	3	5	6	7	8	9	10	[System 4: Etac–MeOH–NH4OH (85+10+5) / Silica]
49	–	–	–	22	–	52	11	12	11	Alprenolol
49	–	–	–	24	25	45	20	34	11	Thebacon
49	–	–	–	26	33	47	15	38	4	Diamorphine
49	–	–	–	28	–	–	6	–	–	Flecainide
49	–	–	–	30	63	51	4	12	6	Quinidine
49	–	–	–	34	–	–	12	–	–	Bupivacaine M/Mepivacaine M
49	–	–	–	47	–	–	–	–	–	Dixyracine
49	–	–	–	52	–	–	–	–	–	Etodroxizine
49	–	–	–	62	–	–	–	–	–	Bisbentiamin
49	–	–	–	63	–	–	0	–	–	Ketanserin M/Ketanserinol
49	–	–	–	71	–	58	2	33	29	Proxyphylline
49	–	–	–	87	–	52	0	1	3	Dobutamine
49	–	0	–	20	–	–	12	–	–	Bupranolol
49	5	5	37	76	77	74	1	43	15	Midazolam M (4-Hydroxy-)
49	7	23	38	84	–	70	–	–	–	Methocarbamol
49	15	28	35	88	89	67	0	30	38	Quazepam M (3-Hydroxy-N-dealkyl-2-oxo-)
49	77	74	68	–	–	–	–	–	–	Thiopental
50	–	–	–	0	–	–	–	–	–	Tropicamide
50	–	–	–	13	–	–	–	–	–	Bromopride
50	–	–	–	13	16	48	26	19	4	Carbinoxamine
50	–	–	–	13	28	42	35	22	6	Tropacocaine
50	–	–	–	14	39	40	9	9	6	Dimethyltryptamine,N,N-/DMT
50	–	–	–	16	–	44	36	22	9	Benzydamine
50	–	–	–	19	–	–	–	–	–	Trimeprazine M
50	–	–	–	20	75	52	6	11	9	Timolol
50	–	0	–	22	80	50	13	15	8	Penbutolol
50	0	0	15	28	62	58	8	19	11	Flurazepam M (Monodeethyl-)
50	7	5	44	71	60	71	2	53	25	Estazolam
50	11	50	22	–	–	–	–	–	–	Ethiazide
50	38	55	43	83	84	–	0	–	–	Salicylamide
50	59	71	63	86	–	–	–	–	–	Sigmodal (5-(2-Bromoallyl)-5-(1-methyl-butyl)barbituric Acid)
50	65	67	56	90	–	–	–	–	–	Propyl p-Hydroxybenzoate
51	–	–	–	17	–	47	1	7	13	Metoclopramide
51	–	–	–	25	60	46	11	31	6	Mianserin M (Nor-)
51	–	–	–	26	51	51	4	17	10	Clomipramine M (8-Hydroxy-)
51	–	–	–	46	61	58	4	16	18	Periciazine
51	–	–	–	54	–	–	0	–	–	Astemizole M (Nor-)
51	–	–	–	82	–	–	–	–	–	Primidone M/PEMA
51	–	0	–	22	–	–	11	–	–	Betaxolol
51	13	46	22	83	–	67	–	–	–	Sulfanilamide
51	19	11	41	78	76	73	0	40	49	Flunitrazepam M (7-Acetamido-)
52	–	–	–	20	–	50	38	44	12	Thenalidine
52	–	–	–	24	–	–	24	–	–	Ketotifen
52	–	–	–	26	–	53	28	37	5	Butaperazine
52	–	–	–	39	–	–	4	–	–	Dibenzepin M (5-Desmethyl-)

Table 8.5. (Continued)

TLC System 4	1	2	3	5	6	7	8	9	10	[System 4: Etac–MeOH–NH4OH (85+10+5) / Silica]
52	–	–	–	72	–	–	–	–	–	Etizolam
52	–	–	–	84	–	–	–	–	–	Ethyl-2-(p-tolyl)malonamide,2-
52	10	10	53	76	77	62	2	50	22	Chlordiazepoxide
52	11	20	40	76	77	73	0	30	47	Clonazepam M (7-Amino-)
52	11	34	32	87	–	82	0	–	–	Chlorphenesin Carbamate
52	15	5	53	72	71	72	5	52	27	Brotizolam
52	15	10	55	59	55	52	3	58	25	Caffeine
53	–	–	–	17	–	–	–	–	–	Lofexidine
53	–	–	–	19	–	54	13	34	12	Emetine
53	–	–	–	27	–	51	30	41	8	Thiethylperazine
53	–	–	–	40	59	66	3	32	21	Pipotiazine
53	–	–	–	46	65	63	7	13	13	Mazindol
53	–	–	–	62	–	–	12	–	–	Benzquinamide M (Desacetyl-)
53	3	4	41	73	72	70	3	52	8	Midazolam M (α-Hydroxy-)
53	13	39	30	–	–	62	1	20	53	Styramate
53	19	50	27	–	–	–	–	–	–	Methyclothiazide
53	38	59	39	86	–	–	–	–	–	Phenolphthalein
53	65	65	69	85	–	–	–	–	–	Hexobarbital
53	72	67	69	–	–	–	–	–	–	Narcobarbital
54	–	–	–	16	–	–	28	–	–	Encainide
54	–	–	–	25	42	49	15	19	9	Clomipramine M (10-Hydroxy-)
54	–	–	–	26	25	44	23	43	12	Acetylcodeine
54	–	–	–	44	–	55	3	45	14	Fenetylline
54	–	–	–	44	–	49	28	28	39	Phenethylamine
54	–	–	–	46	52	60	4	37	18	Mianserin M (8-Hydroxy-)
54	–	–	–	49	–	–	36	–	–	Loxapine
54	–	–	–	57	65	68	10	54	19	Hydroxyzine
54	–	–	–	86	–	49	3	45	17	Phanquinone
54	–	–	44	71	70	63	3	42	46	Pyrazinamide
54	–	0	–	19	–	–	–	–	–	Sulforidazine
54	10	20	30	77	79	71	0	30	46	Nitrazepam M (7-Amino-)
54	11	19	35	78	75	72	0	34	47	Flunitrazepam M (7-Aminonor-)
54	66	65	69	87	–	–	–	–	–	Metharbital
55	–	–	–	12	–	–	42	–	–	Aprindine M (Desethyl-)
55	–	–	–	12	41	40	45	19	4	Ethoheptazine
55	–	–	–	19	30	51	11	20	6	Triprolidine
55	–	–	–	24	–	–	–	–	–	Medrylamine
55	–	–	–	25	78	40	17	4	9	Mexiletine
55	–	–	–	26	26	49	34	37	7	Prochlorperazine
55	–	–	–	30	29	53	33	30	8	Trifluoperazine
55	–	–	–	38	22	54	22	35	14	Dibenzepin
55	–	–	–	41	38	55	12	36	18	Physostigmine
55	–	–	–	64	–	–	–	–	–	Pentoxifylline
55	–	–	–	87	89	64	2	43	57	Mephenesin
55	4	5	29	42	–	57	4	38	17	Clozapine
55	14	36	35	–	–	–	–	–	–	Mephenesin Carbamate

Table 8.5. (Continued)

TLC System 4	1	2	3	5	6	7	8	9	10	[System 4: Etac–MeOH–NH4OH (85+10+5) / Silica]
55	19	38	39	–	–	79	–	–	–	Clopamide
56	–	–	–	–	–	–	–	–	–	Methylbromazepam,1-
56	–	–	–	9	–	–	–	–	–	Aminoacridine
56	–	–	–	14	–	44	35	9	15	Etafedrine
56	–	–	–	19	–	–	–	–	–	Piprinhydrinate
56	–	–	–	22	–	62	–	–	–	Ajmaline
56	–	–	–	47	–	–	–	–	–	Diltiazem
56	–	–	–	51	–	–	–	–	–	Tizanidine
56	–	–	–	68	–	–	1	–	–	Bupivacaine M1 (4-OH)
56	–	–	–	79	75	60	2	56	47	Carbamazepine
56	–	–	–	83	–	–	–	–	–	Ethaverine
56	–	1	–	28	–	–	11	–	–	Gallopamil M (Nor-)
56	3	–	–	60	59	60	3	39	18	Lysergide/LSD
56	9	36	32	63	87	75	0	32	58	Meprobamate
57	–	–	–	–	–	–	–	–	–	Chrysophanol
57	–	–	–	46	–	60	1	60	29	Vinblastine
57	–	–	–	52	–	–	6	–	–	Haloperidol M
57	–	–	–	72	76	70	2	40	33	Harman
57	–	–	–	74	77	77	0	59	50	Ergocristine
57	9	2	51	68	66	69	6	55	22	Adinazolam
57	23	43	42	81	–	–	–	–	–	Sultiame
57	23	51	33	–	–	–	–	–	–	Metolazone
57	25	71	30	–	–	–	–	–	–	Bendroflumethiazide/Bendrofluthiazide
57	38	37	64	81	80	71	1	59	52	Flunitrazepam M (3-Hydroxy-)
58	–	–	–	18	49	46	49	25	9	Clemastine
58	–	–	–	22	–	–	13	–	–	Clenbuterol
58	–	–	–	22	33	51	39	25	14	Mepyramine/Pyrilamine
58	–	–	–	41	67	54	39	33	48	Tranylcypromine
58	–	–	–	54	–	83	0	21	27	Apomorphine
58	–	–	–	60	–	57	3	44	33	Nimorazole
58	–	–	–	61	–	58	8	46	31	Morazone
58	–	–	–	66	–	61	2	31	21	Pyrimethamine
58	–	–	–	69	–	66	1	48	38	Ergoloid
58	–	–	–	71	73	67	2	48	36	Droperidol
58	–	–	–	74	–	62	–	–	–	Methylphenazone,4-
58	–	–	–	83	85	80	0	55	6	Flubendazole
58	62	65	65	–	–	–	–	–	–	Fenimide
58	71	71	70	–	–	–	–	–	–	Enallylpropymal
58	73	72	71	85	–	–	–	–	–	Methohexital
59	–	–	–	–	–	–	–	–	–	Bisacodyl M (-diphenol)
59	–	–	–	8	–	41	57	8	6	Perhexiline
59	–	–	–	15	29	47	43	23	9	Prothipendyl
59	–	–	–	16	74	47	47	19	8	Etilamfetamine
59	–	–	–	24	33	48	18	36	25	Oxomemazine
59	–	–	–	24	–	49	40	43	13	Bamipine
59	–	–	–	30	–	50	33	32	17	Nefopam

Table 8.5. (Continued)

TLC System 4	1	2	3	5	6	7	8	9	10	[System 4: Etac–MeOH–NH4OH (85+10+5) / Silica]
59	–	–	–	39	50	–	21	48	12	Cocaine M (Nor-)
59	–	–	–	68	74	60	1	47	49	Prazosin
59	–	–	–	71	–	65	0	54	34	Bamifylline
59	–	–	–	71	67	59	15	56	29	Nikethamide
59	–	–	–	74	77	78	0	58	50	Ergocryptine
59	18	60	26	–	–	77	–	–	–	Cyclothiazide
59	21	40	30	75	–	–	3	–	–	Paracetamol M/p-Aminophenol
59	46	45	60	82	82	52	6	61	50	Lormetazepam
59	49	48	60	84	84	–	–	–	–	Acecarbromal
60	–	–	–	9	7	45	7	8	13	Disopyramide
60	–	–	–	12	–	48	41	10	9	Doxylamine
60	–	–	–	12	–	50	42	19	8	Aminopromazine
60	–	–	–	19	–	48	41	16	11	Fenfluramine
60	–	–	–	34	40	52	37	34	11	Pethidine/Meperidine
60	–	–	–	72	74	72	7	64	63	Pentetrazol
60	–	–	–	77	–	70	–	–	–	Pargyline
60	–	–	–	80	84	65	0	59	49	Mebendazole
60	–	3	–	62	69	78	3	51	32	Benperidol
60	10	35	35	82	85	–	0	33	56	Mebutamate
60	13	5	53	69	70	72	6	60	19	Midazolam
60	31	51	47	79	–	76	–	–	–	Clorexolone
61	–	–	–	21	–	50	45	25	12	Metixene
61	–	–	–	39	22	54	39	35	13	Nicotine
61	–	–	–	49	43	–	13	59	25	Orientalidine
61	–	–	–	49	43	55	14	57	26	Isothebaine
61	–	–	–	51	72	74	4	49	39	Spiperone
61	–	–	–	73	74	70	1	45	42	Piritramide
61	–	–	–	79	–	–	–	–	–	Bencyclane
61	–	–	–	90	–	–	–	–	–	Trichlorfon (Trichlorphos)
61	–	2	–	29	–	–	30	–	–	Melperone M (FG 5155)
61	19	28	54	81	82	74	2	46	55	Flurazepam M (N(1)-Hydroxyethyl-)
61	30	14	61	76	72	71	3	63	44	Flumazenil/RO 15-1788
62	–	–	–	–	–	–	–	–	–	Apronal
62	–	–	–	–	–	61	2	33	57	Amiphenazole
62	–	–	–	18	–	48	17	24	12	Carbocromen
62	–	–	–	18	35	44	38	30	11	Promazine
62	–	–	–	20	–	–	–	–	–	Mecloxamine
62	–	–	–	22	–	–	–	–	–	Dimetacrine
62	–	–	–	27	–	–	–	–	–	Chlorprothixene M
62	–	–	–	30	33	50	25	51	39	Oxycodone
62	–	–	–	33	83	74	3	14	50	Buphenine
62	–	–	–	43	–	53	–	–	–	Arecoline
62	–	–	–	49	41	57	36	51	24	Phendimetrazine
62	–	–	–	62	81	78	3	32	53	Isoxsuprine
62	–	–	–	66	–	–	0	–	–	Cisapride
62	–	–	–	74	–	–	4	–	–	Oxatomide

Table 8.5. (Continued)

TLC System 4	1	2	3	5	6	7	8	9	10	[System 4: Etac–MeOH–NH4OH (85+10+5) / Silica]
62	–	–	–	77	–	–	–	–	–	Molsidomine
62	–	–	–	78	81	71	4	55	52	Carnidazole
62	–	–	–	81	–	–	–	–	–	Ornidazole
62	–	–	–	82	–	–	–	–	–	Ectylurea
62	–	–	–	85	–	–	1	–	–	Valnoctamine
62	25	10	58	70	68	66	21	–	–	Aminophenazone/Amidopyrine
62	51	47	65	82	82	53	8	59	53	Temazepam/M of Camazepam, Diazepam, Ketazolam, Medazepam
63	–	–	–	14	56	46	14	10	11	Diethyltryptamine,N,N-/DET
63	–	–	–	24	45	51	48	37	13	Doxepin
63	–	–	–	28	–	48	26	24	12	Acepromazine
63	–	–	–	33	–	56	40	34	18	Methadone M (Nor-)
63	–	–	–	78	–	–	–	–	–	Isopropylaminophenazone/Isopyrin
63	–	–	0	22	–	52	41	28	17	Chloropyramine
63	13	18	47	73	69	61	6	41	53	Bromazepam
63	21	21	52	77	74	74	1	55	52	Flunitrazepam M (7-Amino-)
63	22	60	32	–	–	–	–	–	–	Polythiazide
63	26	42	52	89	93	77	0	52	56	Budesonide
63	31	25	55	78	–	58	–	–	–	Methyprylon
64	–	–	–	19	–	–	–	–	–	Etifelmine
64	–	–	–	22	35	52	41	30	14	Isothipendyl
64	–	–	–	28	–	48	45	–	–	Pizotifen
64	–	–	–	30	50	51	45	44	13	Cyproheptadine
64	–	–	–	43	39	57	15	32	16	Tetracaine/Amethocaine
64	–	–	–	53	52	56	9	29	31	Nomifensine
64	–	–	–	56	–	–	1	–	–	Astemizole
64	–	–	–	64	67	75	0	47	29	Ergosine
64	–	–	–	66	70	63	6	38	52	Yohimbine
64	–	–	–	66	–	58	16	58	41	Metyrapone
64	–	–	–	78	–	66	6	45	68	Leucinocaine
64	32	43	54	84	90	75	0	52	61	Clobazam M (Nor-)
64	35	46	53	84	86	68	0	36	55	Nitrazepam
65	–	–	–	–	–	–	–	–	–	Glutethimide M (Amino-)
65	–	–	–	15	67	–	9	15	25	Psilocin-(eth)/CZ-74
65	–	–	–	22	31	55	38	28	14	Thonzylamine
65	–	–	–	24	74	53	15	13	25	Cyclazocine
65	–	–	–	27	48	55	44	33	15	Diphenhydramine
65	–	–	–	27	–	–	52	–	–	Melitracene
65	–	–	–	29	73	55	36	38	13	Noracymethadol
65	–	–	–	30	44	50	36	35	17	Promethazine
65	–	–	–	34	–	–	–	–	–	Aceprometazine
65	–	–	–	36	46	57	27	45	20	Mebhydrolin
65	–	–	–	53	52	62	18	48	42	Levamisole
65	–	–	–	73	–	–	3	–	–	Ketanserin
65	–	–	–	87	–	–	–	–	–	Thibenzazoline
65	–	–	–	92	96	71	41	79	70	Bezitramide

Table 8.5. (Continued)

TLC System 4	1	2	3	5	6	7	8	9	10	[System 4: Etac–MeOH–NH4OH (85+10+5) / Silica]
65	–	1	–	27	–	53	47	44	16	Pecazine/Mepazine
65	15	10	48	–	–	–	–	–	–	Pethidine Intermediate A
65	22	40	50	–	–	–	–	–	–	Phenacemide
65	32	47	53	–	–	–	–	–	–	Aminoglutethimide
65	35	68	–	–	–	77	–	–	–	Tybamate
65	78	68	76	87	–	79	–	–	–	Phenylbutazone
66	–	–	–	6	68	20	32	10	2	Cyclopentamine
66	–	–	–	17	–	–	–	–	–	Butamirate
66	–	–	–	21	24	52	41	26	13	Methapyrilene
66	–	–	–	27	74	54	42	27	17	Eucaine,β-
66	–	–	–	29	–	53	43	35	18	Noxiptiline
66	–	–	–	40	70	57	35	41	23	Methylphenidate
66	–	–	–	43	–	56	13	48	28	Dimetiotazine
66	–	–	–	51	53	69	18	54	46	Tetramisole
66	–	–	–	63	60	65	31	62	48	Mepivacaine
66	–	–	–	64	61	63	10	58	37	Trazodone
66	–	–	–	72	–	55	6	66	46	Pentifylline
66	21	62	27	–	–	–	–	–	–	Cyclopenthiazide
66	38	61	46	89	–	–	–	–	–	Indapamide
66	45	45	52	80	84	–	–	–	–	Acetanilide
66	50	53	59	84	–	70	5	–	–	Ethosuximide
67	–	–	–	19	–	50	51	22	19	Oxeladine
67	–	–	–	20	55	48	42	30	13	Thioridazine
67	–	–	–	21	47	48	48	23	13	Imipramine
67	–	–	–	32	–	53	38	48	15	Phenyltoloxamine
67	–	–	–	35	52	57	42	46	14	Chlorcyclizine
67	–	–	–	42	–	63	25	34	35	Cinchocaine/Dibucaine
67	–	–	–	46	–	77	26	42	22	Meclofenoxate
67	–	–	–	59	–	–	6	–	–	Bromhexine M/Ambroxol
67	9	22	30	66	63	73	4	51	47	Medazepam M (Nor-)
67	35	45	56	85	87	72	0	53	61	Clonazepam
67	45	58	55	–	–	–	–	–	–	Mefruside
68	–	–	–	16	49	48	47	22	14	Pentoxyverine/Carbetapentane
68	–	–	–	22	34	55	44	27	15	Tripelennamine
68	–	–	–	23	49	46	37	28	8	Diphenylpyraline
68	–	–	–	25	49	55	48	33	16	Orphenadrine
68	–	–	–	26	–	51	52	32	15	Tolpropamine
68	–	–	–	28	48	55	45	33	15	Dimenhydrinate
68	–	–	–	30	52	55	34	42	26	Propiomazine
68	–	–	–	35	–	56	41	35	20	Nicametate
68	–	–	–	40	52	57	49	41	16	Cyclizine
68	–	–	–	41	49	63	45	57	21	Phenindamine
68	–	–	–	48	50	58	39	58	23	Mianserin
68	–	–	–	55	–	–	–	–	–	Metamfepramone
68	–	–	–	80	–	61	11	72	52	Trimetozine
68	–	–	–	84	–	–	–	–	–	Flutazolam

Table 8.5. (Continued)

TLC System 4	1	2	3	5	6	7	8	9	10	[System 4: Etac–MeOH–NH4OH (85+10+5) / Silica]
68	–	–	–	84	80	66	1	63	57	Chlormezanone
68	34	46	57	83	87	84	3	56	60	Clorazepic Acid
68	35	46	56	84	85	76	0	58	59	Flunitrazepam M (Nor-)
68	35	51	52	84	–	–	6	–	–	Bromisoval
68	38	37	52	83	–	–	–	–	–	Phenacetin
68	52	53	64	–	88	–	–	–	–	Bemegride
69	–	–	–	–	–	–	–	–	–	Cyclobenzaprine
69	–	–	–	27	48	54	42	43	13	Bromazine/Bromdiphenhydramine
69	–	–	–	27	51	51	50	32	15	Amitriptyline
69	–	–	–	51	–	–	–	–	–	Benzetimide
69	–	–	–	74	74	61	8	65	47	Papaverine
69	–	–	–	75	–	–	–	–	–	Chlorazanil
69	–	–	–	83	79	68	28	67	55	Crotethamide
69	–	7	–	56	–	–	21	–	–	Gallopamil M (D517)
69	34	45	57	82	83	62	3	55	60	Nordazepam/M of Chlordiazepoxide, Clorazepic Acid, Diazepam, Halazepam, Ketazolam, Medazepam, Oxazolam, Prazepam
69	42	52	58	88	90	71	2	55	59	Quazepam M (3-Hydroxy-2-oxo-)
69	52	37	68	78	–	–	–	–	–	Griseofulvin
70	–	–	–	21	–	–	–	–	–	Camylofine
70	–	–	–	23	79	–	41	33	29	Phenylcyclohexylamine,1-
70	–	–	–	25	45	49	45	35	17	Chlorpromazine
70	–	–	–	27	41	51	49	42	16	Dothiepin
70	–	–	–	29	–	53	47	36	17	Chlorphenoxamine
70	–	–	–	34	72	61	16	12	28	Pentazocine
70	–	–	–	44	76	62	8	31	53	Clonidine
70	–	–	–	51	67	71	21	48	44	Dexetimide
70	–	–	–	70	77	70	11	63	36	Lidoflazine
70	–	–	–	74	76	75	13	58	51	Bromolysergide,2-
70	–	–	–	80	–	59	1	50	53	Phenazopyridine
70	–	–	–	81	–	–	–	–	–	Khellin
70	–	–	–	81	–	–	4	–	–	Pirenzepine M (Desamide)
70	–	–	–	86	–	67	0	51	62	Benorilate
71	–	–	–	11	28	54	41	21	9	Diamocaine
71	–	–	–	12	–	3	1	0	1	Hexoprenaline
71	–	–	–	17	–	59	4	35	17	Veratrine
71	–	–	–	25	66	54	54	37	18	Pyrrobutamine
71	–	–	–	36	42	54	5	31	30	Procaine
71	–	–	–	59	57	61	15	64	52	Hydrastine
71	–	–	–	63	78	69	4	59	49	Fluspirilene
71	–	–	–	73	82	71	3	60	40	Pimozide
71	–	–	–	77	78	–	11	73	63	Meconin
71	–	–	–	79	–	68	1	65	68	Nifedipine
71	–	–	–	80	–	–	–	–	–	Pyrithyldione
71	–	–	–	84	–	–	–	–	–	Phenaglycodol
71	–	–	–	86	–	–	4	–	–	Ethinylestradiol

Table 8.5. (Continued)

TLC System 4	1	2	3	5	6	7	8	9	10	[System 4: Etac–MeOH–NH4OH (85+10+5) / Silica]
71	–	4	–	46	–	–	24	–	–	Gallopamil
71	–	5	–	64	–	–	30	–	–	Buspirone
71	3	3	41	52	45	62	30	48	40	Flurazepam
71	35	40	62	–	–	–	–	–	–	Alfadolone Acetate
71	38	53	59	–	–	76	–	–	–	Pheneturide
71	53	54	60	–	–	–	–	–	–	Ethotoin
71	53	54	65	78	95	77	15	68	66	Oxazolam
72	–	–	–	26	54	51	53	34	18	Clomipramine
72	–	–	–	37	–	–	–	–	–	Oxyfedrine
72	–	–	–	42	–	–	–	–	–	Moperone
72	–	–	–	69	–	64	20	70	54	Doxapram
72	–	–	–	74	77	61	37	66	59	Methyl Nicotinate
72	–	–	–	77	–	–	–	–	–	Pyricarbate
72	–	–	–	77	–	–	–	–	–	Tofisopam
72	–	–	–	78	–	–	–	–	–	Enilconazole/Imazalil
72	–	–	–	80	–	66	20	70	55	Propanidid
72	–	–	–	92	–	–	–	–	–	Pentaerithritol Tetranitrate
72	–	–	–	94	–	–	–	–	–	Paraflutizide
72	34	45	60	83	88	75	2	56	58	Flurazepam M (N(1)-Dealkyl-)/Quazepam M (N-Dealkyl-2-oxo-)
72	35	41	57	82	86	73	5	58	56	Delorazepam
72	55	55	70	85	88	71	12	–	52	Prazepam M (3-Hydroxy-)
73	–	–	–	14	50	–	22	46	43	Cryptopine
73	–	–	–	25	–	–	50	–	–	Melperone
73	–	–	–	35	46	66	26	55	38	Pitofenone
73	–	–	–	43	–	–	–	–	–	Nicergoline
73	–	–	–	43	61	59	23	70	42	Verapamil
73	–	–	–	58	50	60	30	58	33	Methylpiperidyl Benzilate/JB-336
73	–	–	–	78	85	67	26	71	52	Etomidate
73	–	–	–	84	–	–	–	–	–	Azintamid
73	–	–	–	84	88	72	2	69	61	Bromocriptine
73	–	–	–	92	92	–	–	–	–	Diethylstilbestrol
73	–	3	–	50	–	–	13	–	–	Bromperidol
73	33	25	65	81	84	76	15	68	58	Metaclazepam M (N-Nor-)
73	35	45	56	85	92	77	5	58	60	Tetrazepam M (Nor-)
73	47	55	60	82	–	75	–	–	–	Phenprobamate
74	–	–	–	20	68	48	62	31	23	Procyclidine
74	–	–	–	25	83	70	29	69	57	Cropropamide
74	–	–	–	34	51	56	51	51	25	Chlorprothixene
74	–	–	–	42	73	67	19	24	45	Levallorphan
74	–	–	–	45	–	–	13	–	–	Terfenadine
74	–	–	–	50	81	68	16	39	49	Phenazocine
74	–	–	–	52	81	70	9	32	22	Loperamide
74	–	–	–	56	–	–	–	–	–	Vincamine
74	–	–	–	62	59	61	35	53	42	Thiopropazate
74	–	–	–	77	81	71	23	65	48	Metomidate

Table 8.5. (Continued)

TLC System 4	1	2	3	5	6	7	8	9	10	[System 4: Etac–MeOH–NH4OH (85+10+5) / Silica]
74	–	–	–	80	–	–	–	–	–	Indolyl Acetate,3-
74	–	–	–	82	–	–	–	–	–	Coumarine
74	–	–	–	86	–	–	–	–	–	Cyprazepam
74	45	45	62	83	80	66	14	64	66	Ketazolam
74	49	62	57	–	–	–	–	–	–	Methylpentynol Carbamate
74	53	58	62	85	93	76	0	62	67	Ethyl Loflazepate
74	54	47	72	80	82	63	10	72	63	Flunitrazepam
74	61	49	65	81	–	71	32	–	–	Propyphenazone
74	62	58	66	–	–	–	–	–	–	Mephenytoin
75	–	–	–	23	–	61	17	35	29	Lobeline
75	–	–	–	32	–	–	–	–	–	Histapyrrodine
75	–	–	–	32	49	54	47	35	22	Triflupromazine
75	–	–	–	38	–	–	43	–	–	Mefenorex
75	–	–	–	42	68	68	11	39	58	Aconitine
75	–	–	–	62	79	77	29	64	60	Prilocaine
75	–	–	–	65	–	76	14	54	61	Butanilicaine
75	–	–	–	78	80	80	9	61	–	Econazole
75	–	–	–	78	–	57	32	27	29	Octacaine
75	–	–	–	80	–	–	–	–	–	Etidocaine
75	–	–	–	84	86	71	20	74	61	Isocarboxazid
75	–	–	–	86	–	–	–	–	–	Mevinphos/Phosdrin
75	–	24	–	25	–	–	58	–	–	Bornaprine
75	36	53	59	85	79	–	4	–	–	Carisoprodol
75	39	45	63	78	84	73	0	66	59	Cloxazolam
75	46	50	65	80	90	74	11	66	62	Haloxazolam
75	53	47	70	84	85	62	8	70	62	Clobazam
75	53	55	64	85	87	–	12	–	–	Carbromal
75	55	32	69	82	83	76	12	73	65	Camazepam
75	56	50	67	78	86	76	24	69	63	Fludiazepam
75	60	59	63	–	–	–	–	–	–	Triclocarban
75	64	50	69	–	79	–	–	–	–	Santonin
75	64	67	59	–	–	–	–	–	–	Tribromoethanol
76	–	–	–	3	–	2	0	–	–	Methanthelinium Bromide
76	–	–	–	20	–	–	63	–	–	Aprindine
76	–	–	–	21	56	55	53	37	27	Piperocaine
76	–	–	–	32	49	57	47	38	46	Levomepromazine/Methotrimeprazine
76	–	–	–	51	75	67	11	27	33	Haloperidol
76	–	–	–	65	–	–	–	–	–	Azaperone
76	–	–	–	66	–	–	–	–	–	Metofenazate
76	–	–	–	66	80	75	36	64	56	Articaine
76	–	–	–	70	82	71	26	64	58	Phenoperidine
76	–	–	–	76	88	73	5	68	61	Ambucetamide
76	–	–	–	80	86	69	8	52	59	Fenyramidol
76	–	–	–	80	85	64	44	69	58	Clomethiazole
76	–	–	–	87	–	64	3	59	55	Ethenzamide
76	–	–	–	90	–	65	8	65	64	Beclamide

Table 8.5. (Continued)

TLC System										[System 4:
4	1	2	3	5	6	7	8	9	10	Etac–MeOH–NH4OH (85+10+5) / Silica]
76	–	–	0	12	–	30	3	2	2	Ethambutol
76	–	–	68	80	–	–	–	–	–	Clotrimazole
76	49	59	58	86	87	76	5	–	–	Ethinamate
76	56	62	63	84	87	67	5	57	66	Benzocaine
76	58	49	72	82	85	75	27	73	59	Diazepam/Ketazolam M/Medazepam M
76	61	60	64	84	96	–	22	64	69	Cholesterol
77	–	–	–	–	–	–	–	–	–	Androsterone
77	–	–	–	–	–	–	–	–	–	Fenproporex
77	–	–	–	16	60	48	59	20	27	Methadone/Levomethadone
77	–	–	–	21	–	54	62	34	30	Fencamfamine
77	–	–	–	32	46	58	54	39	31	Trimeprazine/Alimemazine
77	–	–	–	33	54	58	57	51	39	Diethazine
77	–	–	–	35	30	65	45	47	54	Cocaine
77	–	–	–	38	–	59	56	52	43	Benzoctamine
77	–	–	–	47	–	66	49	–	–	Captodiame
77	–	–	–	48	–	62	36	67	45	Flavoxate
77	–	–	–	52	57	66	43	53	53	Benactyzine
77	–	–	–	76	80	69	2	74	63	Reserpine
77	–	–	–	85	–	–	–	–	–	Kavain
77	–	–	–	87	–	–	–	–	–	Triflubazam
77	2	5	26	55	76	73	13	–	–	Trifluperidol
77	53	46	71	81	81	74	12	70	55	Nimetazepam
77	71	59	72	–	–	75	–	–	–	Phensuximide
78	–	–	–	14	–	70	57	67	56	Azapetine
78	–	–	–	20	–	–	–	–	–	Clobutinol
78	–	–	–	30	–	–	–	–	–	Tramadol
78	–	–	–	30	73	56	63	37	26	Volazocine
78	–	–	–	31	–	–	–	–	–	Suloctidil
78	–	–	–	34	30	70	42	60	55	Cinnamoylcocaine
78	–	–	–	43	–	64	52	41	35	Naftidrofuryl
78	–	–	–	51	80	69	18	39	47	Phenomorphan
78	–	–	–	59	54	75	44	63	55	Ethylpiperidyl Benzilate/JB-318
78	–	–	–	70	77	70	43	74	58	Fentanyl
78	–	–	–	72	75	64	21	74	64	Noscapine
78	–	–	–	74	–	–	–	–	–	Oxydemeton
78	–	–	–	74	–	68	4	65	58	Oxypertine
78	–	–	–	76	73	78	31	69	52	Clemizole
78	–	–	–	78	81	71	21	78	69	Disulfiram
78	–	–	–	78	–	74	52	79	71	Broxaldine
78	–	–	–	79	84	70	36	80	56	Methaqualone
78	–	–	–	87	–	–	–	–	–	Kebuzone
78	–	–	–	87	–	–	–	–	–	Meproscillarine
78	–	–	–	87	–	–	1	–	–	Nimodipine
78	–	–	–	89	–	–	–	–	–	Etofenamate
78	–	55	–	86	–	–	12	–	–	Mexazolam
78	56	40	73	79	83	67	41	74	62	Medazepam

Table 8.5. (Continued)

TLC System 4	1	2	3	5	6	7	8	9	10	[System 4: Etac–MeOH–NH4OH (85+10+5) / Silica]
78	57	65	60	86	–	–	9	–	–	Hexapropymate
78	61	58	64	–	–	–	–	–	–	Estradiol
78	66	51	75	84	–	–	–	–	–	Spironolactone
79	–	–	–	–	–	–	–	–	–	Carbofuran
79	–	–	–	15	65	–	66	25	26	Rolicyclidine/PHP/1-(1-phenylcyclohexyl)pyrrolidine
79	–	–	–	21	78	–	65	27	37	Eticyclidine/PCE/N-Ethyl-1-phenylcyclohexylamine
79	–	–	–	22	–	50	67	32	25	Prolintane
79	–	–	–	30	91	70	48	40	20	Hexetidine
79	–	–	–	33	–	–	–	–	–	Salverine
79	–	–	–	60	66	73	12	56	51	Anileridine
79	–	–	–	68	72	63	37	63	64	Ketamine
79	–	–	–	70	72	68	13	70	61	Etonitazene
79	–	–	–	72	78	73	42	71	60	Dextromoramide/Racemoramide/Levomoramide
79	–	–	–	80	–	69	41	78	67	Tetrabenazine
79	37	51	55	78	–	67	63	–	–	Sulfadoxine
79	47	35	71	82	84	77	40	73	62	Metaclazepam
79	57	49	67	84	89	78	32	69	66	Tetrazepam
79	60	58	63	–	–	–	–	–	–	Ergosterol
80	–	–	–	18	–	52	62	19	26	Terodiline
80	–	–	–	36	56	59	62	54	37	Trimipramine
80	–	–	–	41	–	–	48	–	–	Lorcainide
80	–	–	–	57	–	–	–	–	–	Caroverine
80	–	–	–	69	79	69	42	73	65	Bupivacaine
80	–	–	–	72	69	70	35	71	63	Lidocaine
80	–	–	–	80	–	–	–	–	–	Benfluralin
80	–	–	–	80	–	76	9	68	69	Buprenorphine
80	–	–	–	82	–	–	1	–	–	Nicardipine
80	–	–	–	82	87	74	17	76	66	Bisacodyl
80	–	–	–	87	–	–	2	–	–	Nitrendipine
80	–	–	0	93	48	1	0	0	0	Metformin
80	–	–	68	88	–	49	2	10	4	Procarbazine
80	–	7	56	77	80	73	11	67	37	Miconazole
80	55	48	66	84	87	78	31	69	66	Clotiazepam
80	59	57	70	87	90	78	16	89	71	Quazepam M (2-Oxo-)
80	63	62	70	86	89	75	31	–	–	Glutethimide
80	74	77	73	87	95	–	37	–	–	Cyclandelate
81	–	–	–	19	79	54	59	38	39	Pipradrol
81	–	–	–	41	–	–	–	–	–	Etafenone
81	–	–	–	45	–	68	58	54	55	Isoaminile
81	–	–	–	48	56	69	59	57	47	Butetamate
81	–	–	–	71	–	–	0	–	–	Doxazosin
81	–	–	–	73	–	72	3	77	66	Deserpidine
81	–	–	–	77	79	73	1	75	64	Rescinnamine

Table 8.5. (Continued)

TLC System 4	1	2	3	5	6	7	8	9	10	[System 4: Etac–MeOH–NH4OH (85+10+5) / Silica]
81	–	–	–	81	–	–	–	–	–	Ajmalicine/Raubasine
81	–	–	–	86	–	–	–	–	–	Demeton-S-methyl
81	37	41	54	80	79	72	5	57	57	Nitrazepam M Hy/DAB
81	59	59	70	89	91	78	15	6	71	Halazepam
81	64	55	72	84	89	65	36	74	63	Prazepam
81	65	61	73	85	92	75	29	72	70	Pinazepam
81	73	68	65	–	92	–	–	–	–	Thymol
82	–	–	–	27	77	55	66	44	36	Prozapine
82	–	–	–	30	–	60	56	52	35	Clomifene
82	–	–	–	41	–	–	–	–	–	Tolylfluanid
82	–	–	–	50	63	68	58	55	54	Propoxyphene/Dextropropoxyphene/Levopropoxyphene
82	–	–	–	54	52	69	55	81	55	Piperidolate
82	–	–	–	54	64	72	62	68	55	Amiodarone
82	–	–	–	55	57	–	54	60	62	Eucaine,α-
82	–	–	–	61	–	–	49	–	–	Prenoxdiazine/Tibexin
82	–	–	–	82	–	–	–	–	–	Fominoben
82	–	–	–	82	89	79	11	79	71	Phenothiazine
82	–	–	–	83	–	–	26	–	–	Nifedipine M
82	–	–	–	86	–	–	–	–	–	Nimodipine M
82	–	–	–	86	–	–	34	–	–	Nitrendipine M
82	–	23	–	67	75	73	39	68	60	Fluanisone
82	41	46	55	80	80	74	5	59	61	Clonazepam Hy/DCB
82	48	50	57	87	91	77	7	60	66	Flurazepam Hy/HCFB
83	–	–	–	–	–	–	–	–	–	Chlorobenzoic Acid,o-
83	–	–	–	5	–	–	–	–	–	Dihydroergocristine
83	–	–	–	28	–	–	–	–	–	Benproperine
83	–	–	–	29	82	77	37	12	63	Phenelzine
83	–	–	–	31	55	67	64	47	66	Profenamine/Ethopropazine
83	–	–	–	43	75	68	66	61	59	Trihexyphenidyl/Benzhexol
83	–	–	–	44	76	71	7	30	64	Butacaine
83	–	–	–	45	–	64	68	64	64	Biperiden
83	–	–	–	49	–	64	55	60	51	Adiphenine
83	–	–	–	54	–	62	23	41	36	Oxybuprocaine
83	–	–	–	70	78	–	38	73	59	Fluorofentanyl,p-
83	–	–	–	72	–	–	–	–	–	Piribedil
83	–	–	–	84	–	83	–	–	–	Crotamiton
83	–	–	–	85	–	–	–	–	–	Dichlorvos/DDVP
83	–	–	–	88	92	79	64	84	74	Phenanthrene
83	9	8	41	45	51	67	53	60	51	Flurazepam Hy/DCFB
83	66	66	80	76	86	73	31	77	68	Nimetazepam Hy/MNB
83	70	69	74	81	93	83	11	71	69	Haloxazolam Hy/ABFB
83	71	69	81	77	90	75	48	77	69	Fludiazepam Hy/CFMB
83	78	71	78	87	96	74	27	75	76	Quazepam
84	–	–	–	–	–	–	–	–	–	Drofenine
84	–	–	–	23	69	59	73	35	66	Phencyclidine

Table 8.5. (Continued)

TLC System 4	1	2	3	5	6	7	8	9	10	[System 4: Etac–MeOH–NH4OH (85+10+5) / Silica]
84	–	–	–	38	62	57	62	45	50	Alphacetylmethadol
84	–	–	–	43	85	68	55	68	56	Prenylamine
84	–	–	–	61	–	–	–	–	–	Tilidine
84	–	–	–	64	77	–	45	75	64	Alphamethylfentanyl
84	–	–	–	72	89	76	17	60	60	Penfluridol
84	–	–	–	86	–	–	–	–	–	Visnadin
84	–	–	–	89	–	–	–	–	–	Sulazepam
84	75	66	71	81	–	–	–	–	–	Clofibrate
85	–	–	–	43	–	–	–	–	–	Fendiline
85	–	–	–	55	56	76	62	63	64	Amfepramone/Diethylpropion
85	–	–	–	80	–	–	–	–	–	Moxaverine
85	–	–	–	85	–	–	–	–	–	Floctafenine
85	64	61	66	86	88	77	9	68	69	Bromazepam Hy/ABP
85	71	65	68	90	91	77	5	69	69	Nitrazepam Hy/ANB
85	74	73	80	77	93	79	44	77	73	Quazepam Hy/CFTB
86	–	–	–	23	75	55	68	36	52	Diisopromine
86	–	–	–	32	–	63	40	53	49	Mebeverine
86	–	–	–	33	86	70	57	59	68	Clofazimine
86	–	–	–	38	70	–	73	54	68	Tenocyclidine/TCP/1-[1-(2-Thienyl)cyclohexyl]piperidine
86	–	–	–	39	72	65	28	50	62	Ibogaine/Rotenone
86	–	–	–	78	–	–	–	–	–	Etozolin
86	–	–	–	79	87	76	54	78	65	Cinnarizine
86	–	–	–	85	90	83	53	77	66	Clocinizine
86	–	–	–	86	–	–	–	–	–	Benzyl Alcohol
86	–	–	–	86	–	–	–	–	–	Coumachlor
86	–	–	–	87	–	–	–	–	–	Tritoqualine
86	–	–	–	88	–	72	47	78	67	Fenbutrazate
86	–	–	–	93	97	83	44	78	75	Proclonol
86	–	–	–	94	–	–	–	–	–	Tribenoside
86	7	63	80	89	90	77	20	79	71	Flunitrazepam M Hy/MNFB
86	11	29	46	–	–	68	–	–	–	Butalamine
86	71	66	68	87	91	77	4	69	70	Flunitrazepam M Hy/ANFB
86	71	72	69	–	–	–	–	–	–	Nitroglycerin
86	72	66	70	87	93	77	4	69	70	Clonazepam Hy/ANCB
86	75	73	80	86	91	83	48	77	72	Pinazepam Hy/CPB
86	76	68	74	89	93	78	11	75	70	Flurazepam Hy/ACFB
86	76	69	76	90	91	78	12	76	71	Oxazepam Hy/ACB
86	81	74	82	87	–	–	0	–	–	Ethchlorvynol
87	–	–	–	–	–	–	–	–	–	Clofenotane (o,p′-Isomer)/o,p′-DDT
87	–	–	–	27	72	66	67	33	70	Dipipanone
87	–	–	–	31	78	–	72	47	68	Methyl-1-phenylcyclohexyl)piperidine,1-(4-
87	–	–	–	39	–	–	–	–	–	Pridinol
87	–	–	–	60	–	73	67	70	70	Benzphetamine
87	–	–	–	72	73	–	62	74	64	Phenylcyclohexyl)morpholine,4-(1-/PCM
87	–	–	–	80	88	76	61	79	70	Meclozine

Table 8.5. (Continued)

TLC System 4	1	2	3	5	6	7	8	9	10	[System 4: Etac–MeOH–NH4OH (85+10+5) / Silica]
87	–	–	–	84	97	73	63	76	68	Phenoxybenzamine
87	–	–	–	90	–	76	–	–	–	Mesuximide
87	–	–	–	90	–	72	37	74	67	Famprofazone
87	–	–	–	90	92	74	42	84	70	Diphenoxylate
87	77	70	76	89	93	78	11	76	70	Lorazepam Hy/ADB
87	81	69	81	82	93	75	49	81	7	Lormetazepam Hy/MDB
87	82	74	80	–	–	–	–	–	–	Clofenotane (p,p′-Isomer)/p,p′-DDT
88	–	–	–	63	77	67	36	64	59	Piminodine
88	–	–	–	83	90	82	45	75	64	Flunarizine
88	–	–	–	84	–	75	69	79	71	Bromhexine
88	82	72	82	89	93	79	53	81	71	Diazepam Hy/MACB
88	83	74	82	92	95	80	58	82	74	Prazepam Hy/CCB
89	–	–	–	92	–	–	–	–	–	Methenolone
89	–	–	–	95	–	–	–	–	–	Mofebutazone
89	81	74	81	94	95	82	49	81	75	Halazepam Hy/TCB
90	–	–	–	82	–	–	–	–	–	Lofepramine
90	–	–	–	89	–	–	–	–	–	Heptenophos
92	–	–	–	87	–	–	–	–	–	Chlorfenvinphos
99	–	–	–	80	–	–	–	–	–	Dimethoate
99	–	–	–	83	–	–	–	–	–	Bromophos
99	–	–	–	84	–	–	–	–	–	Bromophos-ethyl
99	–	–	–	84	–	–	–	–	–	Etrimfos
99	–	–	–	84	–	–	–	–	–	Etrimfos
99	–	–	–	84	–	–	–	–	–	Malathion
99	–	–	–	85	–	–	–	–	–	Chlorpyriphos
99	–	–	–	86	–	–	–	–	–	Chlormephos
99	–	–	–	86	–	–	–	–	–	Chlorthiophos
99	–	–	–	86	–	–	–	–	–	Isofenphos
99	–	–	–	89	–	–	–	–	–	Dimefos
99	–	–	–	90	–	–	–	–	–	Phosalone
99	91	92	95	80	88	86	55	99	99	Methidathion
99	91	93	94	81	87	84	40	99	99	Azinphos-methyl

8.6 hR$_f$ Data in TLC System 5 in Ascending Numerical Order vs. Substance Names

TLC System 5	1	2	3	4	6	7	8	9	10	[System 5: MeOH / Silica]
0	–	–	–	–	–	–	–	–	–	Amikacin
0	–	–	–	–	–	–	–	–	–	Busulfan
0	–	–	–	–	–	–	–	–	–	Dihydrostreptomycin
0	–	–	–	–	–	–	–	–	–	Fosfomycin
0	–	–	–	–	–	–	–	–	–	Gentamycin
0	–	–	–	–	–	–	–	–	–	Kanamycin
0	–	–	–	–	–	–	–	–	–	Neomycin
0	–	–	–	–	–	–	–	–	–	Rolitetracycline
0	–	–	–	–	–	–	–	–	–	Sisomicin
0	–	–	–	–	–	–	–	–	–	Streptomycin
0	–	–	–	–	–	–	–	–	–	Tobramycin
0	–	–	0	–	–	1	1	0	0	Pentamidine
0	–	–	–	0	–	–	0	–	–	Atropine Methonitrate
0	–	–	–	0	0	0	0	0	0	Diquat Dibromide
0	–	–	–	0	0	0	0	0	0	Paraquat
0	–	–	–	0	0	1	0	0	0	Suxamethonium Chloride
0	–	–	–	0	1	0	0	0	0	Hexamethonium
0	–	–	–	0	1	0	0	0	0	Pentolinium
0	–	–	–	0	2	0	0	0	0	Decamethonium
0	–	–	–	0	2	0	0	0	0	Gallamine
0	–	–	–	0	11	0	0	0	0	Tubocurarine
0	–	–	–	0	12	0	0	0	0	Homatropine Methylbromide
0	–	–	–	0	15	1	0	0	0	Pancuronium Bromide
0	–	–	–	0	29	0	0	1	0	Cetrimide
0	–	–	–	0	29	0	0	1	0	Cetylpyridinium
0	–	–	–	7	3	13	0	0	0	Histamine
0	–	–	–	50	–	–	–	–	–	Tropicamide
1	–	–	–	–	–	–	–	–	–	Piperidine
1	–	–	–	–	4	5	1	1	0	Piperazine
1	–	–	–	0	–	15	0	1	2	Amidefrine
1	–	–	–	0	25	2	0	0	0	Thiazinamium Metilsulfate
1	–	–	–	1	13	2	0	–	–	Neostigmine
1	–	–	–	8	33	–	4	–	–	Tropine
1	–	–	–	13	–	33	–	–	–	Chlorhexidine
1	–	–	–	26	22	2	38	1	0	Cotarnine
1	–	–	–	37	15	0	46	0	0	Hydrastinine
2	–	–	–	1	18	1	0	0	0	Thiamine Hydrochloride
2	–	–	–	1	36	3	0	1	0	Oxyphenonium Bromide
2	–	–	–	3	–	–	–	–	–	Distigmine Bromide
2	–	–	–	3	–	–	–	–	–	Tiemonium Iodide
2	–	–	–	6	18	2	1	3	0	Oxyphencyclimine
3	–	–	–	0	–	–	–	–	–	Metformin
3	–	–	–	0	–	–	–	–	–	Moroxydine
3	–	–	–	0	76	1	0	0	0	Guanoxan

Table 8.6. (Continued)

TLC System 5	1	2	3	4	6	7	8	9	10	[System 5: MeOH / Silica]
3	–	–	–	1	–	2	–	–	–	Clidinium
3	–	–	–	1	–	2	–	–	–	Hexocyclium Metilsulfate
3	–	–	–	1	–	3	–	–	–	Glycopyrronium Bromide
3	–	–	–	1	18	2	0	0	0	Methscopolamine Bromide
3	–	–	–	1	30	1	0	2	0	Guanethidine
3	–	–	–	2	–	5	0	–	–	Emepronium Bromide
3	–	–	–	2	41	8	0	0	0	Butylscopolammonium Bromide,N-
3	–	–	–	3	41	5	0	5	0	Isopropamide Iodide
3	–	–	–	4	31	4	0	4	0	Propantheline Bromide
3	–	–	–	6	–	–	–	–	–	Buzepide
3	–	–	–	12	–	–	–	–	–	Viquidil
3	–	–	–	13	–	0	0	1	0	Epinephrine
3	–	–	–	14	–	10	0	3	1	Azacyclonol
3	–	–	–	25	55	13	2	2	2	Tolazoline
3	–	–	–	27	–	10	6	6	0	Mequitazine
3	–	–	–	27	52	14	3	6	3	Naphazoline
3	–	–	–	76	–	2	0	–	–	Methanthelinium Bromide
4	–	–	–	–	–	6	4	2	2	Tramazoline
4	–	–	–	0	23	0	0	0	0	Carbachol
4	–	–	–	2	66	2	0	4	0	Benzalkonium Chloride
4	–	–	–	5	39	2	0	4	0	Fenpiverinium
4	–	–	–	8	56	7	38	0	0	Berberine
4	–	–	–	40	10	5	67	3	5	Sparteine
4	–	–	0	46	14	38	14	4	2	Chloroquine
5	–	–	–	–	42	26	11	0	24	Betazole
5	–	–	–	7	44	17	0	0	0	Morphine M (Nor-)
5	–	–	–	24	28	18	5	3	1	Atropine/Hyoscyamine
5	–	–	–	25	7	16	0	17	1	Brucine
5	–	–	–	26	60	13	7	2	2	Tetryzoline
5	–	–	–	30	64	13	7	5	3	Xylometazoline
5	–	–	–	37	70	26	39	13	5	Coniine
5	–	–	–	47	66	31	6	7	3	Antazoline
5	–	–	–	83	–	–	–	–	–	Dihydroergocristine
6	–	–	–	24	63	20	3	10	12	Mescaline
6	–	–	0	24	74	24	0	1	2	Amiloride
6	–	–	–	26	–	23	1	2	11	Tryptamine
6	–	–	–	28	37	27	2	8	1	Dothiepin M (Nor-,Sulfoxide)
6	–	–	–	31	61	26	5	11	14	Dimethoxyphenethylamine,3,4-
6	–	–	–	33	–	32	1	3	2	Phentolamine
6	–	–	–	36	–	13	26	6	2	Benzatropine
6	–	–	–	36	71	15	18	5	2	Maprotiline
6	–	–	–	38	69	19	18	7	2	Protriptyline
6	–	–	–	66	68	20	32	10	2	Cyclopentamine
7	–	–	–	–	–	–	–	–	–	Ethacridine
7	–	–	–	–	–	–	–	–	–	Strychnine M (N(6)-Oxide)
7	–	–	–	18	79	3	0	1	1	Proguanil/Chlorguanide

Table 8.6. (Continued)

TLC System 5	1	2	3	4	6	7	8	9	10	[System 5: MeOH / Silica]
7	–	–	–	23	27	1	5	1	1	Homatropine
7	–	–	–	29	70	33	6	11	12	Trimethoxyamphetamine,3,4,5-
7	–	–	–	32	–	23	26	4	2	Methoxyphenamine
7	–	–	–	35	77	23	17	7	4	Amantadine
7	–	–	–	37	–	45	1	2	3	Hydroxychloroquine
7	–	–	–	40	71	26	19	11	3	Desipramine/Imipramine M
8	–	–	–	–	–	59	–	–	–	Prajmalium Bitartrate
8	–	–	0	0	–	5	–	–	–	Oxytetracycline
8	–	–	–	11	–	25	3	1	18	Synephrine/Oxedrine
8	–	–	–	12	67	33	1	1	0	Phenylephrine
8	–	–	–	23	–	–	–	–	–	Homofenazine M
8	–	–	–	32	11	26	8	19	2	Strychnine
8	–	–	–	32	63	–	7	8	2	Pethidine M (Nor-)
8	–	–	–	32	75	33	1	4	1	Clomipramine M (8-Hydroxy-nor-)
8	–	–	–	43	75	36	24	10	14	Dimethoxy-4-ethylamphetamine,2,5-/DOET
8	–	–	–	45	75	34	24	13	3	Clomipramine M (Nor-)
8	–	–	–	45	75	36	19	10	17	Dimethoxyamphetamine,2,5-
8	–	–	–	59	–	41	57	8	6	Perhexiline
9	–	–	–	–	–	–	–	–	–	Pramiverine
9	–	–	–	3	–	2	–	–	–	Penthienate Bromide
9	–	–	–	27	–	29	3	3	3	Pholedrine
9	–	–	–	28	–	–	–	–	–	Promazine M
9	–	–	–	41	76	51	15	17	16	Dimethoxy-4-methylamphetamine,2,5-/STP/DOM
9	–	–	–	42	63	31	28	13	5	Deoxyephedrine,(+/-)-
9	–	–	–	42	63	31	28	13	5	Methamfetamine
9	–	–	–	43	74	35	23	10	18	Methoxy-4,5-methylenedioxyamphetamine,2-/MMDA-2
9	–	–	–	43	74	73	23	77	69	Methoxyamphetamine,4-/PMA
9	–	–	–	56	–	–	–	–	–	Aminoacridine
9	–	–	–	60	7	45	7	8	13	Disopyramide
10	–	–	–	25	64	30	5	5	1	Ephedrine
10	–	–	–	33	34	35	0	1	1	Bufotenine
10	–	–	–	35	–	–	–	–	–	Prothipendyl M
10	–	–	–	38	–	–	–	–	–	Thioridazine M
10	–	–	–	42	49	35	14	4	3	Dextrorphan/Levorphanol
10	–	–	–	42	76	39	18	12	17	Methylenedioxyamphetamine,3,4-/MDA
10	–	–	–	43	76	36	15	14	17	Bromo-2,5-dimethoxyamphetamine,4-/DOB
10	–	–	–	47	–	42	35	13	6	Dimetindene
10	–	–	–	47	42	33	42	18	6	Dextromethorphan/Racemethorphan/Levo-methorphan
11	–	–	–	–	75	–	–	–	–	Methylenedioxyphentermin,3,4-
11	0	0	0	0	–	–	–	–	–	Levodopa
11	–	–	–	17	–	–	0	–	–	Octopamine
11	–	–	–	25	–	–	–	–	–	Perazine M #4
11	–	–	–	27	14	–	5	13	3	Metopon

Table 8.6. (Continued)

TLC System 5	1	2	3	4	6	7	8	9	10	[System 5: MeOH / Silica]
11	–	–	–	29	19	26	8	13	2	Dihydrocodeine
11	–	–	–	30	–	38	3	6	1	Mesoridazine
11	–	–	–	33	13	25	4	20	4	Hydrocodone
11	–	–	–	42	–	–	–	–	–	Metixene M
11	–	–	–	48	78	46	26	24	12	Phentermine
11	–	–	–	49	–	–	–	–	–	Pentorex
11	–	–	–	71	28	54	41	21	9	Diamocaine
12	–	–	–	–	12	30	4	13	3	Methenamine
12	–	–	–	11	73	55	24	4	38	Methoxamine
12	–	–	–	13	–	–	–	–	–	Norfenefrine
12	–	–	–	16	–	47	10	44	6	Hexobendine
12	–	–	–	18	–	25	2	3	1	Dihydromorphine
12	–	–	–	18	14	23	3	9	2	Hydromorphone
12	–	–	–	29	–	42	25	5	46	Pseudoephedrine M (Nor-)/Cathine
12	–	0	–	30	75	44	4	4	33	Ephedrine M/Phenylpropanolamine
12	–	–	–	34	–	–	1	–	–	Propafenone M (N-Desmethyl)
12	–	–	–	35	–	32	–	–	–	Methylephedrine,N-
12	–	–	–	36	–	–	–	–	–	Tiotixene M
12	–	–	–	36	–	–	–	–	–	Tiotixene M #1
12	–	–	–	36	–	–	–	–	–	Tiotixene M #2
12	–	–	–	43	–	–	–	–	–	Etryptamin
12	–	–	–	43	75	43	20	9	18	Amphetamine/Amphetaminil M artifact
12	–	–	–	46	21	45	35	18	2	Chlorphenamine/Chlorpheniramine
12	–	–	–	48	–	47	17	13	6	Pipazetate
12	–	–	–	49	–	45	33	16	6	Brompheniramine
12	–	–	–	55	–	–	42	–	–	Aprindine M (Desethyl-)
12	–	–	–	55	41	40	45	19	4	Ethoheptazine
12	–	–	–	60	–	48	41	10	9	Doxylamine
12	–	–	–	60	–	50	42	19	8	Aminopromazine
12	–	–	–	71	–	3	1	0	1	Hexoprenaline
12	–	–	0	76	–	30	3	2	2	Ethambutol
13	–	–	–	18	76	42	1	1	24	Metaraminol
13	–	–	–	26	–	–	–	–	–	Thioproperazine M
13	–	–	–	30	–	–	–	–	–	Thiethylperazine M
13	–	–	–	33	71	47	0	3	6	Acebutolol
13	–	–	–	39	–	–	6	–	–	Homofenazine
13	–	–	–	50	–	–	–	–	–	Bromopride
13	–	–	–	50	16	48	26	19	4	Carbinoxamine
13	–	–	–	50	28	42	35	22	6	Tropacocaine
14	–	–	–	20	–	42	1	1	1	Nadolol
14	–	–	–	21	69	40	0	1	3	Isoprenaline
14	–	–	–	22	–	23	1	2	5	Heptaminol
14	–	–	–	22	–	45	0	2	2	Atenolol
14	–	–	–	33	–	–	–	–	–	Carteolol
14	–	–	–	36	–	–	–	–	–	Pecazine M
14	–	–	–	43	59	18	0	0	0	Dopamine

Table 8.6. (Continued)

TLC System 5	1	2	3	4	6	7	8	9	10	[System 5: MeOH / Silica]
14	–	–	–	46	26	45	35	13	3	Pheniramine
14	–	–	–	47	48	39	5	9	9	Psilocin
14	–	–	–	48	–	45	26	21	7	Tofenacin
14	–	–	–	48	77	44	18	17	8	Chlorphentermine
14	–	–	–	50	39	40	9	9	6	Dimethyltryptamine,N,N-/DMT
14	–	–	–	56	–	44	35	9	15	Etafedrine
14	–	–	–	63	56	46	14	10	11	Diethyltryptamine,N,N-/DET
14	–	–	–	73	50	–	22	46	43	Cryptopine
14	–	–	–	78	–	70	57	67	56	Azapetine
15	–	–	–	22	74	41	2	2	3	Etilefrine
15	–	–	–	25	–	36	3	18	2	Pholcodine
15	–	–	–	33	–	–	–	–	–	Celiprolol
15	–	–	–	59	29	47	43	23	9	Prothipendyl
15	–	–	–	65	67	–	9	15	25	Psilocin-(eth)/CZ-74
15	–	–	–	79	65	–	66	25	26	Rolicyclidine/PHP/1-(1-phenylcyclohex-yl)pyrrolidine
16	–	–	–	–	22	–	–	–	–	Narcophin
16	–	–	–	0	11	–	0	0	–	Betaine
16	–	–	–	0	11	17	0	0	0	Cocaine M/Ecgonine
16	–	0	–	20	74	46	1	1	4	Salbutamol/Albuterol
16	–	–	–	40	–	–	–	–	–	Bunitrolol
16	–	–	–	40	–	–	–	–	–	Propiomazine M
16	–	–	–	40	–	–	1	–	–	Tiapride
16	–	–	–	42	–	–	–	–	–	Chlorpromazine M
16	–	–	–	50	–	44	36	22	9	Benzydamine
16	–	–	–	54	–	–	28	–	–	Encainide
16	–	–	–	59	74	47	47	19	8	Etilamfetamine
16	–	–	–	68	49	48	47	22	14	Pentoxyverine/Carbetapentane
16	–	–	–	77	60	48	59	20	27	Methadone/Levomethadone
17	–	–	–	34	–	38	0	–	–	Sulpiride
17	–	–	–	39	33	49	1	5	9	Procainamide
17	–	1	–	43	–	–	13	–	–	Tertatolol
17	–	–	–	48	–	–	3	–	–	Propafenone
17	–	–	–	51	–	47	1	7	13	Metoclopramide
17	–	–	–	53	–	–	–	–	–	Lofexidine
17	–	–	–	66	–	–	–	–	–	Butamirate
17	–	–	–	71	–	59	4	35	17	Veratrine
18	–	–	–	16	–	–	–	–	–	Pirenzepine
18	–	–	–	20	23	37	0	9	1	Morphine
18	–	–	–	21	77	47	1	1	5	Terbutaline
18	–	–	–	39	–	–	0	–	–	Propafenone M (5-OH)
18	–	–	–	40	–	–	–	–	–	Promethazine M #1
18	–	–	–	40	–	–	–	–	–	Promethazine M #2
18	–	–	–	40	–	42	3	34	6	Nicomorphine
18	–	–	–	43	78	49	2	5	8	Pindolol
18	–	–	–	46	–	–	12	–	–	Fluvoxamine

Table 8.6. (Continued)

TLC System 5	1	2	3	4	6	7	8	9	10	[System 5: MeOH / Silica]
18	–	–	–	58	49	46	49	25	9	Clemastine
18	–	–	–	62	–	48	17	24	12	Carbocromen
18	–	–	–	62	35	44	38	30	11	Promazine
18	–	–	–	80	–	52	62	19	26	Terodiline
19	–	–	–	–	–	–	–	–	–	Butaperazine M
19	–	–	–	–	–	–	–	–	–	Triflupromazine M
19	–	–	–	–	36	53	42	25	12	Thenyldiamine
19	–	–	–	8	–	–	0	–	–	Clozapine M (N-Oxyde)
19	–	–	–	30	–	53	1	3	5	Sotalol
19	–	–	–	40	–	–	6	–	–	Dibenzepin M (Nor-)
19	–	–	–	41	–	–	1	–	–	Clozapine M (Nor-)
19	–	–	–	44	22	45	17	28	5	Dothiepin M (Nor-)
19	–	–	–	44	61	49	6	12	5	Cinchonine
19	–	–	–	47	–	43	12	42	10	Nicocodine
19	–	–	–	50	–	–	–	–	–	Trimeprazine M
19	–	–	–	53	–	54	13	34	12	Emetine
19	–	0	–	54	–	–	–	–	–	Sulforidazine
19	–	–	–	55	30	51	11	20	6	Triprolidine
19	–	–	–	56	–	–	–	–	–	Piprinhydrinate
19	–	–	–	60	–	48	41	16	11	Fenfluramine
19	–	–	–	64	–	–	–	–	–	Etifelmine
19	–	–	–	67	–	50	51	22	19	Oxeladine
19	–	–	–	81	79	54	59	38	39	Pipradrol
20	–	–	–	0	0	–	–	–	0	Morphine M (-3-glucuronide)
20	–	–	–	38	–	–	0	–	–	Dibenzepin M (Dinor-,5-desmethyl-)
20	–	–	–	38	–	–	1	–	–	Dibenzepin M (Nor-,5-desmethyl-)
20	0	0	14	40	60	58	2	18	29	Flurazepam M (Dideethyl-)
20	–	–	–	41	23	41	6	23	8	Benzylmorphine
20	–	–	–	43	–	–	3	–	–	Mepindolol
20	–	–	–	43	70	45	3	8	5	Hydroquinidine
20	–	–	–	44	–	–	–	–	–	Metipranolol
20	–	–	–	44	74	49	10	8	9	Metoprolol
20	–	–	–	45	78	48	11	11	13	Oxprenolol
20	–	–	–	46	22	45	11	28	5	Dothiepin M (Sulfoxide)
20	–	–	–	48	–	–	–	–	–	Levomepromazine M
20	–	–	–	48	–	47	27	25	–	Zimeldine
20	–	0	–	49	–	–	12	–	–	Bupranolol
20	–	–	–	50	75	52	6	11	9	Timolol
20	–	–	–	52	–	50	38	44	12	Thenalidine
20	–	–	–	62	–	–	–	–	–	Mecloxamine
20	–	–	–	67	55	48	42	30	13	Thioridazine
20	–	–	–	74	68	48	62	31	23	Procyclidine
20	–	–	–	76	–	–	63	–	–	Aprindine
20	–	–	–	78	–	–	–	–	–	Clobutinol
21	–	–	–	–	38	39	18	24	6	Dimethoxanate
21	–	–	–	18	77	48	1	3	6	Orciprenaline

Table 8.6. (Continued)

TLC System 5	1	2	3	4	6	7	8	9	10	[System 5: MeOH / Silica]
21	–	–	–	35	22	33	6	18	3	Codeine
21	–	–	–	36	–	–	–	–	–	Trifluoperazine M
21	–	–	–	36	26	40	7	22	6	Ethylmorphine
21	–	–	–	39	–	–	3	–	–	Dibenzepin M (Dinor-)
21	–	–	–	40	69	44	2	7	3	Hydroquinine
21	–	–	–	44	–	–	–	–	–	Isothipendyl M
21	–	0	–	44	–	–	12	–	–	Toliprolol
21	–	0	–	47	–	–	1	–	–	Carazolol
21	–	–	–	47	23	48	25	37	3	Perazine
21	–	–	–	49	79	50	6	10	7	Propranolol
21	–	–	–	61	–	50	45	25	12	Metixene
21	–	–	–	66	24	52	41	26	13	Methapyrilene
21	–	–	–	67	47	48	48	23	13	Imipramine
21	–	–	–	70	–	–	–	–	–	Camylofine
21	–	–	–	76	56	55	53	37	27	Piperocaine
21	–	–	–	77	–	54	62	34	30	Fencamfamine
21	–	–	–	79	78	–	65	27	37	Eticyclidine/PCE/N-Ethyl-1-phenylcyclohexyl-amine
22	–	–	–	–	–	–	–	–	–	Clofexamide
22	–	–	–	2	14	21	0	1	3	Cocaine M/N-Benzoylecgonine
22	1	1	2	27	–	49	–	–	–	Mafenide
22	–	–	–	43	–	–	–	–	–	Nifenalol
22	–	–	–	43	22	46	7	34	6	Thioproperazine
22	–	0	–	48	–	–	8	–	–	Bisoprolol
22	–	–	–	49	–	52	11	12	11	Alprenolol
22	–	0	–	50	80	50	13	15	8	Penbutolol
22	–	0	–	51	–	–	11	–	–	Betaxolol
22	–	–	–	56	–	62	–	–	–	Ajmaline
22	–	–	–	58	–	–	13	–	–	Clenbuterol
22	–	–	–	58	33	51	39	25	14	Mepyramine/Pyrilamine
22	–	–	–	62	–	–	–	–	–	Dimetacrine
22	–	–	0	63	–	52	41	28	17	Chloropyramine
22	–	–	–	64	35	52	41	30	14	Isothipendyl
22	–	–	–	65	31	55	38	28	14	Thonzylamine
22	–	–	–	68	34	55	44	27	15	Tripelennamine
22	–	–	–	79	–	50	67	32	25	Prolintane
23	–	–	–	18	16	37	0	9	0	Mianserin M (N-Oxide)
23	–	–	–	38	–	55	3	6	0	Bamethan
23	–	–	–	45	32	45	24	37	5	Thebaine
23	–	–	–	68	49	46	37	28	8	Diphenylpyraline
23	–	–	–	70	79	–	41	33	29	Phenylcyclohexylamine,1-
23	–	–	–	75	–	61	17	35	29	Lobeline
23	–	–	–	84	69	59	73	35	66	Phencyclidine
23	–	–	–	86	75	55	68	36	52	Diisopromine
24	–	–	–	44	55	49	6	8	6	Cinchonidine
24	–	–	–	45	–	–	–	–	–	Aceprometazine M

Table 8.6. (Continued)

TLC System 5	1	2	3	4	6	7	8	9	10	[System 5: MeOH / Silica]
24	–	–	–	46	–	–	11	–	–	Erythromycin
24	–	–	–	47	–	–	2	–	–	Trimethobenzamide
24	–	–	–	49	25	45	20	34	11	Thebacon
24	–	–	–	52	–	–	24	–	–	Ketotifen
24	–	–	–	55	–	–	–	–	–	Medrylamine
24	–	–	–	59	–	49	40	43	13	Bamipine
24	–	–	–	59	33	48	18	36	25	Oxomemazine
24	–	–	–	63	45	51	48	37	13	Doxepin
24	–	–	–	65	74	53	15	13	25	Cyclazocine
25	–	–	–	23	–	–	–	–	–	Xanthinol Nicotinate
25	–	–	–	36	–	42	6	23	6	Viloxazine
25	–	–	–	44	27	46	6	26	8	Morphine-6-acetate/Diamorphine M
25	–	–	–	48	–	54	3	16	15	Fenpipramide
25	–	–	–	51	60	46	11	31	6	Mianserin M (Nor-)
25	–	–	–	54	42	49	15	19	9	Clomipramine M (10-Hydroxy-)
25	–	–	–	55	78	40	17	4	9	Mexiletine
25	–	–	–	68	49	55	48	33	16	Orphenadrine
25	–	–	–	70	45	49	45	35	17	Chlorpromazine
25	–	–	–	71	66	54	54	37	18	Pyrrobutamine
25	–	–	–	73	–	–	50	–	–	Melperone
25	–	–	–	74	83	70	29	69	57	Cropropamide
25	–	24	–	75	–	–	58	–	–	Bornaprine
26	–	–	–	26	–	–	–	–	–	Midodrine
26	–	–	–	37	–	47	2	9	6	Ketobemidone
26	3	1	36	40	15	40	1	48	5	Loprazolam
26	–	–	–	44	24	49	10	40	7	Tiotixene
26	–	–	–	45	51	70	1	24	11	Clomipramine M (2-Hydroxy-)
26	–	–	–	45	65	51	2	11	4	Quinine
26	–	–	–	49	33	47	15	38	4	Diamorphine
26	–	–	–	51	51	51	4	17	10	Clomipramine M (8-Hydroxy-)
26	–	–	–	52	–	53	28	37	5	Butaperazine
26	–	–	–	54	25	44	23	43	12	Acetylcodeine
26	–	–	–	55	26	49	34	37	7	Prochlorperazine
26	–	–	–	68	–	51	52	32	15	Tolpropamine
26	–	–	–	72	54	51	53	34	18	Clomipramine
27	–	0	–	0	–	–	0	–	–	Lisinopril
27	–	–	–	16	20	41	0	9	0	Clomipramine M (N-Oxid)
27	–	–	–	33	36	48	10	37	30	Oxymorphone
27	–	–	–	53	–	51	30	41	8	Thiethylperazine
27	–	–	–	62	–	–	–	–	–	Chlorprothixene M
27	–	–	–	65	–	–	52	–	–	Melitracene
27	–	1	–	65	–	53	47	44	16	Pecazine/Mepazine
27	–	–	–	65	48	55	44	33	15	Diphenhydramine
27	–	–	–	66	74	54	42	27	17	Eucaine,β-
27	–	–	–	69	48	54	42	43	13	Bromazine/Bromdiphenhydramine
27	–	–	–	69	51	51	50	32	15	Amitriptyline

Table 8.6. (Continued)

TLC System 5	1	2	3	4	6	7	8	9	10	[System 5: MeOH / Silica]
27	–	–	–	70	41	51	49	42	16	Dothiepin
27	–	–	–	82	77	55	66	44	36	Prozapine
27	–	–	–	87	72	66	67	33	70	Dipipanone
28	–	–	–	23	–	–	–	–	–	Perphenazine M
28	–	–	–	49	–	–	6	–	–	Flecainide
28	0	0	15	50	62	58	8	19	11	Flurazepam M (Monodeethyl-)
28	–	1	–	56	–	–	11	–	–	Gallopamil M (Nor-)
28	–	–	–	63	–	48	26	24	12	Acepromazine
28	–	–	–	64	–	48	45	–	–	Pizotifen
28	–	–	–	68	48	55	45	33	15	Dimenhydrinate
28	–	–	–	83	–	–	–	–	–	Benproperine
29	–	2	–	61	–	–	30	–	–	Melperone M (FG 5155)
29	–	–	–	65	73	55	36	38	13	Noracymethadol
29	–	–	–	66	–	53	43	35	18	Noxiptiline
29	–	–	–	70	–	53	47	36	17	Chlorphenoxamine
29	–	–	–	83	82	77	37	12	63	Phenelzine
30	–	–	–	0	2	–	0	0	0	Codeine M (-6-glucuronide)
30	–	–	–	43	–	–	1	–	–	Clenbuterol M #1 (Hydroxy-)
30	–	–	–	49	63	51	4	12	6	Quinidine
30	–	–	–	55	29	53	33	30	8	Trifluoperazine
30	–	–	–	59	–	50	33	32	17	Nefopam
30	–	–	–	62	33	50	25	51	39	Oxycodone
30	–	–	–	64	50	51	45	44	13	Cyproheptadine
30	–	–	–	65	44	50	36	35	17	Promethazine
30	–	–	–	68	52	55	34	42	26	Propiomazine
30	–	–	–	78	–	–	–	–	–	Tramadol
30	–	–	–	78	73	56	63	37	26	Volazocine
30	–	–	–	79	91	70	48	40	20	Hexetidine
30	–	–	–	82	–	60	56	52	35	Clomifene
31	–	–	–	29	–	–	–	–	–	Dixyracine M
31	–	–	–	34	–	–	–	–	–	Oxypendyl
31	–	–	–	78	–	–	–	–	–	Suloctidil
31	–	–	–	83	55	67	64	47	66	Profenamine/Ethopropazine
31	–	–	–	87	78	–	72	47	68	Methyl-1-phenylcyclohexyl)piperidine,1-(4-
32	–	–	–	–	–	52	41	37	29	Clofedanol
32	–	1	–	29	–	–	0	–	–	Labetalol
32	–	–	–	33	–	–	–	–	–	Clopenthixol M
32	–	–	–	67	–	53	38	48	15	Phenyltoloxamine
32	–	–	–	75	–	–	–	–	–	Histapyrrodine
32	–	–	–	75	49	54	47	35	22	Triflupromazine
32	–	–	–	76	49	57	47	38	46	Levomepromazine/Methotrimeprazine
32	–	–	–	77	46	58	54	39	31	Trimeprazine/Alimemazine
32	–	–	–	86	–	63	40	53	49	Mebeverine
33	–	–	–	–	32	53	3	25	3	Acetophenazine
33	–	–	–	26	–	–	–	–	–	Fluphenazine M
33	–	–	–	43	61	56	1	12	8	Pipamperone

Table 8.6. (Continued)

TLC System 5	1	2	3	4	6	7	8	9	10	[System 5: MeOH / Silica]
33	–	–	–	62	83	74	3	14	50	Buphenine
33	–	–	–	63	–	56	40	34	18	Methadone M (Nor-)
33	–	–	–	77	54	58	57	51	39	Diethazine
33	–	–	–	79	–	–	–	–	–	Salverine
33	–	–	–	86	86	70	57	59	68	Clofazimine
34	–	–	–	0	14	52	0	3	0	Narceine
34	–	–	–	46	45	50	14	27	14	Phenmetrazine/Phendimetrazine M (Nor-)
34	–	–	–	49	–	–	12	–	–	Bupivacaine M/Mepivacaine M
34	–	–	–	60	40	52	37	34	11	Pethidine/Meperidine
34	–	–	–	65	–	–	–	–	–	Aceprometazine
34	–	–	–	70	72	61	16	12	28	Pentazocine
34	–	–	–	74	51	56	51	51	25	Chlorprothixene
34	–	–	–	78	30	70	42	60	55	Cinnamoylcocaine
35	–	–	–	33	–	–	–	–	–	Periciazine M
35	–	–	–	38	39	54	6	22	7	Opipramol
35	–	–	–	67	52	57	42	46	14	Chlorcyclizine
35	–	–	–	68	–	56	41	35	20	Nicametate
35	–	–	–	73	46	66	26	55	38	Pitofenone
35	–	–	–	77	30	65	45	47	54	Cocaine
36	–	–	–	65	46	57	27	45	20	Mebhydrolin
36	–	–	–	71	42	54	5	31	30	Procaine
36	–	–	–	80	56	59	62	54	37	Trimipramine
37	–	–	–	72	–	–	–	–	–	Oxyfedrine
38	–	–	–	–	–	–	–	–	–	Theodrenaline
38	–	0	–	25	81	76	0	1	4	Fenoterol
38	–	–	–	55	22	54	22	35	14	Dibenzepin
38	–	–	–	75	–	–	43	–	–	Mefenorex
38	–	–	–	77	–	59	56	52	43	Benzoctamine
38	–	–	–	84	62	57	62	45	50	Alphacetylmethadol
38	–	–	–	86	70	–	73	54	68	Tenocyclidine/TCP/1-[1-(2-Thienyl)cyclohexyl] piperidine
39	–	–	–	39	–	–	–	–	–	Imiclopazine/Chlorimiphenine
39	–	–	–	39	–	54	5	27	7	Carfenazine
39	–	–	–	52	–	–	4	–	–	Dibenzepin M (5-Desmethyl-)
39	–	–	–	59	50	–	21	48	12	Cocaine M (Nor-)
39	–	–	–	61	22	54	39	35	13	Nicotine
39	–	–	–	86	72	65	28	50	62	Ibogaine/Rotenone
39	–	–	–	87	–	–	–	–	–	Pridinol
40	–	–	70	30	88	56	0	5	9	Clioquinol
40	–	–	–	37	–	–	–	–	–	Flupentixol M
40	–	–	–	42	40	55	7	29	9	Perphenazine
40	–	–	–	53	59	66	3	32	21	Pipotiazine
40	–	–	–	66	70	57	35	41	23	Methylphenidate
40	–	–	–	68	52	57	49	41	16	Cyclizine
41	–	–	–	43	52	65	0	17	14	Pipamazine
41	–	–	–	55	38	55	12	36	18	Physostigmine

Table 8.6. (Continued)

TLC System 5	1	2	3	4	6	7	8	9	10	[System 5: MeOH / Silica]
41	–	–	–	58	67	54	39	33	48	Tranylcypromine
41	–	–	–	68	49	63	45	57	21	Phenindamine
41	–	–	–	80	–	–	48	–	–	Lorcainide
41	–	–	–	81	–	–	–	–	–	Etafenone
41	–	–	–	82	–	–	–	–	–	Tolylfluanid
42	–	–	–	44	74	60	2	23	–	Tocainide
42	–	0	–	47	–	–	4	–	–	Zopiclone
42	4	5	29	55	–	57	4	38	17	Clozapine
42	–	–	–	67	–	63	25	34	35	Cinchocaine/Dibucaine
42	–	–	–	72	–	–	–	–	–	Moperone
42	–	–	–	74	73	67	19	24	45	Levallorphan
42	–	–	–	75	68	68	11	39	58	Aconitine
43	–	–	–	3	–	0	0	0	0	Levarterenol/Norepinephrine
43	–	–	–	62	–	53	–	–	–	Arecoline
43	–	–	–	64	39	57	15	32	16	Tetracaine/Amethocaine
43	–	–	–	66	–	56	13	48	28	Dimetiotazine
43	–	–	–	73	–	–	–	–	–	Nicergoline
43	–	–	–	73	61	59	23	70	42	Verapamil
43	–	–	–	78	–	64	52	41	35	Naftidrofuryl
43	–	–	–	83	75	68	66	61	59	Trihexyphenidyl/Benzhexol
43	–	–	–	84	85	68	55	68	56	Prenylamine
43	–	–	–	85	–	–	–	–	–	Fendiline
44	–	0	–	0	–	–	0	–	–	Flumazenil M/RO 15-3890
44	–	–	–	3	–	–	0	–	–	Pirenzepine M (N-Desmethyl-)
44	–	–	–	3	–	36	1	3	3	Penicillamine
44	–	–	–	54	–	49	28	28	39	Phenethylamine
44	–	–	–	54	–	55	3	45	14	Fenetylline
44	–	–	–	70	76	62	8	31	53	Clonidine
44	–	–	–	83	76	71	7	30	64	Butacaine
45	–	–	–	–	55	63	5	26	31	Butethamine
45	–	–	–	44	–	56	7	32	11	Clopenthixol/Zuclopenthixol/cis-Ordinol
45	–	–	–	45	49	63	5	23	10	Fluphenazine
45	–	–	20	45	59	55	0	22	12	Trimethoprim
45	–	–	–	74	–	–	13	–	–	Terfenadine
45	–	–	–	81	–	68	58	54	55	Isoaminile
45	–	–	–	83	–	64	68	64	64	Biperiden
45	9	8	41	83	51	67	53	60	51	Flurazepam Hy/DCFB
46	–	–	–	–	–	–	–	–	–	Spiramycin
46	–	–	–	51	61	58	4	16	18	Periciazine
46	–	–	–	53	65	63	7	13	13	Mazindol
46	–	–	–	54	52	60	4	37	18	Mianserin M (8-Hydroxy-)
46	–	–	–	57	–	60	1	60	29	Vinblastine
46	–	–	–	67	–	77	26	42	22	Meclofenoxate
46	–	4	–	71	–	–	24	–	–	Gallopamil
47	–	–	–	49	–	–	–	–	–	Dixyracine
47	–	–	–	56	–	–	–	–	–	Diltiazem

Table 8.6. (Continued)

TLC System 5	1	2	3	4	6	7	8	9	10	[System 5: MeOH / Silica]
47	–	–	–	77	–	66	49	–	–	Captodiame
48	–	–	–	9	–	–	–	–	–	Cimetidine M (Sulfoxide)
48	–	–	–	46	–	–	–	–	–	Thiopropazate M
48	–	–	–	68	50	58	39	58	23	Mianserin
48	–	–	–	77	–	62	36	67	45	Flavoxate
48	–	–	–	81	56	69	59	57	47	Butetamate
49	1	0	28	40	39	63	1	36	2	Adinazolam M (Nor-)
49	–	–	–	48	47	55	6	37	18	Scopolamine
49	–	–	–	54	–	–	36	–	–	Loxapine
49	–	–	–	61	43	–	13	59	25	Orientalidine
49	–	–	–	61	43	55	14	57	26	Isothebaine
49	–	–	–	62	41	57	36	51	24	Phendimetrazine
49	–	–	–	83	–	64	55	60	51	Adiphenine
50	–	–	13	30	65	51	1	8	4	Triamterene
50	–	–	–	46	–	62	6	33	–	Flupentixol
50	–	–	–	47	–	–	–	–	–	Cafedrine
50	–	3	–	73	–	–	13	–	–	Bromperidol
50	–	–	–	74	81	68	16	39	49	Phenazocine
50	–	–	–	82	63	68	58	55	54	Propoxyphene/Dextropropoxyphene/Levopropoxyphene
51	–	–	–	–	–	–	–	–	–	Hydroxyphenethylmorphinan
51	–	–	–	56	–	–	–	–	–	Tizanidine
51	–	–	–	61	72	74	4	49	39	Spiperone
51	–	–	–	66	53	69	18	54	46	Tetramisole
51	–	–	–	69	–	–	–	–	–	Benzetimide
51	–	–	–	70	67	71	21	48	44	Dexetimide
51	–	–	–	76	75	67	11	27	33	Haloperidol
51	–	–	–	78	80	69	18	39	47	Phenomorphan
52	–	–	–	–	–	–	–	–	–	Sarcolysis/Merphalan
52	–	–	–	–	–	68	67	64	54	Dicycloverine/Dicyclomine
52	–	–	–	44	45	53	0	32	12	Pilocarpine
52	–	–	–	49	–	–	–	–	–	Etodroxizine
52	–	–	–	57	–	–	6	–	–	Haloperidol M
52	3	3	41	71	45	62	30	48	40	Flurazepam
52	–	–	–	74	81	70	9	32	22	Loperamide
52	–	–	–	77	57	66	43	53	53	Benactyzine
53	–	–	–	0	8	1	0	0	0	Obidoxime Chloride
53	–	–	–	27	55	54	0	9	12	Cimetidine
53	–	–	–	41	–	–	4	–	–	Cotinine
53	–	–	–	64	52	56	9	29	31	Nomifensine
53	–	–	–	65	52	62	18	48	42	Levamisole
54	–	–	–	43	–	–	17	–	–	Maprotiline M (Nor-)
54	–	–	–	51	–	–	0	–	–	Astemizole M (Nor-)
54	–	–	–	58	–	83	0	21	27	Apomorphine
54	–	–	–	82	52	69	55	81	55	Piperidolate
54	–	–	–	82	64	72	62	68	55	Amiodarone

Table 8.6. (Continued)

TLC System 5	1	2	3	4	6	7	8	9	10	[System 5: MeOH / Silica]
54	–	–	–	83	–	62	23	41	36	Oxybuprocaine
55	–	–	18	29	49	47	1	11	20	Isoniazid
55	–	–	–	68	–	–	–	–	–	Metamfepramone
55	2	5	26	77	76	73	13	–	–	Trifluperidol
55	–	–	–	82	57	–	54	60	62	Eucaine,α-
55	–	–	–	85	56	76	62	63	64	Amfepramone/Diethylpropion
56	–	–	–	35	–	53	0	41	11	Demecolcine
56	–	–	–	64	–	–	1	–	–	Astemizole
56	–	7	–	69	–	–	21	–	–	Gallopamil M (D517)
56	–	–	–	74	–	–	–	–	–	Vincamine
57	–	–	–	32	59	59	1	23	29	Nalorphine
57	–	–	–	36	51	60	0	19	7	Lysergamide
57	–	–	–	54	65	68	10	54	19	Hydroxyzine
57	–	–	–	80	–	–	–	–	–	Caroverine
58	–	–	–	25	–	–	–	–	–	Orazamide
58	–	–	–	36	–	–	0	–	–	Nifenazone
58	–	–	–	42	–	60	1	28	16	Dihydroergotamine
58	–	–	–	73	50	60	30	58	33	Methylpiperidyl Benzilate/JB-336
59	–	4	–	34	54	53	1	31	21	Theobromine
59	–	–	–	38	–	65	–	–	–	Dropropizine
59	15	10	55	52	55	52	3	58	25	Caffeine
59	–	–	–	67	–	–	6	–	–	Bromhexine M/Ambroxol
59	–	–	–	71	57	61	15	64	52	Hydrastine
59	–	–	–	78	54	75	44	63	55	Ethylpiperidyl Benzilate/JB-318
60	–	–	–	0	–	–	–	–	–	Protheobromine
60	–	–	–	2	75	49	1	1	1	Methyldopa
60	–	–	–	40	–	–	0	–	–	Dihydroergotamine M (8′-OH)
60	3	–	–	56	59	60	3	39	18	Lysergide/LSD
60	–	–	–	58	–	57	3	44	33	Nimorazole
60	–	–	–	79	66	73	12	56	51	Anileridine
60	–	–	–	87	–	73	67	70	70	Benzphetamine
61	–	–	–	38	–	52	10	7	15	Oxetacaine
61	–	–	–	47	–	–	–	–	–	Ampyrone
61	–	–	–	58	–	58	8	46	31	Morazone
61	–	–	–	82	–	–	49	–	–	Prenoxdiazine/Tibexin
61	–	–	–	84	–	–	–	–	–	Tilidine
62	–	–	–	–	60	70	43	55	41	Pramoxine
62	1	1	25	31	50	64	0	–	–	Bromazepam M (N(py)-Oxide)
62	–	–	–	33	60	57	0	12	8	Ergometrine
62	–	–	–	49	–	–	–	–	–	Bisbentiamin
62	–	–	–	53	–	–	12	–	–	Benzquinamide M (Desacetyl-)
62	–	3	–	60	69	78	3	51	32	Benperidol
62	–	–	–	62	81	78	3	32	53	Isoxsuprine
62	–	–	–	74	59	61	35	53	42	Thiopropazate
62	–	–	–	75	79	77	29	64	60	Prilocaine
63	39	31	62	2	–	–	–	–	–	Nalidixic Acid

Table 8.6. (Continued)

TLC System 5	1	2	3	4	6	7	8	9	10	[System 5: MeOH / Silica]
63	–	–	–	49	–	–	0	–	–	Ketanserin M/Ketanserinol
63	9	36	32	56	87	75	0	32	58	Meprobamate
63	–	–	–	66	60	65	31	62	48	Mepivacaine
63	–	–	–	71	78	69	4	59	49	Fluspirilene
63	–	–	–	88	77	67	36	64	59	Piminodine
64	–	–	–	27	–	–	–	–	–	Fosazepam
64	–	–	–	55	–	–	–	–	–	Pentoxifylline
64	–	–	–	64	67	75	0	47	29	Ergosine
64	–	–	–	66	61	63	10	58	37	Trazodone
64	–	5	–	71	–	–	30	–	–	Buspirone
64	–	–	–	84	77	–	45	75	64	Alphamethylfentanyl
65	–	–	–	–	68	63	0	22	28	Harmine
65	–	–	–	46	14	52	2	52	21	Endralazine Mesilate
65	–	–	–	75	–	76	14	54	61	Butanilicaine
65	–	–	–	76	–	–	–	–	–	Azaperone
66	–	6	–	38	–	–	0	–	–	Etofylline
66	18	14	50	45	66	65	4	–	–	Phenazone/Antipyrine
66	–	–	–	58	–	61	2	31	21	Pyrimethamine
66	–	–	–	62	–	–	0	–	–	Cisapride
66	–	–	–	64	–	58	16	58	41	Metyrapone
66	–	–	–	64	70	63	6	38	52	Yohimbine
66	9	22	30	67	63	73	4	51	47	Medazepam M (Nor-)
66	–	–	–	76	–	–	–	–	–	Metofenazate
66	–	–	–	76	80	75	36	64	56	Articaine
67	–	–	–	–	–	–	–	–	–	Undecylenic Acid
67	7	2	40	47	66	67	1	57	14	Alprazolam
67	–	23	–	82	75	73	39	68	60	Fluanisone
68	–	–	–	–	64	70	2	25	4	Nialamide
68	3	3	33	40	60	69	0	32	42	Chlordiazepoxide M (Nor-)
68	–	–	–	40	66	54	0	21	27	Nicotinamide
68	–	–	–	44	64	63	1	34	22	Ergotamine
68	5	2	41	44	65	60	1	40	16	Triazolam
68	–	–	–	47	–	–	2	–	–	Benzquinamide M (Nor-)
68	–	–	–	56	–	–	1	–	–	Bupivacaine M1 (4-OH)
68	9	2	51	57	66	69	6	55	22	Adinazolam
68	–	–	–	59	74	60	1	47	49	Prazosin
68	–	–	–	79	72	63	37	63	64	Ketamine
69	–	–	–	–	–	–	–	–	–	Eprazinone
69	–	–	–	–	–	–	–	–	–	Fedrilate
69	–	–	0	–	–	5	0	1	1	Cytarabine
69	–	–	–	33	63	55	0	37	12	Colchicine
69	–	–	–	38	–	–	0	–	–	Benzquinamide M (Desacetyl-,N-deethyl)
69	–	–	–	41	–	62	0	14	12	Methylergometrine
69	–	–	–	41	66	76	0	41	8	Domperidone
69	–	–	–	58	–	66	1	48	38	Ergoloid
69	13	5	53	60	70	72	6	60	19	Midazolam

Table 8.6. (Continued)

TLC System 5	1	2	3	4	6	7	8	9	10	[System 5: MeOH / Silica]
69	–	–	–	72	–	64	20	70	54	Doxapram
69	–	–	–	80	79	69	42	73	65	Bupivacaine
70	0	0	0	0	16	58	0	0	0	Lysergic Acid
70	–	–	–	25	59	48	0	12	12	Diprophylline/Propyphylline
70	–	–	0	38	–	55	0	4	8	Ambazone
70	–	–	–	42	69	69	1	23	17	Iproniazid
70	25	10	58	62	68	66	21	–	–	Aminophenazone/Amidopyrine
70	–	–	–	70	77	70	11	63	36	Lidoflazine
70	–	–	–	76	82	71	26	64	58	Phenoperidine
70	–	–	–	78	77	70	43	74	58	Fentanyl
70	–	–	–	79	72	68	13	70	61	Etonitazene
70	–	–	–	83	78	–	38	73	59	Fluorofentanyl,p-
71	3	2	5	5	–	–	–	–	–	Etacrynic Acid
71	–	–	–	49	–	58	2	33	29	Proxyphylline
71	7	5	44	50	60	71	2	53	25	Estazolam
71	–	–	44	54	70	63	3	42	46	Pyrazinamide
71	–	–	–	58	73	67	2	48	36	Droperidol
71	–	–	–	59	–	65	0	54	34	Bamifylline
71	–	–	–	59	67	59	15	56	29	Nikethamide
71	–	–	–	81	–	–	0	–	–	Doxazosin
72	1	0	4	0	–	58	–	–	–	Nicotinic Acid
72	–	–	–	48	–	–	–	–	–	Ketoconazole
72	–	–	–	52	–	–	–	–	–	Etizolam
72	15	5	53	52	71	72	5	52	27	Brotizolam
72	–	–	–	57	76	70	2	40	33	Harman
72	–	–	–	60	74	72	7	64	63	Pentetrazol
72	–	–	–	66	–	55	6	66	46	Pentifylline
72	–	–	–	78	75	64	21	74	64	Noscapine
72	–	–	–	79	78	73	42	71	60	Dextromoramide/Racemoramide/Levomoramide
72	–	–	–	80	69	70	35	71	63	Lidocaine
72	–	–	–	83	–	–	–	–	–	Piribedil
72	–	–	–	84	89	76	17	60	60	Penfluridol
72	–	–	–	87	73	–	62	74	64	Phenylcyclohexyl)morpholine,4-(1-/PCM
73	–	–	–	–	–	59	0	0	0	Isoetarine
73	3	2	26	26	69	67	1	25	7	Alprazolam M (4-Hydroxy-)
73	–	–	–	37	–	–	–	–	–	Clonazepam M (7-Amino-3-hydroxy-)
73	0	6	26	39	61	61	0	24	28	Bromazepam M (3-Hydroxy-)
73	3	4	41	53	72	70	3	52	8	Midazolam M (α-Hydroxy-)
73	–	–	–	61	74	70	1	45	42	Piritramide
73	13	18	47	63	69	61	6	41	53	Bromazepam
73	–	–	–	65	–	–	3	–	–	Ketanserin
73	–	–	–	71	82	71	3	60	40	Pimozide
73	–	–	–	81	–	72	3	77	66	Deserpidine
74	–	–	–	–	–	–	–	–	–	Chlorbenzoxamine
74	–	–	–	–	–	–	–	–	–	Cyclophosphamide

Table 8.6. (Continued)

TLC System 5	1	2	3	4	6	7	8	9	10	[System 5: MeOH / Silica]
74	–	9	–	11	66	75	1	30	11	Theophylline
74	–	–	–	47	–	65	9	66	63	Naloxone
74	–	–	–	48	69	56	4	17	22	Nicotinyl Alcohol
74	–	–	–	57	77	77	0	59	50	Ergocristine
74	–	–	–	58	–	62	–	–	–	Methylphenazone,4-
74	–	–	–	59	77	78	0	58	50	Ergocryptine
74	–	–	–	62	–	–	4	–	–	Oxatomide
74	–	–	–	69	74	61	8	65	47	Papaverine
74	–	–	–	70	76	75	13	58	51	Bromolysergide,2-
74	–	–	–	72	77	61	37	66	59	Methyl Nicotinate
74	–	–	–	78	–	–	–	–	–	Oxydemeton
74	–	–	–	78	–	68	4	65	58	Oxypertine
75	–	–	–	–	–	–	–	–	–	Lincomycin
75	–	–	–	–	–	–	–	–	–	Thiotepa
75	–	–	–	–	–	67	7	54	53	Tiabendazole
75	46	57	54	6	–	–	–	–	–	Ibuprofen
75	–	–	–	15	67	59	0	8	5	Pyridoxine Hydrochloride
75	–	–	–	21	–	–	–	–	–	Allopurinol
75	3	1	29	24	71	63	0	29	8	Triazolam M (4-Hydroxy-)
75	1	6	7	25	–	65	–	–	–	Sulfaguanidine
75	4	3	39	42	74	71	4	49	21	Triazolam M (α-Hydroxy-)
75	63	47	68	42	75	72	1	–	–	Carbimazole
75	–	–	32	46	70	58	2	36	40	Metronidazole
75	21	40	30	59	–	–	3	–	–	Paracetamol M/p-Aminophenol
75	–	–	–	69	–	–	–	–	–	Chlorazanil
76	–	–	–	–	–	–	–	–	–	Mercaptopurine,6-
76	–	–	–	–	–	–	–	–	–	Pyritinol
76	5	4	37	28	76	68	1	35	13	Brotizolam M (6-Hydroxy-)
76	4	4	37	37	76	72	1	44	18	Alprazolam M (α-Hydroxy-)
76	8	23	28	41	–	–	–	–	–	Primidone
76	5	5	37	49	77	74	1	43	15	Midazolam M (4-Hydroxy-)
76	10	10	53	52	77	62	2	50	22	Chlordiazepoxide
76	11	20	40	52	77	73	0	30	47	Clonazepam M (7-Amino-)
76	30	14	61	61	72	71	3	63	44	Flumazenil/RO 15-1788
76	–	–	–	76	88	73	5	68	61	Ambucetamide
76	–	–	–	77	80	69	2	74	63	Reserpine
76	–	–	–	78	73	78	31	69	52	Clemizole
76	66	66	80	83	86	73	31	77	68	Nimetazepam Hy/MNB
77	–	–	–	–	–	–	–	–	–	Nitrofural/Nitrofurazone
77	–	–	–	–	–	66	1	38	57	Protionamide
77	–	–	–	31	–	–	–	–	–	Vinclozolin
77	15	32	26	45	–	–	0	–	–	Paracetamol
77	10	20	30	54	79	71	0	30	46	Nitrazepam M (7-Amino-)
77	–	–	–	60	–	70	–	–	–	Pargyline
77	–	–	–	62	–	–	–	–	–	Molsidomine
77	21	21	52	63	74	74	1	55	52	Flunitrazepam M (7-Amino-)

Table 8.6. (Continued)

TLC System 5	1	2	3	4	6	7	8	9	10	[System 5: MeOH / Silica]
77	–	–	–	71	78	–	11	73	63	Meconin
77	–	–	–	72	–	–	–	–	–	Pyricarbate
77	–	–	–	72	–	–	–	–	–	Tofisopam
77	–	–	–	74	81	71	23	65	48	Metomidate
77	–	7	56	80	80	73	11	67	37	Miconazole
77	–	–	–	81	79	73	1	75	64	Rescinnamine
77	71	69	81	83	90	75	48	77	69	Fludiazepam Hy/CFMB
77	74	73	80	85	93	79	44	77	73	Quazepam Hy/CFTB
78	0	0	4	0	–	–	–	–	–	Hydroxyhippuric Acid,o-
78	–	13	–	0	–	–	–	–	–	Ibuprofen M (2-4′-(2-Carboxypropyl)phenyl-propionic Acid)
78	18	30	31	9	–	–	–	–	–	Acetylsalicylic Acid
78	4	34	11	34	–	–	–	–	–	Hydrochlorothiazide
78	2	5	33	38	81	75	4	25	14	Midazolam M (α,4-Dihydroxy-)
78	7	6	41	45	78	72	2	46	31	Brotizolam M (α-Hydroxy-)
78	19	11	41	51	76	73	0	40	49	Flunitrazepam M (7-Acetamido-)
78	11	19	35	54	75	72	0	34	47	Flunitrazepam M (7-Aminonor-)
78	–	–	–	62	81	71	4	55	52	Carnidazole
78	–	–	–	63	–	–	–	–	–	Isopropylaminophenazone/Isopyrin
78	31	25	55	63	–	58	–	–	–	Methyprylon
78	–	–	–	64	–	66	6	45	68	Leucinocaine
78	52	37	68	69	–	–	–	–	–	Griseofulvin
78	53	54	65	71	95	77	15	68	66	Oxazolam
78	–	–	–	72	–	–	–	–	–	Enilconazole/Imazalil
78	–	–	–	73	85	67	26	71	52	Etomidate
78	–	–	–	75	–	57	32	27	29	Octacaine
78	–	–	–	75	80	80	9	61	–	Econazole
78	39	45	63	75	84	73	0	66	59	Cloxazolam
78	56	50	67	75	86	76	24	69	63	Fludiazepam
78	–	–	–	78	–	74	52	79	71	Broxaldine
78	–	–	–	78	81	71	21	78	69	Disulfiram
78	37	51	55	79	–	67	63	–	–	Sulfadoxine
78	–	–	–	86	–	–	–	–	–	Etozolin
79	–	–	–	–	–	–	–	–	–	Viminol
79	26	54	41	5	–	65	–	–	–	Sulfamethoxazole
79	9	20	27	8	–	66	–	–	–	Sulfathiazole
79	23	45	44	13	–	62	–	–	–	Sulfadimidine
79	7	11	30	42	82	73	0	22	45	Clonazepam M (7-Acetamido-)
79	–	–	–	56	75	60	2	56	47	Carbamazepine
79	31	51	47	60	–	76	–	–	–	Clorexolone
79	–	–	–	61	–	–	–	–	–	Bencyclane
79	–	–	–	71	–	68	1	65	68	Nifedipine
79	56	40	73	78	83	67	41	74	62	Medazepam
79	–	–	–	78	84	70	36	80	56	Methaqualone
79	–	–	–	86	87	76	54	78	65	Cinnarizine
80	–	–	–	–	–	–	–	–	–	Sulfaperine

Table 8.6. (Continued)

TLC System 5	1	2	3	4	6	7	8	9	10	[System 5: MeOH / Silica]
80	–	–	–	–	–	–	–	–	–	Trofosfamide
80	–	–	–	–	83	80	58	77	69	Phenadoxone
80	–	–	–	0	1	5	0	0	0	Psilocybine
80	–	–	–	1	–	–	–	–	–	Proscillaridine
80	5	16	27	5	–	64	–	–	–	Sulfisomidine
80	–	–	71	6	–	53	3	8	4	Azathioprine
80	23	41	42	8	–	65	–	–	–	Sulfamerazine
80	–	–	–	10	88	5	62	0	0	Phencyclidine interm./1-Piperidino-1-cyclohexanecarbonitrile/PCC
80	–	–	–	34	–	9	1	1	1	Oxymetazoline
80	–	–	–	41	–	62	1	52	59	Thiamazole
80	7	12	24	44	83	71	0	20	49	Nitrazepam M (7-Acetamido-)
80	–	–	–	60	84	65	0	59	49	Mebendazole
80	45	45	52	66	84	–	–	–	–	Acetanilide
80	–	–	–	68	–	61	11	72	52	Trimetozine
80	–	–	–	70	–	59	1	50	53	Phenazopyridine
80	–	–	–	71	–	–	–	–	–	Pyrithyldione
80	–	–	–	72	–	66	20	70	55	Propanidid
80	–	–	–	74	–	–	–	–	–	Indolyl Acetate,3-
80	54	47	72	74	82	63	10	72	63	Flunitrazepam
80	–	–	–	75	–	–	–	–	–	Etidocaine
80	46	50	65	75	90	74	11	66	62	Haloxazolam
80	–	–	68	76	–	–	–	–	–	Clotrimazole
80	–	–	–	76	85	64	44	69	58	Clomethiazole
80	–	–	–	76	86	69	8	52	59	Fenyramidol
80	–	–	–	79	–	69	41	78	67	Tetrabenazine
80	–	–	–	80	–	–	–	–	–	Benfluralin
80	–	–	–	80	–	76	9	68	69	Buprenorphine
80	37	41	54	81	79	72	5	57	57	Nitrazepam M Hy/DAB
80	41	46	55	82	80	74	5	59	61	Clonazepam Hy/DCB
80	–	–	–	85	–	–	–	–	–	Moxaverine
80	–	–	–	87	88	76	61	79	70	Meclozine
80	–	–	–	99	–	–	–	–	–	Dimethoate
80	91	92	95	99	88	86	55	99	99	Methidathion
81	–	–	–	–	–	–	–	–	–	Chlorobenzoic Acid,m-
81	–	–	–	–	–	–	–	–	–	Dimethylaminobenzaldehyde,p-
81	–	–	14	–	–	76	0	1	8	Sulfacarbamid
81	–	–	–	0	–	–	0	–	–	Strophanthin
81	22	39	38	4	–	64	–	–	–	Sulfadiazine
81	25	52	33	6	–	65	–	–	–	Sulfafurazole
81	–	–	–	28	–	–	0	–	–	Clindamycin
81	–	–	40	36	–	–	–	–	–	Sulfaguanole
81	–	11	–	36	81	60	0	23	40	Pemoline
81	11	17	40	39	–	–	2	–	–	Guaifenesin
81	15	24	42	41	83	63	0	35	51	Demoxepam/Chlordiazepoxide M
81	–	65	–	43	–	–	–	–	–	Crotylbarbital

Table 8.6. (Continued)

TLC System 5	1	2	3	4	6	7	8	9	10	[System 5: MeOH / Silica]
81	62	65	55	44	–	–	–	–	–	Methyl p-Hydroxybenzoate/Methyl Paraben
81	–	9	–	46	–	–	0	–	–	Metaclazepam M (Dinor-)
81	–	3	–	46	78	67	1	38	40	Glafenine
81	23	43	42	57	–	–	–	–	–	Sultiame
81	38	37	64	57	80	71	1	59	52	Flunitrazepam M (3-Hydroxy-)
81	19	28	54	61	82	74	2	46	55	Flurazepam M (N(1)-Hydroxyethyl-)
81	–	–	–	62	–	–	–	–	–	Ornidazole
81	–	–	–	70	–	–	–	–	–	Khellin
81	–	–	–	70	–	–	4	–	–	Pirenzepine M (Desamide)
81	33	25	65	73	84	76	15	68	58	Metaclazepam M (N-Nor-)
81	61	49	65	74	–	71	32	–	–	Propyphenazone
81	53	46	71	77	81	74	12	70	55	Nimetazepam
81	–	–	–	81	–	–	–	–	–	Ajmalicine/Raubasine
81	70	69	74	83	93	83	11	71	69	Haloxazolam Hy/ABFB
81	75	66	71	84	–	–	–	–	–	Clofibrate
81	91	93	94	99	87	84	40	99	99	Azinphos-methyl
82	33	38	44	6	–	–	–	–	–	Naproxen
82	0	0	7	8	72	75	0	2	0	Cinchophen
82	24	43	47	12	–	60	–	–	–	Sulfametoxydiazine
82	–	–	–	24	21	–	6	28	7	Morphine-3-acetate
82	–	–	–	38	–	–	–	–	–	Methylprimidone,4-
82	23	41	42	43	82	52	1	36	28	Lorazepam/Lormetazepam M
82	–	–	–	44	87	68	0	37	42	Dipyridamole
82	22	35	42	45	82	56	0	40	51	Oxazepam/M of Chlordiazepoxide, Clorazepic Acid, Camazepam, Diazepam, Halazepam, Ketazolam, Medazepam, Oxazolam, Prazepam
82	–	–	–	51	–	–	–	–	–	Primidone M/PEMA
82	46	45	60	59	82	52	6	61	50	Lormetazepam
82	10	35	35	60	85	–	0	33	56	Mebutamate
82	–	–	–	62	–	–	–	–	–	Ectylurea
82	51	47	65	62	82	53	8	59	53	Temazepam/M of Camazepam, Diazepam, Ketazolam, Medazepam
82	34	45	57	69	83	62	3	55	60	Nordazepam/M of Chlordiazepoxide, Clorazepic Acid, Diazepam, Halazepam, Ketazolam, Medazepam, Oxazolam, Prazepam
82	35	41	57	72	86	73	5	58	56	Delorazepam
82	47	55	60	73	–	75	–	–	–	Phenprobamate
82	–	–	–	74	–	–	–	–	–	Coumarine
82	55	32	69	75	83	76	12	73	65	Camazepam
82	58	49	72	76	85	75	27	73	59	Diazepam/Ketazolam M/Medazepam M
82	47	35	71	79	84	77	40	73	62	Metaclazepam
82	–	–	–	80	–	–	1	–	–	Nicardipine
82	–	–	–	80	87	74	17	76	66	Bisacodyl
82	–	–	–	82	–	–	–	–	–	Fominoben
82	–	–	–	82	89	79	11	79	71	Phenothiazine

Table 8.6. (Continued)

TLC System 5	1	2	3	4	6	7	8	9	10	[System 5: MeOH / Silica]
82	81	69	81	87	93	75	49	81	7	Lormetazepam Hy/MDB
82	–	–	–	90	–	–	–	–	–	Lofepramine
83	–	–	–	–	–	–	–	–	–	Rifampin
83	–	–	–	–	90	75	6	63	70	Butyl Aminobenzoate
83	–	–	–	3	–	–	–	–	–	Thioctic Acid
83	12	23	27	4	–	65	–	–	–	Sulfamethizole
83	–	–	–	5	–	76	0	65	29	Glymidine
83	16	13	38	5	63	–	–	–	–	Indometacin
83	19	39	43	9	–	65	–	–	–	Sulfamethoxypyridazine
83	–	–	–	34	–	–	–	–	–	Etiroxate
83	–	–	–	42	–	–	–	–	–	Isovaleryl-1,3-indanedione,2-
83	38	55	43	50	84	–	0	–	–	Salicylamide
83	13	46	22	51	–	67	–	–	–	Sulfanilamide
83	–	–	–	56	–	–	–	–	–	Ethaverine
83	–	–	–	58	85	80	0	55	6	Flubendazole
83	38	37	52	68	–	–	–	–	–	Phenacetin
83	34	46	57	68	87	84	3	56	60	Clorazepic Acid
83	–	–	–	69	79	68	28	67	55	Crotethamide
83	34	45	60	72	88	75	2	56	58	Flurazepam M (N(1)-Dealkyl-)/Quazepam M (N-Dealkyl-2-oxo-)
83	45	45	62	74	80	66	14	64	66	Ketazolam
83	–	–	–	82	–	–	26	–	–	Nifedipine M
83	–	–	–	88	90	82	45	75	64	Flunarizine
83	–	–	–	99	–	–	–	–	–	Bromophos
84	0	0	0	0	–	–	–	–	–	Barbituric Acid
84	–	1	–	0	–	–	–	–	–	Ibuprofen M (2-4′-(2-Hydroxy-2-methyl-propyl)phenylpropionic Acid)
84	–	22	–	3	–	–	–	–	–	Flurbiprofen M #1
84	4	31	18	5	–	85	–	–	–	Acetazolamide
84	33	40	50	6	–	–	–	–	–	Chlorambucil
84	2	30	33	6	–	–	–	–	–	Nitrofurantoin
84	–	–	–	9	–	–	0	–	–	Gliclazide
84	–	20	–	10	–	–	–	–	–	Flurbiprofen M #2
84	–	–	–	18	–	–	–	–	–	Flufenamic Acid
84	41	61	57	32	–	–	–	–	–	Barbital
84	7	23	38	49	–	70	–	–	–	Methocarbamol
84	–	–	–	52	–	–	–	–	–	Ethyl-2-(p-tolyl)malonamide,2-
84	49	48	60	59	84	–	–	–	–	Acecarbromal
84	35	46	53	64	86	68	0	36	55	Nitrazepam
84	32	43	54	64	90	75	0	52	61	Clobazam M (Nor-)
84	50	53	59	66	–	70	5	–	–	Ethosuximide
84	–	–	–	68	–	–	–	–	–	Flutazolam
84	35	51	52	68	–	–	6	–	–	Bromisoval
84	–	–	–	68	80	66	1	63	57	Chlormezanone
84	35	46	56	68	85	76	0	58	59	Flunitrazepam M (Nor-)
84	–	–	–	71	–	–	–	–	–	Phenaglycodol

Table 8.6. (Continued)

TLC System 5	1	2	3	4	6	7	8	9	10	[System 5: MeOH / Silica]
84	–	–	–	73	–	–	–	–	–	Azintamid
84	–	–	–	73	88	72	2	69	61	Bromocriptine
84	53	47	70	75	85	62	8	70	62	Clobazam
84	–	–	–	75	86	71	20	74	61	Isocarboxazid
84	56	62	63	76	87	67	5	57	66	Benzocaine
84	61	60	64	76	96	–	22	64	69	Cholesterol
84	66	51	75	78	–	–	–	–	–	Spironolactone
84	57	49	67	79	89	78	32	69	66	Tetrazepam
84	55	48	66	80	87	78	31	69	66	Clotiazepam
84	64	55	72	81	89	65	36	74	63	Prazepam
84	–	–	–	83	–	83	–	–	–	Crotamiton
84	–	–	–	87	97	73	63	76	68	Phenoxybenzamine
84	–	–	–	88	–	75	69	79	71	Bromhexine
84	–	–	–	99	–	–	–	–	–	Bromophos-ethyl
84	–	–	–	99	–	–	–	–	–	Etrimfos
84	–	–	–	99	–	–	–	–	–	Etrimfos
84	–	–	–	99	–	–	–	–	–	Malathion
85	–	–	–	–	–	–	–	–	–	Sulfadicramide
85	–	–	–	–	–	–	–	–	–	Sulfaloxic Acid
85	0	0	2	0	–	66	–	–	–	Sulfasalazine
85	0	0	2	2	59	84	0	1	2	Metamizol/Dipyrone
85	–	28	–	5	–	–	–	–	–	Pirprofen
85	13	20	30	5	–	–	–	–	–	Tolmetin
85	–	–	–	5	–	–	0	–	–	Glisoxepide
85	27	25	41	6	–	–	–	–	–	Ketoprofen
85	–	–	–	11	–	–	–	–	–	Trospium Chloride
85	–	–	–	14	–	–	–	–	–	Chlorotheophylline,8-
85	47	65	53	28	–	–	–	–	–	Phenobarbital
85	1	5	28	33	–	–	–	–	–	Digoxin
85	38	35	59	41	–	–	–	–	–	Etamivan
85	65	65	69	53	–	–	–	–	–	Hexobarbital
85	73	72	71	58	–	–	–	–	–	Methohexital
85	–	–	–	62	–	–	1	–	–	Valnoctamine
85	35	45	56	67	87	72	0	53	61	Clonazepam
85	55	55	70	72	88	71	12	–	52	Prazepam M (3-Hydroxy-)
85	35	45	56	73	92	77	5	58	60	Tetrazepam M (Nor-)
85	53	58	62	74	93	76	0	62	67	Ethyl Loflazepate
85	36	53	59	75	79	–	4	–	–	Carisoprodol
85	53	55	64	75	87	–	12	–	–	Carbromal
85	–	–	–	77	–	–	–	–	–	Kavain
85	65	61	73	81	92	75	29	72	70	Pinazepam
85	–	–	–	83	–	–	–	–	–	Dichlorvos/DDVP
85	–	–	–	85	–	–	–	–	–	Floctafenine
85	–	–	–	86	90	83	53	77	66	Clocinizine
85	–	–	–	99	–	–	–	–	–	Chlorpyriphos
86	–	–	–	–	–	–	–	–	–	Dimpylate/Diazinon

Table 8.6. (Continued)

TLC System										
5	1	2	3	4	6	7	8	9	10	[System 5: MeOH / Silica]
86	–	–	–	4	–	–	–	–	–	Nafcillin
86	–	5	–	4	–	–	–	–	–	Tiaprofenic Acid
86	1	7	7	6	–	–	0	–	–	Furosemide
86	52	50	66	7	–	–	0	–	–	Tolazamide
86	–	–	–	8	–	–	–	–	–	Bumadizon
86	7	1	24	10	–	–	–	–	–	Salicylic Acid
86	31	51	48	10	–	65	–	–	–	Sulfadimethoxine
86	–	–	–	23	–	–	–	–	–	Amino-5-bromo-3-hydroxyphenyl)(2-pyridyl) ketone,(2-
86	49	64	56	36	–	–	1	–	–	Reposal
86	11	31	34	36	90	69	0	31	48	Chloramphenicol
86	48	65	57	40	–	–	–	–	–	Aprobarbital
86	50	65	58	41	–	–	–	–	–	Butobarbital/Butethal
86	70	67	70	41	–	–	–	–	–	Methylphenobarbital
86	33	55	53	41	–	–	–	–	–	Phenytoin
86	–	28	–	45	–	–	0	–	–	Hydrocortisone
86	59	71	63	50	–	–	–	–	–	Sigmodal (5-(2-Bromoallyl)-5-(1-methylbutyl) barbituric Acid)
86	38	59	39	53	–	–	–	–	–	Phenolphthalein
86	–	–	–	54	–	49	3	45	17	Phanquinone
86	–	–	–	70	–	67	0	51	62	Benorilate
86	–	–	–	71	–	–	4	–	–	Ethinylestradiol
86	–	–	–	74	–	–	–	–	–	Cyprazepam
86	–	–	–	75	–	–	–	–	–	Mevinphos/Phosdrin
86	49	59	58	76	87	76	5	–	–	Ethinamate
86	57	65	60	78	–	–	9	–	–	Hexapropymate
86	–	55	–	78	–	–	12	–	–	Mexazolam
86	63	62	70	80	89	75	31	–	–	Glutethimide
86	–	–	–	81	–	–	–	–	–	Demeton-S-methyl
86	–	–	–	82	–	–	–	–	–	Nimodipine M
86	–	–	–	82	–	–	34	–	–	Nitrendipine M
86	–	–	–	84	–	–	–	–	–	Visnadin
86	64	61	66	85	88	77	9	68	69	Bromazepam Hy/ABP
86	–	–	–	86	–	–	–	–	–	Benzyl Alcohol
86	–	–	–	86	–	–	–	–	–	Coumachlor
86	75	73	80	86	91	83	48	77	72	Pinazepam Hy/CPB
86	–	–	–	99	–	–	–	–	–	Chlormephos
86	–	–	–	99	–	–	–	–	–	Chlorthiophos
86	–	–	–	99	–	–	–	–	–	Isofenphos
87	–	–	–	–	–	–	–	–	–	Carbutamide
87	–	–	–	–	–	–	–	–	–	Sulfathiourea
87	–	0	–	2	–	–	0	–	–	Clenbuterol M #2
87	–	–	–	2	46	88	0	10	10	Dimenhydrinate
87	1	10	6	4	–	–	–	–	–	Bumetanide
87	14	10	34	4	–	–	–	–	–	Sulindac
87	17	42	28	4	–	70	–	–	–	Sulfacetamid

Table 8.6. (Continued)

TLC System 5	1	2	3	4	6	7	8	9	10	[System 5: MeOH / Silica]
87	13	3	24	5	–	–	–	–	–	Probenecid
87	–	8	–	5	–	–	–	–	–	Propantheline M/Xanthanoic Acid
87	–	19	–	6	–	–	0	–	–	Clenbuterol M #3
87	23	49	33	6	–	–	2	–	–	Sulfasomizole
87	1	1	4	7	–	–	–	–	–	Saccharin
87	38	43	49	10	88	72	0	–	–	Chlorpropamide
87	41	48	54	11	–	–	–	–	–	Mefenamic Acid
87	52	68	57	29	–	–	–	–	–	Brallobarbital
87	50	66	56	34	–	–	–	–	–	Allobarbital
87	7	47	13	36	–	–	–	–	–	Hydroflumethiazide
87	–	27	–	41	–	–	0	–	–	Methylprednisolon
87	54	67	57	44	–	–	1	–	–	Butalbital
87	–	–	–	46	71	34	28	16	4	Amitriptyline M/Nortriptyline
87	–	–	–	49	–	52	0	1	3	Dobutamine
87	11	34	32	52	–	82	0	–	–	Chlorphenesin Carbamate
87	66	65	69	54	–	–	–	–	–	Metharbital
87	–	–	–	55	89	64	2	43	57	Mephenesin
87	–	–	–	65	–	–	–	–	–	Thibenzazoline
87	78	68	76	65	–	79	–	–	–	Phenylbutazone
87	–	–	–	76	–	64	3	59	55	Ethenzamide
87	–	–	–	77	–	–	–	–	–	Triflubazam
87	–	–	–	78	–	–	–	–	–	Kebuzone
87	–	–	–	78	–	–	–	–	–	Meproscillarine
87	–	–	–	78	–	–	1	–	–	Nimodipine
87	–	–	–	80	–	–	2	–	–	Nitrendipine
87	59	57	70	80	90	78	16	89	71	Quazepam M (2-Oxo-)
87	74	77	73	80	95	–	37	–	–	Cyclandelate
87	48	50	57	82	91	77	7	60	66	Flurazepam Hy/HCFB
87	78	71	78	83	96	74	27	75	76	Quazepam
87	–	–	–	86	–	–	–	–	–	Tritoqualine
87	81	74	82	86	–	–	0	–	–	Ethchlorvynol
87	71	66	68	86	91	77	4	69	70	Flunitrazepam M Hy/ANFB
87	72	66	70	86	93	77	4	69	70	Clonazepam Hy/ANCB
87	–	–	–	92	–	–	–	–	–	Chlorfenvinphos
88	–	–	–	–	–	–	–	–	–	Anhydrotetracycline
88	–	–	–	–	–	–	–	–	–	Doxycycline
88	–	–	–	–	–	–	–	–	–	Gitoformate
88	–	–	–	–	–	–	–	–	–	Minocycline
88	–	–	–	–	–	5	–	–	–	Tetracycline
88	6	23	19	2	69	–	–	–	–	Gentisic Acid
88	12	12	–	4	–	–	–	–	–	Zomepirac
88	–	–	–	6	–	–	–	–	–	Lonazolac
88	–	–	–	8	–	68	53	5	67	Azapropazone
88	3	3	15	11	–	–	–	–	–	Niflumic Acid
88	51	55	62	12	88	76	–	–	–	Tolbutamide
88	15	60	23	14	–	80	–	–	–	Trichlormethiazide

Table 8.6. (Continued)

TLC System 5	1	2	3	4	6	7	8	9	10	[System 5: MeOH / Silica]
88	51	38	71	17	–	–	–	–	–	Piroxicam
88	–	–	–	26	–	–	–	–	–	Phenytoin M (4-Hydroxyphenyl-)
88	54	61	56	33	90	85	0	51	47	Chlorzoxazone
88	3	10	42	36	–	–	–	–	–	Digitoxin
88	50	64	59	38	–	–	–	–	–	Heptabarbital
88	50	64	58	40	–	–	–	–	–	Cyclobarbital
88	4	40	23	42	83	–	–	–	–	Chlortalidone
88	52	66	58	44	–	–	–	–	–	Amobarbital
88	53	68	57	45	–	–	–	–	–	Butallylonal
88	55	68	62	45	–	–	–	–	–	Secobarbital
88	64	66	57	46	–	–	–	–	–	Ethyl p-Hydroxybenzoate
88	50	64	57	48	–	–	–	–	–	Secbutabarbital
88	15	28	35	49	89	67	0	30	38	Quazepam M (3-Hydroxy-N-dealkyl-2-oxo-)
88	42	52	58	69	90	71	2	55	59	Quazepam M (3-Hydroxy-2-oxo-)
88	–	–	68	80	–	49	2	10	4	Procarbazine
88	–	–	–	83	92	79	64	84	74	Phenanthrene
88	–	–	–	86	–	72	47	78	67	Fenbutrazate
89	–	–	–	–	–	–	–	–	–	Sulfaproxyline
89	–	–	–	5	–	–	–	–	–	Bezafibrate
89	–	–	–	6	–	–	–	–	–	Lanatoside C
89	29	51	43	9	–	69	–	–	–	Sulfaphenazole
89	8	5	18	16	–	–	–	–	–	Diflunisal
89	50	65	57	34	–	–	–	–	–	Vinbarbital
89	38	64	66	39	–	–	–	–	–	Vinylbital
89	–	20	–	39	–	–	24	–	–	Methylenedioxymethamphetamine,3,4-/MDMA/XTC
89	26	42	52	63	93	77	0	52	56	Budesonide
89	38	61	46	66	–	–	–	–	–	Indapamide
89	–	–	–	78	–	–	–	–	–	Etofenamate
89	59	59	70	81	91	78	15	6	71	Halazepam
89	–	–	–	84	–	–	–	–	–	Sulazepam
89	7	63	80	86	90	77	20	79	71	Flunitrazepam M Hy/MNFB
89	76	68	74	86	93	78	11	75	70	Flurazepam Hy/ACFB
89	77	70	76	87	93	78	11	76	70	Lorazepam Hy/ADB
89	82	72	82	88	93	79	53	81	71	Diazepam Hy/MACB
89	–	–	–	90	–	–	–	–	–	Heptenophos
89	–	–	–	99	–	–	–	–	–	Dimefos
90	52	62	57	9	–	77	0	–	–	Oxyphenbutazone/Phenylbutazone M (Hydroxy-)
90	30	30	57	11	–	–	0	–	–	Glibenclamide/Glyburide
90	25	27	47	12	–	–	–	–	–	Diclofenac
90	–	–	–	23	–	–	–	–	–	Hydroxybenzoic Acid, p-
90	0	72	0	34	–	–	–	–	–	Thiobarbital
90	50	65	59	39	–	–	–	–	–	Cyclopentobarbital
90	55	66	59	45	–	–	–	–	–	Pentobarbital
90	–	–	–	47	–	90	2	67	70	Buclosamide

Table 8.6. (Continued)

TLC System 5	1	2	3	4	6	7	8	9	10	[System 5: MeOH / Silica]
90	65	67	56	50	–	–	–	–	–	Propyl p-Hydroxybenzoate
90	–	–	–	61	–	–	–	–	–	Trichlorfon (Trichlorphos)
90	–	–	–	76	–	65	8	65	64	Beclamide
90	71	65	68	85	91	77	5	69	69	Nitrazepam Hy/ANB
90	76	69	76	86	91	78	12	76	71	Oxazepam Hy/ACB
90	–	–	–	87	–	72	37	74	67	Famprofazone
90	–	–	–	87	–	76	–	–	–	Mesuximide
90	–	–	–	87	92	74	42	84	70	Diphenoxylate
90	–	–	–	99	–	–	–	–	–	Phosalone
91	–	–	–	–	–	–	–	–	–	Fenticlor
91	–	–	–	23	–	91	–	–	–	Niclosamide
91	50	66	56	32	–	–	–	–	–	Ibomal/Propallylonal
91	58	67	57	34	–	–	–	–	–	Dichlorophen
92	40	60	54	5	–	–	–	–	–	Glibornuride
92	52	48	60	16	92	–	0	–	–	Acenocoumarol
92	–	–	–	19	–	–	–	–	–	Feprazone
92	58	68	60	44	–	–	–	–	–	Nealbarbital
92	53	67	60	46	–	–	–	–	–	Talbutal
92	–	–	–	65	96	71	41	79	70	Bezitramide
92	–	–	–	72	–	–	–	–	–	Pentaerithritol Tetranitrate
92	–	–	–	73	92	–	–	–	–	Diethylstilbestrol
92	83	74	82	88	95	80	58	82	74	Prazepam Hy/CCB
92	–	–	–	89	–	–	–	–	–	Methenolone
93	–	–	–	–	–	–	–	–	–	Gliquidone
93	38	64	36	13	–	–	0	–	–	Xipamide
93	–	50	–	17	–	–	0	–	–	Chlorphenprocoumon,p-
93	62	58	61	19	–	–	–	–	–	Phenprocoumon
93	–	–	0	80	48	1	0	0	0	Metformin
93	–	–	–	86	97	83	44	78	75	Proclonol
94	–	–	–	24	–	–	–	–	–	Benzbromarone
94	32	34	44	30	98	–	0	–	–	Hexachlorophene
94	–	–	–	72	–	–	–	–	–	Paraflutizide
94	–	–	–	86	–	–	–	–	–	Tribenoside
94	81	74	81	89	95	82	49	81	75	Halazepam Hy/TCB
95	–	–	–	89	–	–	–	–	–	Mofebutazone
96	–	–	–	26	98	95	0	10	90	Closantel
99	–	–	–	17	–	33	54	4	63	Pseudoephedrine
99	–	–	–	18	–	–	3	–	–	Lorcainide M

8.7 hR$_f$ Data in TLC System 6 in Ascending Numerical Order vs. Substance Names

TLC System 6	1	2	3	4	5	7	8	9	10	[System 6: MeOH-BuOH (60+40) 0.1M NaBr / Silica]
0	–	–	–	0	0	0	0	0	0	Diquat Dibromide
0	–	–	–	0	0	0	0	0	0	Paraquat
0	–	–	–	0	0	1	0	0	0	Suxamethonium Chloride
0	–	–	–	0	20	–	–	–	0	Morphine M (-3-glucuronide)
1	–	–	–	–	–	–	–	–	–	Carminic Acid
1	–	–	–	0	0	0	0	0	0	Hexamethonium
1	–	–	–	0	0	0	0	0	0	Pentolinium
1	–	–	–	0	80	5	0	0	0	Psilocybine
2	–	–	–	0	0	0	0	0	0	Decamethonium
2	–	–	–	0	0	0	0	0	0	Gallamine
2	–	–	–	0	30	–	0	0	0	Codeine M (-6-glucuronide)
3	–	–	–	7	0	13	0	0	0	Histamine
4	–	–	–	–	1	5	1	1	0	Piperazine
7	–	–	–	25	5	16	0	17	1	Brucine
7	–	–	–	60	9	45	7	8	13	Disopyramide
8	–	–	–	0	53	1	0	0	0	Obidoxime Chloride
10	–	–	–	40	4	5	67	3	5	Sparteine
11	–	–	–	0	0	0	0	0	0	Tubocurarine
11	–	–	–	0	16	–	0	0	–	Betaine
11	–	–	–	0	16	17	0	0	0	Cocaine M/Ecgonine
11	–	–	–	32	8	26	8	19	2	Strychnine
12	–	–	–	–	12	30	4	13	3	Methenamine
12	–	–	–	0	0	0	0	0	0	Homatropine Methylbromide
13	–	–	–	1	1	2	0	–	–	Neostigmine
13	–	–	–	33	11	25	4	20	4	Hydrocodone
14	–	–	–	0	34	52	0	3	0	Narceine
14	–	–	–	2	22	21	0	1	3	Cocaine M/N-Benzoylecgonine
14	–	–	–	18	12	23	3	9	2	Hydromorphone
14	–	–	–	27	11	–	5	13	3	Metopon
14	–	–	0	46	4	38	14	4	2	Chloroquine
14	–	–	–	46	65	52	2	52	21	Endralazine Mesilate
15	–	–	–	0	0	1	0	0	0	Pancuronium Bromide
15	–	–	–	37	1	0	46	0	0	Hydrastinine
15	3	1	36	40	26	40	1	48	5	Loprazolam
16	0	0	0	0	70	58	0	0	0	Lysergic Acid
16	–	–	–	18	23	37	0	9	0	Mianserin M (N-Oxide)
16	–	–	–	50	13	48	26	19	4	Carbinoxamine
18	–	–	–	1	2	1	0	0	0	Thiamine Hydrochloride
18	–	–	–	1	3	2	0	0	0	Methscopolamine Bromide
18	–	–	–	6	2	2	1	3	0	Oxyphencyclimine
19	–	–	–	29	11	26	8	13	2	Dihydrocodeine
20	–	–	–	16	27	41	0	9	0	Clomipramine M (N-Oxid)
21	–	–	–	24	82	–	6	28	7	Morphine-3-acetate
21	–	–	–	46	12	45	35	18	2	Chlorphenamine/Chlorpheniramine

Table 8.7. (Continued)

TLC System 6	1	2	3	4	5	7	8	9	10	[System 6: MeOH-BuOH (60+40) 0.1M NaBr / Silica]
22	–	–	–	–	16	–	–	–	–	Narcophin
22	–	–	–	26	1	2	38	1	0	Cotarnine
22	–	–	–	35	21	33	6	18	3	Codeine
22	–	–	–	43	22	46	7	34	6	Thioproperazine
22	–	–	–	44	19	45	17	28	5	Dothiepin M (Nor-)
22	–	–	–	46	20	45	11	28	5	Dothiepin M (Sulfoxide)
22	–	–	–	55	38	54	22	35	14	Dibenzepin
22	–	–	–	61	39	54	39	35	13	Nicotine
23	–	–	–	0	4	0	0	0	0	Carbachol
23	–	–	–	20	18	37	0	9	1	Morphine
23	–	–	–	41	20	41	6	23	8	Benzylmorphine
23	–	–	–	47	21	48	25	37	3	Perazine
24	–	–	–	44	26	49	10	40	7	Tiotixene
24	–	–	–	66	21	52	41	26	13	Methapyrilene
25	–	–	–	0	1	2	0	0	0	Thiazinamium Metilsulfate
25	–	–	–	49	24	45	20	34	11	Thebacon
25	–	–	–	54	26	44	23	43	12	Acetylcodeine
26	–	–	–	36	21	40	7	22	6	Ethylmorphine
26	–	–	–	46	14	45	35	13	3	Pheniramine
26	–	–	–	55	26	49	34	37	7	Prochlorperazine
27	–	–	–	23	7	1	5	1	1	Homatropine
27	–	–	–	44	25	46	6	26	8	Morphine-6-acetate/Diamorphine M
28	–	–	–	24	5	18	5	3	1	Atropine/Hyoscyamine
28	–	–	–	50	13	42	35	22	6	Tropacocaine
28	–	–	–	71	11	54	41	21	9	Diamocaine
29	–	–	–	0	0	0	0	1	0	Cetrimide
29	–	–	–	0	0	0	0	1	0	Cetylpyridinium
29	–	–	–	55	30	53	33	30	8	Trifluoperazine
29	–	–	–	59	15	47	43	23	9	Prothipendyl
30	–	–	–	1	3	1	0	2	0	Guanethidine
30	–	–	–	55	19	51	11	20	6	Triprolidine
30	–	–	–	77	35	65	45	47	54	Cocaine
30	–	–	–	78	34	70	42	60	55	Cinnamoylcocaine
31	–	–	–	4	3	4	0	4	0	Propantheline Bromide
31	–	–	–	65	22	55	38	28	14	Thonzylamine
32	–	–	–	–	33	53	3	25	3	Acetophenazine
32	–	–	–	45	23	45	24	37	5	Thebaine
33	–	–	–	8	1	–	4	–	–	Tropine
33	–	–	–	39	17	49	1	5	9	Procainamide
33	–	–	–	49	26	47	15	38	4	Diamorphine
33	–	–	–	58	22	51	39	25	14	Mepyramine/Pyrilamine
33	–	–	–	59	24	48	18	36	25	Oxomemazine
33	–	–	–	62	30	50	25	51	39	Oxycodone
34	–	–	–	33	10	35	0	1	1	Bufotenine
34	–	–	–	68	22	55	44	27	15	Tripelennamine
35	–	–	–	62	18	44	38	30	11	Promazine

Table 8.7. (Continued)

TLC System 6	1	2	3	4	5	7	8	9	10	[System 6: MeOH-BuOH (60+40) 0.1M NaBr / Silica]
35	–	–	–	64	22	52	41	30	14	Isothipendyl
36	–	–	–	–	19	53	42	25	12	Thenyldiamine
36	–	–	–	1	2	3	0	1	0	Oxyphenonium Bromide
36	–	–	–	33	27	48	10	37	30	Oxymorphone
37	–	–	–	28	6	27	2	8	1	Dothiepin M (Nor-,Sulfoxide)
38	–	–	–	–	21	39	18	24	6	Dimethoxanate
38	–	–	–	55	41	55	12	36	18	Physostigmine
39	–	–	–	5	4	2	0	4	0	Fenpiverinium
39	–	–	–	38	35	54	6	22	7	Opipramol
39	1	0	28	40	49	63	1	36	2	Adinazolam M (Nor-)
39	–	–	–	50	14	40	9	9	6	Dimethyltryptamine,N,N-/DMT
39	–	–	–	64	43	57	15	32	16	Tetracaine/Amethocaine
40	–	–	–	42	40	55	7	29	9	Perphenazine
40	–	–	–	60	34	52	37	34	11	Pethidine/Meperidine
41	–	–	–	2	3	8	0	0	0	Butylscopolammonium Bromide,N-
41	–	–	–	3	3	5	0	5	0	Isopropamide Iodide
41	–	–	–	55	12	40	45	19	4	Ethoheptazine
41	–	–	–	62	49	57	36	51	24	Phendimetrazine
41	–	–	–	70	27	51	49	42	16	Dothiepin
42	–	–	–	–	5	26	11	0	24	Betazole
42	–	–	–	47	10	33	42	18	6	Dextromethorphan/Racemethorphan/Levomethorphan
42	–	–	–	54	25	49	15	19	9	Clomipramine M (10-Hydroxy-)
42	–	–	–	71	36	54	5	31	30	Procaine
43	–	–	–	61	49	–	13	59	25	Orientalidine
43	–	–	–	61	49	55	14	57	26	Isothebaine
44	–	–	–	7	5	17	0	0	0	Morphine M (Nor-)
44	–	–	–	65	30	50	36	35	17	Promethazine
45	–	–	–	44	52	53	0	32	12	Pilocarpine
45	–	–	–	46	34	50	14	27	14	Phenmetrazine/Phendimetrazine M (Nor-)
45	–	–	–	63	24	51	48	37	13	Doxepin
45	–	–	–	70	25	49	45	35	17	Chlorpromazine
45	3	3	41	71	52	62	30	48	40	Flurazepam
46	–	–	–	2	87	88	0	10	10	Dimenhydrinate
46	–	–	–	65	36	57	27	45	20	Mebhydrolin
46	–	–	–	73	35	66	26	55	38	Pitofenone
46	–	–	–	77	32	58	54	39	31	Trimeprazine/Alimemazine
47	–	–	–	48	49	55	6	37	18	Scopolamine
47	–	–	–	67	21	48	48	23	13	Imipramine
48	–	–	–	47	14	39	5	9	9	Psilocin
48	–	–	–	65	27	55	44	33	15	Diphenhydramine
48	–	–	–	68	28	55	45	33	15	Dimenhydrinate
48	–	–	–	69	27	54	42	43	13	Bromazine/Bromdiphenhydramine
48	–	–	0	80	93	1	0	0	0	Metformin
49	–	–	18	29	55	47	1	11	20	Isoniazid
49	–	–	–	42	10	35	14	4	3	Dextrorphan/Levorphanol

Table 8.7. (Continued)

TLC System 6	1	2	3	4	5	7	8	9	10	[System 6: MeOH-BuOH (60+40) 0.1M NaBr / Silica]
49	–	–	–	45	45	63	5	23	10	Fluphenazine
49	–	–	–	58	18	46	49	25	9	Clemastine
49	–	–	–	68	16	48	47	22	14	Pentoxyverine/Carbetapentane
49	–	–	–	68	23	46	37	28	8	Diphenylpyraline
49	–	–	–	68	25	55	48	33	16	Orphenadrine
49	–	–	–	68	41	63	45	57	21	Phenindamine
49	–	–	–	75	32	54	47	35	22	Triflupromazine
49	–	–	–	76	32	57	47	38	46	Levomepromazine/Methotrimeprazine
50	1	1	25	31	62	64	0	–	–	Bromazepam M (N(py)-Oxide)
50	–	–	–	59	39	–	21	48	12	Cocaine M (Nor-)
50	–	–	–	64	30	51	45	44	13	Cyproheptadine
50	–	–	–	68	48	58	39	58	23	Mianserin
50	–	–	–	73	14	–	22	46	43	Cryptopine
50	–	–	–	73	58	60	30	58	33	Methylpiperidyl Benzilate/JB-336
51	–	–	–	36	57	60	0	19	7	Lysergamide
51	–	–	–	45	26	70	1	24	11	Clomipramine M (2-Hydroxy-)
51	–	–	–	51	26	51	4	17	10	Clomipramine M (8-Hydroxy-)
51	–	–	–	69	27	51	50	32	15	Amitriptyline
51	–	–	–	74	34	56	51	51	25	Chlorprothixene
51	9	8	41	83	45	67	53	60	51	Flurazepam Hy/DCFB
52	–	–	–	27	3	14	3	6	3	Naphazoline
52	–	–	–	43	41	65	0	17	14	Pipamazine
52	–	–	–	54	46	60	4	37	18	Mianserin M (8-Hydroxy-)
52	–	–	–	64	53	56	9	29	31	Nomifensine
52	–	–	–	65	53	62	18	48	42	Levamisole
52	–	–	–	67	35	57	42	46	14	Chlorcyclizine
52	–	–	–	68	30	55	34	42	26	Propiomazine
52	–	–	–	68	40	57	49	41	16	Cyclizine
52	–	–	–	82	54	69	55	81	55	Piperidolate
53	–	–	–	66	51	69	18	54	46	Tetramisole
54	–	4	–	34	59	53	1	31	21	Theobromine
54	–	–	–	72	26	51	53	34	18	Clomipramine
54	–	–	–	77	33	58	57	51	39	Diethazine
54	–	–	–	78	59	75	44	63	55	Ethylpiperidyl Benzilate/JB-318
55	–	–	–	–	45	63	5	26	31	Butethamine
55	–	–	–	25	3	13	2	2	2	Tolazoline
55	–	–	–	27	53	54	0	9	12	Cimetidine
55	–	–	–	44	24	49	6	8	6	Cinchonidine
55	15	10	55	52	59	52	3	58	25	Caffeine
55	–	–	–	67	20	48	42	30	13	Thioridazine
55	–	–	–	83	31	67	64	47	66	Profenamine/Ethopropazine
56	–	–	–	8	4	7	38	0	0	Berberine
56	–	–	–	63	14	46	14	10	11	Diethyltryptamine,N,N-/DET
56	–	–	–	76	21	55	53	37	27	Piperocaine
56	–	–	–	80	36	59	62	54	37	Trimipramine
56	–	–	–	81	48	69	59	57	47	Butetamate

Table 8.7. (Continued)

TLC System 6	1	2	3	4	5	7	8	9	10	[System 6: MeOH-BuOH (60+40) 0.1M NaBr / Silica]
56	–	–	–	85	55	76	62	63	64	Amfepramone/Diethylpropion
57	–	–	–	71	59	61	15	64	52	Hydrastine
57	–	–	–	77	52	66	43	53	53	Benactyzine
57	–	–	–	82	55	–	54	60	62	Eucaine,α-
59	0	0	2	2	85	84	0	1	2	Metamizol/Dipyrone
59	–	–	–	25	70	48	0	12	12	Diprophylline/Propyphylline
59	–	–	–	32	57	59	1	23	29	Nalorphine
59	–	–	–	43	14	18	0	0	0	Dopamine
59	–	–	20	45	45	55	0	22	12	Trimethoprim
59	–	–	–	53	40	66	3	32	21	Pipotiazine
59	3	–	–	56	60	60	3	39	18	Lysergide/LSD
59	–	–	–	74	62	61	35	53	42	Thiopropazate
60	–	–	–	–	62	70	43	55	41	Pramoxine
60	–	–	–	26	5	13	7	2	2	Tetryzoline
60	–	–	–	33	62	57	0	12	8	Ergometrine
60	0	0	14	40	20	58	2	18	29	Flurazepam M (Dideethyl-)
60	3	3	33	40	68	69	0	32	42	Chlordiazepoxide M (Nor-)
60	7	5	44	50	71	71	2	53	25	Estazolam
60	–	–	–	51	25	46	11	31	6	Mianserin M (Nor-)
60	–	–	–	66	63	65	31	62	48	Mepivacaine
60	–	–	–	77	16	48	59	20	27	Methadone/Levomethadone
61	–	–	–	31	6	26	5	11	14	Dimethoxyphenethylamine,3,4-
61	0	6	26	39	73	61	0	24	28	Bromazepam M (3-Hydroxy-)
61	–	–	–	43	33	56	1	12	8	Pipamperone
61	–	–	–	44	19	49	6	12	5	Cinchonine
61	–	–	–	51	46	58	4	16	18	Periciazine
61	–	–	–	66	64	63	10	58	37	Trazodone
61	–	–	–	73	43	59	23	70	42	Verapamil
62	0	0	15	50	28	58	8	19	11	Flurazepam M (Monodeethyl-)
62	–	–	–	84	38	57	62	45	50	Alphacetylmethadol
63	16	13	38	5	83	–	–	–	–	Indometacin
63	–	–	–	24	6	20	3	10	12	Mescaline
63	–	–	–	32	8	–	7	8	2	Pethidine M (Nor-)
63	–	–	–	33	69	55	0	37	12	Colchicine
63	–	–	–	42	9	31	28	13	5	Deoxyephedrine,(+/-)-
63	–	–	–	42	9	31	28	13	5	Methamfetamine
63	–	–	–	49	30	51	4	12	6	Quinidine
63	9	22	30	67	66	73	4	51	47	Medazepam M (Nor-)
63	–	–	–	82	50	68	58	55	54	Propoxyphene/Dextropropoxyphene/Levopropoxyphene
64	–	–	–	–	68	70	2	25	4	Nialamide
64	–	–	–	25	10	30	5	5	1	Ephedrine
64	–	–	–	30	5	13	7	5	3	Xylometazoline
64	–	–	–	44	68	63	1	34	22	Ergotamine
64	–	–	–	82	54	72	62	68	55	Amiodarone
65	–	–	13	30	50	51	1	8	4	Triamterene

Table 8.7. (Continued)

TLC System 6	1	2	3	4	5	7	8	9	10	[System 6: MeOH-BuOH (60+40) 0.1M NaBr / Silica]
65	5	2	41	44	68	60	1	40	16	Triazolam
65	–	–	–	45	26	51	2	11	4	Quinine
65	–	–	–	53	46	63	7	13	13	Mazindol
65	–	–	–	54	57	68	10	54	19	Hydroxyzine
65	–	–	–	79	15	–	66	25	26	Rolicyclidine/PHP/1-(1-phenylcyclohex-yl)pyrrolidine
66	–	–	–	2	4	2	0	4	0	Benzalkonium Chloride
66	–	9	–	11	74	75	1	30	11	Theophylline
66	–	–	–	40	68	54	0	21	27	Nicotinamide
66	–	–	–	41	69	76	0	41	8	Domperidone
66	18	14	50	45	66	65	4	–	–	Phenazone/Antipyrine
66	–	–	–	47	5	31	6	7	3	Antazoline
66	7	2	40	47	67	67	1	57	14	Alprazolam
66	9	2	51	57	68	69	6	55	22	Adinazolam
66	–	–	–	71	25	54	54	37	18	Pyrrobutamine
66	–	–	–	79	60	73	12	56	51	Anileridine
67	–	–	–	12	8	33	1	1	0	Phenylephrine
67	–	–	–	15	75	59	0	8	5	Pyridoxine Hydrochloride
67	–	–	–	58	41	54	39	33	48	Tranylcypromine
67	–	–	–	59	71	59	15	56	29	Nikethamide
67	–	–	–	64	64	75	0	47	29	Ergosine
67	–	–	–	65	15	–	9	15	25	Psilocin-(eth)/CZ-74
67	–	–	–	70	51	71	21	48	44	Dexetimide
68	–	–	–	–	65	63	0	22	28	Harmine
68	25	10	58	62	70	66	21	–	–	Aminophenazone/Amidopyrine
68	–	–	–	66	6	20	32	10	2	Cyclopentamine
68	–	–	–	74	20	48	62	31	23	Procyclidine
68	–	–	–	75	42	68	11	39	58	Aconitine
69	6	23	19	2	88	–	–	–	–	Gentisic Acid
69	–	–	–	21	14	40	0	1	3	Isoprenaline
69	3	2	26	26	73	67	1	25	7	Alprazolam M (4-Hydroxy-)
69	–	–	–	38	6	19	18	7	2	Protriptyline
69	–	–	–	40	21	44	2	7	3	Hydroquinine
69	–	–	–	42	70	69	1	23	17	Iproniazid
69	–	–	–	48	74	56	4	17	22	Nicotinyl Alcohol
69	–	3	–	60	62	78	3	51	32	Benperidol
69	13	18	47	63	73	61	6	41	53	Bromazepam
69	–	–	–	80	72	70	35	71	63	Lidocaine
69	–	–	–	84	23	59	73	35	66	Phencyclidine
70	–	–	–	29	7	33	6	11	12	Trimethoxyamphetamine,3,4,5-
70	–	–	–	37	5	26	39	13	5	Coniine
70	–	–	–	43	20	45	3	8	5	Hydroquinidine
70	–	–	32	46	75	58	2	36	40	Metronidazole
70	–	–	44	54	71	63	3	42	46	Pyrazinamide
70	13	5	53	60	69	72	6	60	19	Midazolam
70	–	–	–	64	66	63	6	38	52	Yohimbine

Table 8.7. (Continued)

TLC System 6	1	2	3	4	5	7	8	9	10	[System 6: MeOH-BuOH (60+40) 0.1M NaBr / Silica]
70	–	–	–	66	40	57	35	41	23	Methylphenidate
70	–	–	–	86	38	–	73	54	68	Tenocyclidine/TCP/1-[1-(2-Thienyl)cyclohexyl] piperidine
71	3	1	29	24	75	63	0	29	8	Triazolam M (4-Hydroxy-)
71	–	–	–	33	13	47	0	3	6	Acebutolol
71	–	–	–	36	6	15	18	5	2	Maprotiline
71	–	–	–	40	7	26	19	11	3	Desipramine/Imipramine M
71	–	–	–	46	87	34	28	16	4	Amitriptyline M/Nortriptyline
71	15	5	53	52	72	72	5	52	27	Brotizolam
72	0	0	7	8	82	75	0	2	0	Cinchophen
72	3	4	41	53	73	70	3	52	8	Midazolam M (α-Hydroxy-)
72	–	–	–	61	51	74	4	49	39	Spiperone
72	30	14	61	61	76	71	3	63	44	Flumazenil/RO 15-1788
72	–	–	–	70	34	61	16	12	28	Pentazocine
72	–	–	–	79	68	63	37	63	64	Ketamine
72	–	–	–	79	70	68	13	70	61	Etonitazene
72	–	–	–	86	39	65	28	50	62	Ibogaine/Rotenone
72	–	–	–	87	27	66	67	33	70	Dipipanone
73	–	–	–	11	12	55	24	4	38	Methoxamine
73	–	–	–	58	71	67	2	48	36	Droperidol
73	–	–	–	65	29	55	36	38	13	Noracymethadol
73	–	–	–	74	42	67	19	24	45	Levallorphan
73	–	–	–	78	30	56	63	37	26	Volazocine
73	–	–	–	78	76	78	31	69	52	Clemizole
73	–	–	–	87	72	–	62	74	64	Phenylcyclohexyl)morpholine,4-(1-/PCM
74	–	0	–	20	16	46	1	1	4	Salbutamol/Albuterol
74	–	–	–	22	15	41	2	2	3	Etilefrine
74	–	–	0	24	6	24	0	1	2	Amiloride
74	4	3	39	42	75	71	4	49	21	Triazolam M (α-Hydroxy-)
74	–	–	–	43	9	35	23	10	18	Methoxy-4,5-methylenedioxyamphetamine,2-/MMDA-2
74	–	–	–	43	9	73	23	77	69	Methoxyamphetamine,4-/PMA
74	–	–	–	44	20	49	10	8	9	Metoprolol
74	–	–	–	44	42	60	2	23	–	Tocainide
74	–	–	–	59	16	47	47	19	8	Etilamfetamine
74	–	–	–	59	68	60	1	47	49	Prazosin
74	–	–	–	60	72	72	7	64	63	Pentetrazol
74	–	–	–	61	73	70	1	45	42	Piritramide
74	21	21	52	63	77	74	1	55	52	Flunitrazepam M (7-Amino-)
74	–	–	–	65	24	53	15	13	25	Cyclazocine
74	–	–	–	66	27	54	42	27	17	Eucaine,β-
74	–	–	–	69	74	61	8	65	47	Papaverine
75	–	–	–	–	11	–	–	–	–	Methylenedioxyphentermin,3,4-
75	–	–	–	2	60	49	1	1	1	Methyldopa
75	–	0	–	30	12	44	4	4	33	Ephedrine M/Phenylpropanolamine
75	–	–	–	32	8	33	1	4	1	Clomipramine M (8-Hydroxy-nor-)

Table 8.7. (Continued)

TLC System 6	1	2	3	4	5	7	8	9	10	[System 6: MeOH-BuOH (60+40) 0.1M NaBr / Silica]
75	63	47	68	42	75	72	1	–	–	Carbimazole
75	–	–	–	43	8	36	24	10	14	Dimethoxy-4-ethylamphetamine,2,5-/DOET
75	–	–	–	43	12	43	20	9	18	Amphetamine/Amphetaminil M artifact
75	–	–	–	45	8	34	24	13	3	Clomipramine M (Nor-)
75	–	–	–	45	8	36	19	10	17	Dimethoxyamphetamine,2,5-
75	–	–	–	50	20	52	6	11	9	Timolol
75	11	19	35	54	78	72	0	34	47	Flunitrazepam M (7-Aminonor-)
75	–	–	–	56	79	60	2	56	47	Carbamazepine
75	–	–	–	76	51	67	11	27	33	Haloperidol
75	–	–	–	78	72	64	21	74	64	Noscapine
75	–	23	–	82	67	73	39	68	60	Fluanisone
75	–	–	–	83	43	68	66	61	59	Trihexyphenidyl/Benzhexol
75	–	–	–	86	23	55	68	36	52	Diisopromine
76	–	–	–	0	3	1	0	0	0	Guanoxan
76	–	–	–	18	13	42	1	1	24	Metaraminol
76	5	4	37	28	76	68	1	35	13	Brotizolam M (6-Hydroxy-)
76	4	4	37	37	76	72	1	44	18	Alprazolam M (α-Hydroxy-)
76	–	–	–	41	9	51	15	17	16	Dimethoxy-4-methylamphetamine,2,5- /STP/DOM
76	–	–	–	42	10	39	18	12	17	Methylenedioxyamphetamine,3,4-/MDA
76	–	–	–	43	10	36	15	14	17	Bromo-2,5-dimethoxyamphetamine,4-/DOB
76	19	11	41	51	78	73	0	40	49	Flunitrazepam M (7-Acetamido-)
76	–	–	–	57	72	70	2	40	33	Harman
76	–	–	–	70	44	62	8	31	53	Clonidine
76	–	–	–	70	74	75	13	58	51	Bromolysergide,2-
76	2	5	26	77	55	73	13	–	–	Trifluperidol
76	–	–	–	83	44	71	7	30	64	Butacaine
77	–	–	–	18	21	48	1	3	6	Orciprenaline
77	–	–	–	21	18	47	1	1	5	Terbutaline
77	–	–	–	35	7	23	17	7	4	Amantadine
77	–	–	–	48	14	44	18	17	8	Chlorphentermine
77	5	5	37	49	76	74	1	43	15	Midazolam M (4-Hydroxy-)
77	10	10	53	52	76	62	2	50	22	Chlordiazepoxide
77	11	20	40	52	76	73	0	30	47	Clonazepam M (7-Amino-)
77	–	–	–	57	74	77	0	59	50	Ergocristine
77	–	–	–	59	74	78	0	58	50	Ergocryptine
77	–	–	–	70	70	70	11	63	36	Lidoflazine
77	–	–	–	72	74	61	37	66	59	Methyl Nicotinate
77	–	–	–	78	70	70	43	74	58	Fentanyl
77	–	–	–	82	27	55	66	44	36	Prozapine
77	–	–	–	84	64	–	45	75	64	Alphamethylfentanyl
77	–	–	–	88	63	67	36	64	59	Piminodine
78	–	–	–	43	18	49	2	5	8	Pindolol
78	–	–	–	45	20	48	11	11	13	Oxprenolol
78	7	6	41	45	78	72	2	46	31	Brotizolam M (α-Hydroxy-)
78	–	3	–	46	81	67	1	38	40	Glafenine

Table 8.7. (Continued)

TLC System 6	1	2	3	4	5	7	8	9	10	[System 6: MeOH-BuOH (60+40) 0.1M NaBr / Silica]
78	–	–	–	48	11	46	26	24	12	Phentermine
78	–	–	–	55	25	40	17	4	9	Mexiletine
78	–	–	–	71	63	69	4	59	49	Fluspirilene
78	–	–	–	71	77	–	11	73	63	Meconin
78	–	–	–	79	21	–	65	27	37	Eticyclidine/PCE/N-Ethyl-1-phenylcyclohexyl-amine
78	–	–	–	79	72	73	42	71	60	Dextromoramide/Racemoramide/Levomorami-de
78	–	–	–	83	70	–	38	73	59	Fluorofentanyl,p-
78	–	–	–	87	31	–	72	47	68	Methyl-1-phenylcyclohexyl)piperidine,1-(4-
79	–	–	–	18	7	3	0	1	1	Proguanil/Chlorguanide
79	–	–	–	49	21	50	6	10	7	Propranolol
79	10	20	30	54	77	71	0	30	46	Nitrazepam M (7-Amino-)
79	–	–	–	69	83	68	28	67	55	Crotethamide
79	–	–	–	70	23	–	41	33	29	Phenylcyclohexylamine,1-
79	64	50	69	75	–	–	–	–	–	Santonin
79	–	–	–	75	62	77	29	64	60	Prilocaine
79	36	53	59	75	85	–	4	–	–	Carisoprodol
79	–	–	–	80	69	69	42	73	65	Bupivacaine
79	–	–	–	81	19	54	59	38	39	Pipradrol
79	–	–	–	81	77	73	1	75	64	Rescinnamine
79	37	41	54	81	80	72	5	57	57	Nitrazepam M Hy/DAB
80	–	0	–	50	22	50	13	15	8	Penbutolol
80	38	37	64	57	81	71	1	59	52	Flunitrazepam M (3-Hydroxy-)
80	–	–	–	68	84	66	1	63	57	Chlormezanone
80	45	45	62	74	83	66	14	64	66	Ketazolam
80	–	–	–	75	78	80	9	61	–	Econazole
80	–	–	–	76	66	75	36	64	56	Articaine
80	–	–	–	77	76	69	2	74	63	Reserpine
80	–	–	–	78	51	69	18	39	47	Phenomorphan
80	–	7	56	80	77	73	11	67	37	Miconazole
80	41	46	55	82	80	74	5	59	61	Clonazepam Hy/DCB
81	–	0	–	25	38	76	0	1	4	Fenoterol
81	–	11	–	36	81	60	0	23	40	Pemoline
81	2	5	33	38	78	75	4	25	14	Midazolam M (α,4-Dihydroxy-)
81	–	–	–	62	62	78	3	32	53	Isoxsuprine
81	–	–	–	62	78	71	4	55	52	Carnidazole
81	–	–	–	74	50	68	16	39	49	Phenazocine
81	–	–	–	74	52	70	9	32	22	Loperamide
81	–	–	–	74	77	71	23	65	48	Metomidate
81	53	46	71	77	81	74	12	70	55	Nimetazepam
81	–	–	–	78	78	71	21	78	69	Disulfiram
82	7	11	30	42	79	73	0	22	45	Clonazepam M (7-Acetamido-)
82	23	41	42	43	82	52	1	36	28	Lorazepam/Lormetazepam M

Table 8.7. (Continued)

TLC System 6	1	2	3	4	5	7	8	9	10	[System 6: MeOH-BuOH (60+40) 0.1M NaBr / Silica]
82	22	35	42	45	82	56	0	40	51	Oxazepam/M of Chlordiazepoxide, Clorazepic Acid, Camazepam, Diazepam, Halazepam, Ketazolam, Medazepam, Oxazolam, Prazepam
82	46	45	60	59	82	52	6	61	50	Lormetazepam
82	19	28	54	61	81	74	2	46	55	Flurazepam M (N(1)-Hydroxyethyl-)
82	51	47	65	62	82	53	8	59	53	Temazepam/M of Camazepam, Diazepam, Ketazolam, Medazepam
82	–	–	–	71	73	71	3	60	40	Pimozide
82	54	47	72	74	80	63	10	72	63	Flunitrazepam
82	–	–	–	76	70	71	26	64	58	Phenoperidine
82	–	–	–	83	29	77	37	12	63	Phenelzine
83	–	–	–	–	80	80	58	77	69	Phenadoxone
83	15	24	42	41	81	63	0	35	51	Demoxepam/Chlordiazepoxide M
83	4	40	23	42	88	–	–	–	–	Chlortalidone
83	7	12	24	44	80	71	0	20	49	Nitrazepam M (7-Acetamido-)
83	–	–	–	62	33	74	3	14	50	Buphenine
83	34	45	57	69	82	62	3	55	60	Nordazepam/M of Chlordiazepoxide, Clorazepic Acid, Diazepam, Halazepam, Ketazolam, Medazepam, Oxazolam, Prazepam
83	–	–	–	74	25	70	29	69	57	Cropropamide
83	55	32	69	75	82	76	12	73	65	Camazepam
83	56	40	73	78	79	67	41	74	62	Medazepam
84	38	55	43	50	83	–	0	–	–	Salicylamide
84	49	48	60	59	84	–	–	–	–	Acecarbromal
84	–	–	–	60	80	65	0	59	49	Mebendazole
84	45	45	52	66	80	–	–	–	–	Acetanilide
84	33	25	65	73	81	76	15	68	58	Metaclazepam M (N-Nor-)
84	39	45	63	75	78	73	0	66	59	Cloxazolam
84	–	–	–	78	79	70	36	80	56	Methaqualone
84	47	35	71	79	82	77	40	73	62	Metaclazepam
85	–	–	–	58	83	80	0	55	6	Flubendazole
85	10	35	35	60	82	–	0	33	56	Mebutamate
85	35	46	56	68	84	76	0	58	59	Flunitrazepam M (Nor-)
85	–	–	–	73	78	67	26	71	52	Etomidate
85	53	47	70	75	84	62	8	70	62	Clobazam
85	–	–	–	76	80	64	44	69	58	Clomethiazole
85	58	49	72	76	82	75	27	73	59	Diazepam/Ketazolam M/Medazepam M
85	–	–	–	84	43	68	55	68	56	Prenylamine
86	–	–	–	–	–	–	–	–	–	Methyltestosterone,17-
86	35	46	53	64	84	68	0	36	55	Nitrazepam
86	35	41	57	72	82	73	5	58	56	Delorazepam
86	56	50	67	75	78	76	24	69	63	Fludiazepam
86	–	–	–	75	84	71	20	74	61	Isocarboxazid
86	–	–	–	76	80	69	8	52	59	Fenyramidol
86	66	66	80	83	76	73	31	77	68	Nimetazepam Hy/MNB

Table 8.7. (Continued)

TLC System 6	1	2	3	4	5	7	8	9	10	[System 6: MeOH-BuOH (60+40) 0.1M NaBr / Silica]
86	–	–	–	86	33	70	57	59	68	Clofazimine
87	–	–	–	–	–	–	–	–	–	Piperonal (precursor of MDA, MDE, MDMA)
87	–	–	–	44	82	68	0	37	42	Dipyridamole
87	9	36	32	56	63	75	0	32	58	Meprobamate
87	35	45	56	67	85	72	0	53	61	Clonazepam
87	34	46	57	68	83	84	3	56	60	Clorazepic Acid
87	53	55	64	75	85	–	12	–	–	Carbromal
87	56	62	63	76	84	67	5	57	66	Benzocaine
87	49	59	58	76	86	76	5	–	–	Ethinamate
87	–	–	–	80	82	74	17	76	66	Bisacodyl
87	55	48	66	80	84	78	31	69	66	Clotiazepam
87	–	–	–	86	79	76	54	78	65	Cinnarizine
87	91	93	94	99	81	84	40	99	99	Azinphos-methyl
88	–	–	–	–	–	–	–	–	–	Testosterone
88	–	–	–	10	80	5	62	0	0	Phencyclidine interm./1-Piperidino-1-cyclohexanecarbonitrile/PCC
88	38	43	49	10	87	72	0	–	–	Chlorpropamide
88	51	55	62	12	88	76	–	–	–	Tolbutamide
88	–	–	70	30	40	56	0	5	9	Clioquinol
88	52	53	64	68	–	–	–	–	–	Bemegride
88	34	45	60	72	83	75	2	56	58	Flurazepam M (N(1)-Dealkyl-)/Quazepam M (N-Dealkyl-2-oxo-)
88	55	55	70	72	85	71	12	–	52	Prazepam M (3-Hydroxy-)
88	–	–	–	73	84	72	2	69	61	Bromocriptine
88	–	–	–	76	76	73	5	68	61	Ambucetamide
88	64	61	66	85	86	77	9	68	69	Bromazepam Hy/ABP
88	–	–	–	87	80	76	61	79	70	Meclozine
88	91	92	95	99	80	86	55	99	99	Methidathion
89	–	–	–	–	–	–	–	–	–	Diphenylamine
89	–	–	–	–	–	–	–	–	–	Progesterone
89	15	28	35	49	88	67	0	30	38	Quazepam M (3-Hydroxy-N-dealkyl-2-oxo-)
89	–	–	–	55	87	64	2	43	57	Mephenesin
89	57	49	67	79	84	78	32	69	66	Tetrazepam
89	63	62	70	80	86	75	31	–	–	Glutethimide
89	64	55	72	81	84	65	36	74	63	Prazepam
89	–	–	–	82	82	79	11	79	71	Phenothiazine
89	–	–	–	84	72	76	17	60	60	Penfluridol
90	–	–	–	–	–	–	–	–	–	Methyl Salicylate
90	–	–	–	–	–	–	–	–	–	Phenyl Salicylate
90	–	–	–	–	83	75	6	63	70	Butyl Aminobenzoate
90	54	61	56	33	88	85	0	51	47	Chlorzoxazone
90	11	31	34	36	86	69	0	31	48	Chloramphenicol
90	32	43	54	64	84	75	0	52	61	Clobazam M (Nor-)
90	42	52	58	69	88	71	2	55	59	Quazepam M (3-Hydroxy-2-oxo-)
90	46	50	65	75	80	74	11	66	62	Haloxazolam
90	59	57	70	80	87	78	16	89	71	Quazepam M (2-Oxo-)

Table 8.7. (Continued)

TLC System 6	1	2	3	4	5	7	8	9	10	[System 6: MeOH-BuOH (60+40) 0.1M NaBr / Silica]
90	71	69	81	83	77	75	48	77	69	Fludiazepam Hy/CFMB
90	–	–	–	86	85	83	53	77	66	Clocinizine
90	7	63	80	86	89	77	20	79	71	Flunitrazepam M Hy/MNFB
90	–	–	–	88	83	82	45	75	64	Flunarizine
91	–	–	–	–	–	–	–	–	–	Resorcinol
91	–	–	–	79	30	70	48	40	20	Hexetidine
91	59	59	70	81	89	78	15	6	71	Halazepam
91	48	50	57	82	87	77	7	60	66	Flurazepam Hy/HCFB
91	71	65	68	85	90	77	5	69	69	Nitrazepam Hy/ANB
91	75	73	80	86	86	83	48	77	72	Pinazepam Hy/CPB
91	71	66	68	86	87	77	4	69	70	Flunitrazepam M Hy/ANFB
91	76	69	76	86	90	78	12	76	71	Oxazepam Hy/ACB
92	52	48	60	16	92	–	0	–	–	Acenocoumarol
92	35	45	56	73	85	77	5	58	60	Tetrazepam M (Nor-)
92	–	–	–	73	92	–	–	–	–	Diethylstilbestrol
92	73	68	65	81	–	–	–	–	–	Thymol
92	65	61	73	81	85	75	29	72	70	Pinazepam
92	–	–	–	83	88	79	64	84	74	Phenanthrene
92	–	–	–	87	90	74	42	84	70	Diphenoxylate
93	–	–	–	–	–	–	–	–	–	Camphor
93	26	42	52	63	89	77	0	52	56	Budesonide
93	53	58	62	74	85	76	0	62	67	Ethyl Loflazepate
93	70	69	74	83	81	83	11	71	69	Haloxazolam Hy/ABFB
93	74	73	80	85	77	79	44	77	73	Quazepam Hy/CFTB
93	72	66	70	86	87	77	4	69	70	Clonazepam Hy/ANCB
93	76	68	74	86	89	78	11	75	70	Flurazepam Hy/ACFB
93	81	69	81	87	82	75	49	81	7	Lormetazepam Hy/MDB
93	77	70	76	87	89	78	11	76	70	Lorazepam Hy/ADB
93	82	72	82	88	89	79	53	81	71	Diazepam Hy/MACB
95	53	54	65	71	78	77	15	68	66	Oxazolam
95	74	77	73	80	87	–	37	–	–	Cyclandelate
95	83	74	82	88	92	80	58	82	74	Prazepam Hy/CCB
95	81	74	81	89	94	82	49	81	75	Halazepam Hy/TCB
96	–	–	–	65	92	71	41	79	70	Bezitramide
96	61	60	64	76	84	–	22	64	69	Cholesterol
96	78	71	78	83	87	74	27	75	76	Quazepam
97	–	–	–	86	93	83	44	78	75	Proclonol
97	–	–	–	87	84	73	63	76	68	Phenoxybenzamine
98	–	–	–	26	96	95	0	10	90	Closantel
98	32	34	44	30	94	–	0	–	–	Hexachlorophene

8.8 hR$_f$ Data in TLC System 7 in Ascending Numerical Order vs. Substance Names

TLC System 7	1	2	3	4	5	6	8	9	10	[System 7: MeOH–NH4OH (100+1.5) / Silica-KOH]
0	–	–	–	–	–	–	0	0	0	Diminazene Aceturate
0	–	–	–	0	0	0	0	0	0	Diquat Dibromide
0	–	–	–	0	0	0	0	0	0	Paraquat
0	–	–	–	0	0	1	0	0	0	Hexamethonium
0	–	–	–	0	0	1	0	0	0	Pentolinium
0	–	–	–	0	0	2	0	0	0	Decamethonium
0	–	–	–	0	0	2	0	0	0	Gallamine
0	–	–	–	0	0	11	0	0	0	Tubocurarine
0	–	–	–	0	0	12	0	0	0	Homatropine Methylbromide
0	–	–	–	0	0	29	0	1	0	Cetrimide
0	–	–	–	0	0	29	0	1	0	Cetylpyridinium
0	–	–	–	0	4	23	0	0	0	Carbachol
0	–	–	–	3	43	–	0	0	0	Levarterenol/Norepinephrine
0	–	–	–	13	3	–	0	1	0	Epinephrine
0	–	–	–	37	1	15	46	0	0	Hydrastinine
1	–	–	–	–	–	–	0	1	3	Benserazide
1	–	–	–	–	–	–	0	0	0	Betanidine
1	–	–	–	–	–	–	0	0	0	Debrisoquine
1	–	–	–	–	–	–	0	0	0	Hydroxystilbamidine
1	–	–	–	–	–	–	0	1	0	Pentapiperide
1	–	–	–	–	–	–	0	0	1	Stilbamidine
1	–	–	–	–	–	–	1	1	1	Picloxydine
1	–	–	–	–	–	–	1	1	1	Propamidine
1	–	–	–	–	–	–	33	40	16	Alphameprodine
1	–	–	0	–	0	–	1	0	0	Pentamidine
1	–	–	–	0	0	0	0	0	0	Suxamethonium Chloride
1	–	–	–	0	0	15	0	0	0	Pancuronium Bromide
1	–	–	–	0	3	76	0	0	0	Guanoxan
1	–	–	–	0	53	8	0	0	0	Obidoxime Chloride
1	–	–	–	1	2	18	0	0	0	Thiamine Hydrochloride
1	–	–	–	1	3	30	0	2	0	Guanethidine
1	–	–	–	23	7	27	5	1	1	Homatropine
1	–	–	0	80	93	48	0	0	0	Metformin
2	–	–	–	–	–	–	0	0	0	Aminophylline
2	–	–	–	–	–	–	0	0	0	Buformin
2	–	–	–	–	–	–	0	0	0	Clamoxyquine
2	–	–	–	0	1	25	0	0	0	Thiazinamium Metilsulfate
2	–	–	–	1	1	13	0	–	–	Neostigmine
2	–	–	–	1	3	–	–	–	–	Clidinium
2	–	–	–	1	3	–	–	–	–	Hexocyclium Metilsulfate
2	–	–	–	1	3	18	0	0	0	Methscopolamine Bromide
2	–	–	–	2	4	66	0	4	0	Benzalkonium Chloride
2	–	–	–	3	9	–	–	–	–	Penthienate Bromide
2	–	–	–	5	4	39	0	4	0	Fenpiverinium

Table 8.8. (Continued)

TLC System										[System 7:
7	1	2	3	4	5	6	8	9	10	MeOH–NH4OH (100+1.5) / Silica-KOH]
2	–	–	–	6	2	18	1	3	0	Oxyphencyclimine
2	–	–	–	26	1	22	38	1	0	Cotarnine
2	–	–	–	76	3	–	0	–	–	Methanthelinium Bromide
3	–	–	–	–	–	–	0	1	1	Chlorproguanil
3	–	–	–	–	–	–	0	0	0	Guanoclor
3	–	–	–	–	–	–	0	0	0	Phenformin
3	–	–	–	1	2	36	0	1	0	Oxyphenonium Bromide
3	–	–	–	1	3	–	–	–	–	Glycopyrronium Bromide
3	–	–	–	18	7	79	0	1	1	Proguanil/Chlorguanide
3	–	–	–	71	12	–	1	0	1	Hexoprenaline
4	–	–	–	–	–	–	1	1	1	Acefylline Piperazine
4	–	–	–	4	3	31	0	4	0	Propantheline Bromide
5	–	–	–	–	1	4	1	1	0	Piperazine
5	–	–	0	–	69	–	0	1	1	Cytarabine
5	–	–	–	–	88	–	–	–	–	Tetracycline
5	–	–	0	0	8	–	–	–	–	Oxytetracycline
5	–	–	–	0	80	1	0	0	0	Psilocybine
5	–	–	–	2	3	–	0	–	–	Emepronium Bromide
5	–	–	–	3	3	41	0	5	0	Isopropamide Iodide
5	–	–	–	10	80	88	62	0	0	Phencyclidine interm./1-Piperidino-1-cyclohexanecarbonitrile/PCC
5	–	–	–	40	4	10	67	3	5	Sparteine
6	–	–	–	–	4	–	4	2	2	Tramazoline
7	–	–	–	8	4	56	38	0	0	Berberine
8	–	–	–	2	3	41	0	0	0	Butylscopolammonium Bromide,N-
9	–	–	–	–	–	–	2	1	1	Dextrorphan M/Levorphanol M (Nor-)
9	–	–	–	34	80	–	1	1	1	Oxymetazoline
10	–	–	–	8	–	–	–	–	–	Betahistine
10	–	–	–	14	3	–	0	3	1	Azacyclonol
10	–	–	–	27	3	–	6	6	0	Mequitazine
13	–	–	–	–	–	–	0	5	0	Codeine M (Nor-)
13	–	–	–	–	–	–	24	4	1	Deptropine
13	–	–	–	7	0	3	0	0	0	Histamine
13	–	–	–	25	3	55	2	2	2	Tolazoline
13	–	–	–	26	5	60	7	2	2	Tetryzoline
13	–	–	–	30	5	64	7	5	3	Xylometazoline
13	–	–	–	36	6	–	26	6	2	Benzatropine
14	–	–	–	27	3	52	3	6	3	Naphazoline
15	–	–	–	0	1	–	0	1	2	Amidefrine
15	–	–	–	36	6	71	18	5	2	Maprotiline
16	–	–	–	–	–	–	51	2	4	Mecamylamine
16	–	–	–	25	5	7	0	17	1	Brucine
17	–	–	–	0	16	11	0	0	0	Cocaine M/Ecgonine
17	–	–	–	7	5	44	0	0	0	Morphine M (Nor-)
18	–	–	–	–	–	–	1	3	0	Codeine M (N-Oxide)
18	–	–	–	24	5	28	5	3	1	Atropine/Hyoscyamine

Table 8.8. (Continued)

TLC System 7	1	2	3	4	5	6	8	9	10	[System 7: MeOH–NH4OH (100+1.5) / Silica-KOH]
18	–	–	–	43	14	59	0	0	0	Dopamine
19	–	–	–	–	–	–	13	5	15	Primaquine
19	–	–	–	38	6	69	18	7	2	Protriptyline
20	–	–	–	24	6	63	3	10	12	Mescaline
20	–	–	–	66	6	68	32	10	2	Cyclopentamine
21	–	–	–	2	22	14	0	1	3	Cocaine M/N-Benzoylecgonine
22	–	–	–	–	–	–	5	4	1	Trimetazidine
22	–	–	–	–	–	–	18	7	2	Apoatropine
22	–	–	–	–	–	–	28	11	25	Octamylamine
23	–	–	–	18	12	14	3	9	2	Hydromorphone
23	–	–	–	22	14	–	1	2	5	Heptaminol
23	–	–	–	26	6	–	1	2	11	Tryptamine
23	–	–	–	32	7	–	26	4	2	Methoxyphenamine
23	–	–	–	35	7	77	17	7	4	Amantadine
24	–	–	–	–	–	–	68	3	10	Pempidine
24	–	–	0	24	6	74	0	1	2	Amiloride
25	–	–	–	–	–	–	1	0	10	Nordefrin Hydrochloride
25	–	–	–	–	–	–	34	8	2	Mephentermine
25	–	–	–	11	8	–	3	1	18	Synephrine/Oxedrine
25	–	–	–	18	12	–	2	3	1	Dihydromorphine
25	–	–	–	33	11	13	4	20	4	Hydrocodone
26	–	–	–	–	–	–	34	–	–	Propylhexedrine
26	–	–	–	–	5	42	11	0	24	Betazole
26	–	–	–	29	11	19	8	13	2	Dihydrocodeine
26	–	–	–	31	6	61	5	11	14	Dimethoxyphenethylamine,3,4-
26	–	–	–	32	8	11	8	19	2	Strychnine
26	–	–	–	37	5	70	39	13	5	Coniine
26	–	–	–	40	7	71	19	11	3	Desipramine/Imipramine M
27	–	–	–	28	6	37	2	8	1	Dothiepin M (Nor-,Sulfoxide)
28	–	–	–	–	–	–	49	3	3	Conessine
29	–	–	–	–	–	–	32	15	6	Methdilazine
29	–	–	–	27	9	–	3	3	3	Pholedrine
30	–	–	–	–	12	12	4	13	3	Methenamine
30	–	–	–	25	10	64	5	5	1	Ephedrine
30	–	–	0	76	12	–	3	2	2	Ethambutol
31	–	–	–	–	–	–	10	5	4	Metazocine
31	–	–	–	42	9	63	28	13	5	Deoxyephedrine,(+/-)-
31	–	–	–	42	9	63	28	13	5	Methamfetamine
31	–	–	–	47	5	66	6	7	3	Antazoline
32	–	–	–	33	6	–	1	3	2	Phentolamine
32	–	–	–	35	12	–	–	–	–	Methylephedrine,N-
33	–	–	–	–	–	–	1	7	24	Tuaminoheptane
33	–	–	–	–	–	–	18	17	8	Desomorphine
33	–	–	–	–	–	–	34	5	3	Pentaquine
33	–	–	–	12	8	67	1	1	0	Phenylephrine
33	–	–	–	13	1	–	–	–	–	Chlorhexidine

Table 8.8. (Continued)

TLC System 7	1	2	3	4	5	6	8	9	10	[System 7: MeOH–NH4OH (100+1.5) / Silica-KOH]
33	–	–	–	17	99	–	54	4	63	Pseudoephedrine
33	–	–	–	29	7	70	6	11	12	Trimethoxyamphetamine,3,4,5-
33	–	–	–	32	8	75	1	4	1	Clomipramine M (8-Hydroxy-nor-)
33	–	–	–	35	21	22	6	18	3	Codeine
33	–	–	–	47	10	42	42	18	6	Dextromethorphan/Racemethorphan/Levomethorphan
34	–	–	–	–	–	–	14	9	2	Racemorphan
34	–	–	–	45	8	75	24	13	3	Clomipramine M (Nor-)
34	–	–	–	46	87	71	28	16	4	Amitriptyline M/Nortriptyline
35	–	–	–	–	–	–	2	2	11	Hydroxyamphetamine/Amphetamine M
35	–	–	–	–	–	–	5	12	4	Neopine
35	–	–	–	33	10	34	0	1	1	Bufotenine
35	–	–	–	42	10	49	14	4	3	Dextrorphan/Levorphanol
35	–	–	–	43	9	74	23	10	18	Methoxy-4,5-methylenedioxyamphetamine,2-/MMDA-2
36	–	–	–	3	44	–	1	3	3	Penicillamine
36	–	–	–	25	15	–	3	18	2	Pholcodine
36	–	–	–	43	8	75	24	10	14	Dimethoxy-4-ethylamphetamine,2,5-/DOET
36	–	–	–	43	10	76	15	14	17	Bromo-2,5-dimethoxyamphetamine,4-/DOB
36	–	–	–	45	8	75	19	10	17	Dimethoxyamphetamine,2,5-
37	–	–	–	–	–	–	9	9	3	Diphenazoline
37	–	–	–	–	–	–	50	11	35	Rimantadine
37	–	–	–	18	23	16	0	9	0	Mianserin M (N-Oxide)
37	–	–	–	20	18	23	0	9	1	Morphine
38	–	–	–	–	–	–	13	11	17	Bromo-STP/Bromo-2,5-dimethoxy-4-methylamphetamine
38	–	–	–	30	11	–	3	6	1	Mesoridazine
38	–	–	–	34	17	–	0	–	–	Sulpiride
38	–	–	0	46	4	14	14	4	2	Chloroquine
39	–	–	–	–	21	38	18	24	6	Dimethoxanate
39	–	–	–	42	10	76	18	12	17	Methylenedioxyamphetamine,3,4-/MDA
39	–	–	–	47	14	48	5	9	9	Psilocin
40	–	–	–	–	–	–	1	10	2	Cytisine
40	–	–	–	–	–	–	5	6	5	Hordenine
40	–	–	–	–	–	–	17	8	8	Phenglutarimide
40	–	–	–	21	14	69	0	1	3	Isoprenaline
40	–	–	–	36	21	26	7	22	6	Ethylmorphine
40	3	1	36	40	26	15	1	48	5	Loprazolam
40	–	–	–	50	14	39	9	9	6	Dimethyltryptamine,N,N-/DMT
40	–	–	–	55	12	41	45	19	4	Ethoheptazine
40	–	–	–	55	25	78	17	4	9	Mexiletine
41	–	–	–	16	27	20	0	9	0	Clomipramine M (N-Oxid)
41	–	–	–	22	15	74	2	2	3	Etilefrine
41	–	–	–	41	20	23	6	23	8	Benzylmorphine
41	–	–	–	59	8	–	57	8	6	Perhexiline
42	–	–	–	–	–	–	1	2	24	Ethylnorepinephrine

Table 8.8. (Continued)

TLC System 7	1	2	3	4	5	6	8	9	10	[System 7: MeOH–NH4OH (100+1.5) / Silica-KOH]
42	–	–	–	–	–	–	8	6	8	Profadol
42	–	–	–	–	–	–	39	21	7	Tigloidine
42	–	–	–	18	13	76	1	1	24	Metaraminol
42	–	–	–	20	14	–	1	1	1	Nadolol
42	–	–	–	29	12	–	25	5	46	Pseudoephedrine M (Nor-)/Cathine
42	–	–	–	36	25	–	6	23	6	Viloxazine
42	–	–	–	40	18	–	3	34	6	Nicomorphine
42	–	–	–	47	10	–	35	13	6	Dimetindene
42	–	–	–	50	13	28	35	22	6	Tropacocaine
43	–	–	–	–	–	–	0	40	40	Furaltadone
43	–	–	–	–	–	–	5	4	10	Tacrine
43	–	–	–	–	–	–	6	21	2	Dehydroemetine
43	–	–	–	–	–	–	15	5	9	Mepacrine
43	–	–	–	–	–	–	30	22	12	Methopromazine
43	–	–	–	43	12	75	20	9	18	Amphetamine/Amphetaminil M artifact
43	–	–	–	47	19	–	12	42	10	Nicocodine
44	–	–	–	–	–	–	0	47	59	Furazolidone
44	–	–	–	–	–	–	1	1	2	Cycloserine
44	–	–	–	–	–	–	12	15	6	Methyldesorphine
44	–	0	–	30	12	75	4	4	33	Ephedrine M/Phenylpropanolamine
44	–	–	–	40	21	69	2	7	3	Hydroquinine
44	–	–	–	48	14	77	18	17	8	Chlorphentermine
44	–	–	–	50	16	–	36	22	9	Benzydamine
44	–	–	–	54	26	25	23	43	12	Acetylcodeine
44	–	–	–	56	14	–	35	9	15	Etafedrine
44	–	–	–	62	18	35	38	30	11	Promazine
45	–	–	–	–	–	–	0	1	4	Practolol
45	–	–	–	22	14	–	0	2	2	Atenolol
45	–	–	–	37	7	–	1	2	3	Hydroxychloroquine
45	–	–	–	43	20	70	3	8	5	Hydroquinidine
45	–	–	–	44	19	22	17	28	5	Dothiepin M (Nor-)
45	–	–	–	45	23	32	24	37	5	Thebaine
45	–	–	–	46	12	21	35	18	2	Chlorphenamine/Chlorpheniramine
45	–	–	–	46	14	26	35	13	3	Pheniramine
45	–	–	–	46	20	22	11	28	5	Dothiepin M (Sulfoxide)
45	–	–	–	48	14	–	26	21	7	Tofenacin
45	–	–	–	49	12	–	33	16	6	Brompheniramine
45	–	–	–	49	24	25	20	34	11	Thebacon
45	–	–	–	60	9	7	7	8	13	Disopyramide
46	–	–	–	–	–	–	0	0	0	Pseudomorphine
46	–	–	–	–	–	–	18	13	12	Eucatropine
46	–	–	–	–	–	–	45	30	10	Proheptazine
46	–	0	–	20	16	74	1	1	4	Salbutamol/Albuterol
46	–	–	–	43	22	22	7	34	6	Thioproperazine
46	–	–	–	44	25	27	6	26	8	Morphine-6-acetate/Diamorphine M
46	–	–	–	48	11	78	26	24	12	Phentermine

Table 8.8. (Continued)

TLC System 7	1	2	3	4	5	6	8	9	10	[System 7: MeOH–NH4OH (100+1.5) / Silica-KOH]
46	–	–	–	51	25	60	11	31	6	Mianserin M (Nor-)
46	–	–	–	58	18	49	49	25	9	Clemastine
46	–	–	–	63	14	56	14	10	11	Diethyltryptamine,N,N-/DET
46	–	–	–	68	23	49	37	28	8	Diphenylpyraline
47	–	–	–	–	–	–	49	34	16	Iprindole/Glycophene
47	–	–	–	16	12	–	10	44	6	Hexobendine
47	–	–	–	21	18	77	1	1	5	Terbutaline
47	–	–	18	29	55	49	1	11	20	Isoniazid
47	–	–	–	33	13	71	0	3	6	Acebutolol
47	–	–	–	37	26	–	2	9	6	Ketobemidone
47	–	–	–	48	12	–	17	13	6	Pipazetate
47	–	–	–	48	20	–	27	25	–	Zimeldine
47	–	–	–	49	26	33	15	38	4	Diamorphine
47	–	–	–	51	17	–	1	7	13	Metoclopramide
47	–	–	–	59	15	29	43	23	9	Prothipendyl
47	–	–	–	59	16	74	47	19	8	Etilamfetamine
48	–	–	–	–	–	–	28	11	18	Butaxamine
48	–	–	–	18	21	77	1	3	6	Orciprenaline
48	–	–	–	25	70	59	0	12	12	Diprophylline/Propyphylline
48	–	–	–	33	27	36	10	37	30	Oxymorphone
48	–	–	–	45	20	78	11	11	13	Oxprenolol
48	–	–	–	47	21	23	25	37	3	Perazine
48	–	–	–	50	13	16	26	19	4	Carbinoxamine
48	–	–	–	59	24	33	18	36	25	Oxomemazine
48	–	–	–	60	12	–	41	10	9	Doxylamine
48	–	–	–	60	19	–	41	16	11	Fenfluramine
48	–	–	–	62	18	–	17	24	12	Carbocromen
48	–	–	–	63	28	–	26	24	12	Acepromazine
48	–	–	–	64	28	–	45	–	–	Pizotifen
48	–	–	–	67	20	55	42	30	13	Thioridazine
48	–	–	–	67	21	47	48	23	13	Imipramine
48	–	–	–	68	16	49	47	22	14	Pentoxyverine/Carbetapentane
48	–	–	–	74	20	68	62	31	23	Procyclidine
48	–	–	–	77	16	60	59	20	27	Methadone/Levomethadone
49	–	–	–	–	–	–	26	14	14	Dimethylphenylthiazolanimin (Dimethyl-5-phenyl-2-thiazolidinimine,3,4-
49	–	–	–	2	60	75	1	1	1	Methyldopa
49	1	1	2	27	22	–	–	–	–	Mafenide
49	–	–	–	39	17	33	1	5	9	Procainamide
49	–	–	–	43	18	78	2	5	8	Pindolol
49	–	–	–	44	19	61	6	12	5	Cinchonine
49	–	–	–	44	20	74	10	8	9	Metoprolol
49	–	–	–	44	24	55	6	8	6	Cinchonidine
49	–	–	–	44	26	24	10	40	7	Tiotixene
49	–	–	–	54	25	42	15	19	9	Clomipramine M (10-Hydroxy-)
49	–	–	–	54	44	–	28	28	39	Phenethylamine

Table 8.8. (Continued)

TLC System 7	1	2	3	4	5	6	8	9	10	[System 7: MeOH–NH4OH (100+1.5) / Silica-KOH]
49	–	–	–	54	86	–	3	45	17	Phanquinone
49	–	–	–	55	26	26	34	37	7	Prochlorperazine
49	–	–	–	59	24	–	40	43	13	Bamipine
49	–	–	–	70	25	45	45	35	17	Chlorpromazine
49	–	–	68	80	88	–	2	10	4	Procarbazine
50	–	–	–	–	–	–	29	25	11	Betaprodine
50	–	–	–	–	–	–	30	35	11	Alphaprodine
50	–	–	–	–	–	–	30	53	16	Myrophine
50	–	–	–	46	34	45	14	27	14	Phenmetrazine/Phendimetrazine M (Nor-)
50	–	–	–	49	21	79	6	10	7	Propranolol
50	–	0	–	50	22	80	13	15	8	Penbutolol
50	–	–	–	52	20	–	38	44	12	Thenalidine
50	–	–	–	59	30	–	33	32	17	Nefopam
50	–	–	–	60	12	–	42	19	8	Aminopromazine
50	–	–	–	61	21	–	45	25	12	Metixene
50	–	–	–	62	30	33	25	51	39	Oxycodone
50	–	–	–	65	30	44	36	35	17	Promethazine
50	–	–	–	67	19	–	51	22	19	Oxeladine
50	–	–	–	79	22	–	67	32	25	Prolintane
51	–	–	–	–	–	–	0	6	3	Broxyquinoline
51	–	–	–	–	–	–	0	3	0	Minoxidil
51	–	–	–	–	–	–	38	11	64	Hydralazine
51	–	–	13	30	50	65	1	8	4	Triamterene
51	–	–	–	41	9	76	15	17	16	Dimethoxy-4-methylamphetamine,2,5-/STP/DOM
51	–	–	–	45	26	65	2	11	4	Quinine
51	–	–	–	49	30	63	4	12	6	Quinidine
51	–	–	–	51	26	51	4	17	10	Clomipramine M (8-Hydroxy-)
51	–	–	–	53	27	–	30	41	8	Thiethylperazine
51	–	–	–	55	19	30	11	20	6	Triprolidine
51	–	–	–	58	22	33	39	25	14	Mepyramine/Pyrilamine
51	–	–	–	63	24	45	48	37	13	Doxepin
51	–	–	–	64	30	50	45	44	13	Cyproheptadine
51	–	–	–	68	26	–	52	32	15	Tolpropamine
51	–	–	–	69	27	51	50	32	15	Amitriptyline
51	–	–	–	70	27	41	49	42	16	Dothiepin
51	–	–	–	72	26	54	53	34	18	Clomipramine
52	–	–	–	–	–	–	5	24	21	Amopyroquine
52	–	–	–	–	–	–	17	26	5	Diethylcarbamazine
52	–	–	–	–	–	–	30	30	17	Dimazole/Diamthazole
52	–	–	–	–	–	–	31	44	19	Moxisylyte
52	–	–	–	–	32	–	41	37	29	Clofedanol
52	–	–	–	0	34	14	0	3	0	Narceine
52	–	–	–	38	61	–	10	7	15	Oxetacaine
52	23	41	42	43	82	82	1	36	28	Lorazepam/Lormetazepam M
52	–	–	–	46	65	14	2	52	21	Endralazine Mesilate

Table 8.8. (Continued)

TLC System 7	1	2	3	4	5	6	8	9	10	[System 7: MeOH–NH4OH (100+1.5) / Silica-KOH]
52	–	–	–	49	22	–	11	12	11	Alprenolol
52	–	–	–	49	87	–	0	1	3	Dobutamine
52	–	–	–	50	20	75	6	11	9	Timolol
52	15	10	55	52	59	55	3	58	25	Caffeine
52	46	45	60	59	82	82	6	61	50	Lormetazepam
52	–	–	–	60	34	40	37	34	11	Pethidine/Meperidine
52	–	–	0	63	22	–	41	28	17	Chloropyramine
52	–	–	–	64	22	35	41	30	14	Isothipendyl
52	–	–	–	66	21	24	41	26	13	Methapyrilene
52	–	–	–	80	18	–	62	19	26	Terodiline
53	–	–	–	–	–	–	1	19	8	Cephaeline
53	–	–	–	–	19	36	42	25	12	Thenyldiamine
53	–	–	–	–	33	32	3	25	3	Acetophenazine
53	–	–	71	6	80	–	3	8	4	Azathioprine
53	–	–	–	30	19	–	1	3	5	Sotalol
53	–	4	–	34	59	54	1	31	21	Theobromine
53	–	–	–	35	56	–	0	41	11	Demecolcine
53	–	–	–	44	52	45	0	32	12	Pilocarpine
53	–	–	–	52	26	–	28	37	5	Butaperazine
53	–	–	–	55	30	29	33	30	8	Trifluoperazine
53	–	–	–	62	43	–	–	–	–	Arecoline
53	51	47	65	62	82	82	8	59	53	Temazepam/M of Camazepam, Diazepam, Ketazolam, Medazepam
53	–	–	–	65	24	74	15	13	25	Cyclazocine
53	–	1	–	65	27	–	47	44	16	Pecazine/Mepazine
53	–	–	–	66	29	–	43	35	18	Noxiptiline
53	–	–	–	67	32	–	38	48	15	Phenyltoloxamine
53	–	–	–	70	29	–	47	36	17	Chlorphenoxamine
54	–	–	–	–	–	–	1	44	53	Niridazole
54	–	–	–	–	–	–	8	49	32	Morinamide
54	–	–	–	–	–	–	12	39	17	Etamiphyllin
54	–	–	–	–	–	–	46	42	21	Methaphenilene
54	–	–	–	–	–	–	50	32	17	Embramine
54	–	–	–	27	53	55	0	9	12	Cimetidine
54	–	–	–	38	35	39	6	22	7	Opipramol
54	–	–	–	39	39	–	5	27	7	Carfenazine
54	–	–	–	40	68	66	0	21	27	Nicotinamide
54	–	–	–	48	25	–	3	16	15	Fenpipramide
54	–	–	–	53	19	–	13	34	12	Emetine
54	–	–	–	55	38	22	22	35	14	Dibenzepin
54	–	–	–	58	41	67	39	33	48	Tranylcypromine
54	–	–	–	61	39	22	39	35	13	Nicotine
54	–	–	–	66	27	74	42	27	17	Eucaine,β-
54	–	–	–	69	27	48	42	43	13	Bromazine/Bromdiphenhydramine
54	–	–	–	71	11	28	41	21	9	Diamocaine
54	–	–	–	71	25	66	54	37	18	Pyrrobutamine

Table 8.8. (Continued)

TLC System 7	1	2	3	4	5	6	8	9	10	[System 7: MeOH–NH4OH (100+1.5) / Silica-KOH]
54	–	–	–	71	36	42	5	31	30	Procaine
54	–	–	–	75	32	49	47	35	22	Triflupromazine
54	–	–	–	77	21	–	62	34	30	Fencamfamine
54	–	–	–	81	19	79	59	38	39	Pipradrol
55	–	–	–	11	12	73	24	4	38	Methoxamine
55	–	–	–	15	–	–	36	2	1	Dihydralazine
55	–	–	–	33	69	63	0	37	12	Colchicine
55	–	–	–	38	23	–	3	6	0	Bamethan
55	–	–	0	38	70	–	0	4	8	Ambazone
55	–	–	–	42	40	40	7	29	9	Perphenazine
55	–	–	20	45	45	59	0	22	12	Trimethoprim
55	–	–	–	48	49	47	6	37	18	Scopolamine
55	–	–	–	54	44	–	3	45	14	Fenetylline
55	–	–	–	55	41	38	12	36	18	Physostigmine
55	–	–	–	61	49	43	14	57	26	Isothebaine
55	–	–	–	65	22	31	38	28	14	Thonzylamine
55	–	–	–	65	27	48	44	33	15	Diphenhydramine
55	–	–	–	65	29	73	36	38	13	Noracymethadol
55	–	–	–	66	72	–	6	66	46	Pentifylline
55	–	–	–	68	22	34	44	27	15	Tripelennamine
55	–	–	–	68	25	49	48	33	16	Orphenadrine
55	–	–	–	68	28	48	45	33	15	Dimenhydrinate
55	–	–	–	68	30	52	34	42	26	Propiomazine
55	–	–	–	76	21	56	53	37	27	Piperocaine
55	–	–	–	82	27	77	66	44	36	Prozapine
55	–	–	–	86	23	75	68	36	52	Diisopromine
56	–	–	–	–	–	–	0	11	12	Metopimazine
56	–	–	–	–	–	–	6	19	17	Piperacetazine
56	–	–	–	–	–	–	39	44	16	Properidine
56	–	–	70	30	40	88	0	5	9	Clioquinol
56	–	–	–	43	33	61	1	12	8	Pipamperone
56	–	–	–	44	45	–	7	32	11	Clopenthixol/Zuclopenthixol/cis-Ordinol
56	22	35	42	45	82	82	0	40	51	Oxazepam/M of Chlordiazepoxide, Clorazepic Acid, Camazepam, Diazepam, Halazepam, Ketazolam, Medazepam, Oxazolam, Prazepam
56	–	–	–	48	74	69	4	17	22	Nicotinyl Alcohol
56	–	–	–	63	33	–	40	34	18	Methadone M (Nor-)
56	–	–	–	64	53	52	9	29	31	Nomifensine
56	–	–	–	66	43	–	13	48	28	Dimetiotazine
56	–	–	–	68	35	–	41	35	20	Nicametate
56	–	–	–	74	34	51	51	51	25	Chlorprothixene
56	–	–	–	78	30	73	63	37	26	Volazocine
57	–	–	–	–	–	–	1	43	16	Inositol Nicotinate
57	–	–	–	–	–	–	2	54	34	Tozalinone
57	–	–	–	–	–	–	18	15	20	Clorprenaline

Table 8.8. (Continued)

TLC System 7	1	2	3	4	5	6	8	9	10	[System 7: MeOH–NH4OH (100+1.5) / Silica-KOH]
57	–	–	–	–	–	–	27	39	26	Cyclopentolate
57	–	–	–	33	62	60	0	12	8	Ergometrine
57	4	5	29	55	42	–	4	38	17	Clozapine
57	–	–	–	58	60	–	3	44	33	Nimorazole
57	–	–	–	62	49	41	36	51	24	Phendimetrazine
57	–	–	–	64	43	39	15	32	16	Tetracaine/Amethocaine
57	–	–	–	65	36	46	27	45	20	Mebhydrolin
57	–	–	–	66	40	70	35	41	23	Methylphenidate
57	–	–	–	67	35	52	42	46	14	Chlorcyclizine
57	–	–	–	68	40	52	49	41	16	Cyclizine
57	–	–	–	75	78	–	32	27	29	Octacaine
57	–	–	–	76	32	49	47	38	46	Levomepromazine/Methotrimeprazine
57	–	–	–	84	38	62	62	45	50	Alphacetylmethadol
58	–	–	–	–	–	–	3	33	28	Propoxycaine
58	–	–	–	–	–	–	34	57	37	Betameprodine
58	–	–	–	–	–	–	35	49	24	Allylprodine
58	–	–	–	–	–	–	41	41	17	Trimeperidine
58	–	–	–	–	–	–	55	36	25	Cyclomethycaine
58	–	–	–	–	–	–	56	43	38	Alphamethadol
58	0	0	0	0	70	16	0	0	0	Lysergic Acid
58	1	0	4	0	72	–	–	–	–	Nicotinic Acid
58	19	43	31	1	–	–	–	–	–	Aminobenzoic Acid,p-
58	0	0	14	40	20	60	2	18	29	Flurazepam M (Dideethyl-)
58	–	–	32	46	75	70	2	36	40	Metronidazole
58	–	–	–	49	71	–	2	33	29	Proxyphylline
58	0	0	15	50	28	62	8	19	11	Flurazepam M (Monodeethyl-)
58	–	–	–	51	46	61	4	16	18	Periciazine
58	–	–	–	58	61	–	8	46	31	Morazone
58	31	25	55	63	78	–	–	–	–	Methyprylon
58	–	–	–	64	66	–	16	58	41	Metyrapone
58	–	–	–	68	48	50	39	58	23	Mianserin
58	–	–	–	77	32	46	54	39	31	Trimeprazine/Alimemazine
58	–	–	–	77	33	54	57	51	39	Diethazine
59	–	–	–	–	–	–	0	21	17	Rolicyprine
59	–	–	–	–	–	–	5	23	37	Chloroprocaine
59	–	–	–	–	–	–	15	48	24	Dimefline
59	–	–	–	–	–	–	37	40	42	Aletamine
59	–	–	–	–	–	–	41	59	23	Clotiapine
59	–	–	–	–	–	–	57	42	28	Ethylmethylthiambutene
59	–	–	–	–	–	–	61	48	38	Butriptyline
59	–	–	–	–	8	–	–	–	–	Prajmalium Bitartrate
59	–	–	–	–	73	–	0	0	0	Isoetarine
59	–	–	–	15	75	67	0	8	5	Pyridoxine Hydrochloride
59	–	–	–	32	57	59	1	23	29	Nalorphine
59	–	–	–	59	71	67	15	56	29	Nikethamide
59	–	–	–	70	80	–	1	50	53	Phenazopyridine

Table 8.8. (Continued)

TLC System 7	1	2	3	4	5	6	8	9	10	[System 7: MeOH–NH4OH (100+1.5) / Silica-KOH]
59	–	–	–	71	17	–	4	35	17	Veratrine
59	–	–	–	73	43	61	23	70	42	Verapamil
59	–	–	–	77	38	–	56	52	43	Benzoctamine
59	–	–	–	80	36	56	62	54	37	Trimipramine
59	–	–	–	84	23	69	73	35	66	Phencyclidine
60	–	–	–	–	–	–	9	42	18	Etoxeridine
60	–	–	–	–	–	–	31	63	50	Thurfyl Nicotinate
60	–	–	–	–	–	–	34	47	36	Ethomoxane
60	–	–	–	–	–	–	49	40	25	Dyclonine
60	24	43	47	12	82	–	–	–	–	Sulfametoxydiazine
60	–	–	–	36	57	51	0	19	7	Lysergamide
60	–	11	–	36	81	81	0	23	40	Pemoline
60	–	–	–	42	58	–	1	28	16	Dihydroergotamine
60	–	–	–	44	42	74	2	23	–	Tocainide
60	5	2	41	44	68	65	1	40	16	Triazolam
60	–	–	–	54	46	52	4	37	18	Mianserin M (8-Hydroxy-)
60	3	–	–	56	60	59	3	39	18	Lysergide/LSD
60	–	–	–	56	79	75	2	56	47	Carbamazepine
60	–	–	–	57	46	–	1	60	29	Vinblastine
60	–	–	–	59	68	74	1	47	49	Prazosin
60	–	–	–	73	58	50	30	58	33	Methylpiperidyl Benzilate/JB-336
60	–	–	–	82	30	–	56	52	35	Clomifene
61	–	–	–	–	–	–	42	70	61	Nicofuranose
61	–	–	–	–	–	–	55	46	28	Quinisocaine
61	–	–	–	–	–	–	56	45	51	Difenidol
61	–	–	–	–	–	–	60	52	42	Etymemazine
61	0	6	26	39	73	61	0	24	28	Bromazepam M (3-Hydroxy-)
61	–	–	–	58	66	–	2	31	21	Pyrimethamine
61	–	–	–	62	–	–	2	33	57	Amiphenazole
61	13	18	47	63	73	69	6	41	53	Bromazepam
61	–	–	–	68	80	–	11	72	52	Trimetozine
61	–	–	–	69	74	74	8	65	47	Papaverine
61	–	–	–	70	34	72	16	12	28	Pentazocine
61	–	–	–	71	59	57	15	64	52	Hydrastine
61	–	–	–	72	74	77	37	66	59	Methyl Nicotinate
61	–	–	–	74	62	59	35	53	42	Thiopropazate
61	–	–	–	75	23	–	17	35	29	Lobeline
62	–	–	–	–	–	–	8	40	37	Amodiaquine
62	–	–	–	–	–	–	26	41	35	Proxymetacaine
62	–	–	–	–	–	–	27	45	11	Morpheridine
62	–	–	–	–	–	–	50	41	21	Dimethylthiambutene
62	23	45	44	13	79	–	–	–	–	Sulfadimidine
62	–	–	–	41	69	–	0	14	12	Methylergometrine
62	–	–	–	41	80	–	1	52	59	Thiamazole
62	–	–	–	46	50	–	6	33	–	Flupentixol
62	10	10	53	52	76	77	2	50	22	Chlordiazepoxide

Table 8.8. (Continued)

TLC System 7	1	2	3	4	5	6	8	9	10	[System 7: MeOH–NH4OH (100+1.5) / Silica-KOH]
62	13	39	30	53	–	–	1	20	53	Styramate
62	–	–	–	56	22	–	–	–	–	Ajmaline
62	–	–	–	58	74	–	–	–	–	Methylphenazone,4-
62	–	–	–	65	53	52	18	48	42	Levamisole
62	34	45	57	69	82	83	3	55	60	Nordazepam/M of Chlordiazepoxide, Clorazepic Acid, Diazepam, Halazepam, Ketazolam, Medazepam, Oxazolam, Prazepam
62	–	–	–	70	44	76	8	31	53	Clonidine
62	3	3	41	71	52	45	30	48	40	Flurazepam
62	53	47	70	75	84	85	8	70	62	Clobazam
62	–	–	–	77	48	–	36	67	45	Flavoxate
62	–	–	–	83	54	–	23	41	36	Oxybuprocaine
63	–	–	–	–	–	–	42	71	60	Benzyl Nicotinate
63	–	–	–	–	–	–	45	69	62	Butoxyethyl Nicotinate,2-
63	–	–	–	–	–	–	47	55	43	Methadone Intermediate/Pre-Methadone
63	–	–	–	–	–	–	59	67	61	Piperoxan
63	–	–	–	–	45	55	5	26	31	Butethamine
63	–	–	–	–	65	68	0	22	28	Harmine
63	3	1	29	24	75	71	0	29	8	Triazolam M (4-Hydroxy-)
63	1	0	28	40	49	39	1	36	2	Adinazolam M (Nor-)
63	15	24	42	41	81	83	0	35	51	Demoxepam/Chlordiazepoxide M
63	–	–	–	44	68	64	1	34	22	Ergotamine
63	–	–	–	45	45	49	5	23	10	Fluphenazine
63	–	–	–	53	46	65	7	13	13	Mazindol
63	–	–	44	54	71	70	3	42	46	Pyrazinamide
63	–	–	–	64	66	70	6	38	52	Yohimbine
63	–	–	–	66	64	61	10	58	37	Trazodone
63	–	–	–	67	42	–	25	34	35	Cinchocaine/Dibucaine
63	–	–	–	68	41	49	45	57	21	Phenindamine
63	54	47	72	74	80	82	10	72	63	Flunitrazepam
63	–	–	–	79	68	72	37	63	64	Ketamine
63	–	–	–	86	32	–	40	53	49	Mebeverine
64	–	–	–	–	–	–	0	30	33	Norharman
64	–	–	–	–	–	–	48	59	43	Benethamine
64	–	–	–	–	–	–	55	38	36	Betacetylmethadol
64	22	39	38	4	81	–	–	–	–	Sulfadiazine
64	5	16	27	5	80	–	–	–	–	Sulfisomidine
64	1	1	25	31	62	50	0	–	–	Bromazepam M (N(py)-Oxide)
64	–	–	–	55	87	89	2	43	57	Mephenesin
64	–	–	–	72	69	–	20	70	54	Doxapram
64	–	–	–	76	80	85	44	69	58	Clomethiazole
64	–	–	–	76	87	–	3	59	55	Ethenzamide
64	–	–	–	78	43	–	52	41	35	Naftidrofuryl
64	–	–	–	78	72	75	21	74	64	Noscapine
64	–	–	–	83	45	–	68	64	64	Biperiden
64	–	–	–	83	49	–	55	60	51	Adiphenine

Table 8.8. (Continued)

TLC System 7	1	2	3	4	5	6	8	9	10	[System 7: MeOH–NH4OH (100+1.5) / Silica-KOH]
65	–	–	–	–	–	–	0	36	55	Ethionamide
65	–	–	–	–	–	–	0	56	67	Ethoxazene
65	–	–	–	–	–	–	1	21	15	Methysergide
65	–	–	–	–	–	–	1	3	6	Protokylol
65	–	–	–	–	–	–	3	68	69	Metisazone
65	–	–	–	–	–	–	7	69	36	Benzquinamide
65	–	–	–	–	–	–	60	58	52	Trifluomeprazine
65	12	23	27	4	83	–	–	–	–	Sulfamethizole
65	26	54	41	5	79	–	–	–	–	Sulfamethoxazole
65	25	52	33	6	81	–	–	–	–	Sulfafurazole
65	23	41	42	8	80	–	–	–	–	Sulfamerazine
65	19	39	43	9	83	–	–	–	–	Sulfamethoxypyridazine
65	31	51	48	10	86	–	–	–	–	Sulfadimethoxine
65	1	6	7	25	75	–	–	–	–	Sulfaguanidine
65	–	–	–	38	59	–	–	–	–	Dropropizine
65	–	–	–	43	41	52	0	17	14	Pipamazine
65	18	14	50	45	66	66	4	–	–	Phenazone/Antipyrine
65	–	–	–	47	74	–	9	66	63	Naloxone
65	–	–	–	59	71	–	0	54	34	Bamifylline
65	–	–	–	60	80	84	0	59	49	Mebendazole
65	–	–	–	66	63	60	31	62	48	Mepivacaine
65	–	–	–	76	90	–	8	65	64	Beclamide
65	–	–	–	77	35	30	45	47	54	Cocaine
65	64	55	72	81	84	89	36	74	63	Prazepam
65	–	–	–	86	39	72	28	50	62	Ibogaine/Rotenone
66	–	–	–	–	–	–	0	53	23	Naftazone
66	–	–	–	–	–	–	1	62	48	Ergotoxine
66	–	–	–	–	–	–	2	58	41	Pipobroman
66	–	–	–	–	–	–	8	69	60	Clonitazene
66	–	–	–	–	–	–	16	74	67	Diloxanide
66	–	–	–	–	–	–	57	67	64	Amolanone
66	–	–	–	–	–	–	65	39	38	Alverine
66	–	–	–	–	–	–	67	61	60	Cycrimine
66	–	–	–	–	77	–	1	38	57	Protionamide
66	0	0	2	0	85	–	–	–	–	Sulfasalazine
66	9	20	27	8	79	–	–	–	–	Sulfathiazole
66	–	–	–	53	40	59	3	32	21	Pipotiazine
66	–	–	–	58	69	–	1	48	38	Ergoloid
66	25	10	58	62	70	68	21	–	–	Aminophenazone/Amidopyrine
66	–	–	–	64	78	–	6	45	68	Leucinocaine
66	–	–	–	68	84	80	1	63	57	Chlormezanone
66	–	–	–	72	80	–	20	70	55	Propanidid
66	–	–	–	73	35	46	26	55	38	Pitofenone
66	45	45	62	74	83	80	14	64	66	Ketazolam
66	–	–	–	77	47	–	49	–	–	Captodiame
66	–	–	–	77	52	57	43	53	53	Benactyzine

Table 8.8. (Continued)

TLC System 7	1	2	3	4	5	6	8	9	10	[System 7: MeOH–NH4OH (100+1.5) / Silica-KOH]
66	–	–	–	87	27	72	67	33	70	Dipipanone
67	–	–	–	–	–	–	7	64	69	Isobutyl 4-Aminobenzoate
67	–	–	–	–	–	–	18	77	68	Octaverine
67	–	–	–	–	–	–	42	70	59	Clorgiline
67	–	–	–	–	–	–	52	68	62	Tiletamine
67	–	–	–	–	–	–	58	62	49	Diethylaminoethyl Diphenylpropionate
67	–	–	–	–	75	–	7	54	53	Tiabendazole
67	14	35	34	8	–	–	–	–	–	Sulfaethidole
67	40	50	54	8	–	–	–	–	–	Sulfalene
67	16	42	34	24	–	–	–	–	–	Sulfapyridine
67	3	2	26	26	73	69	1	25	7	Alprazolam M (4-Hydroxy-)
67	–	3	–	46	81	78	1	38	40	Glafenine
67	7	2	40	47	67	66	1	57	14	Alprazolam
67	15	28	35	49	88	89	0	30	38	Quazepam M (3-Hydroxy-N-dealkyl-2-oxo-)
67	13	46	22	51	83	–	–	–	–	Sulfanilamide
67	–	–	–	58	71	73	2	48	36	Droperidol
67	–	–	–	70	86	–	0	51	62	Benorilate
67	–	–	–	73	78	85	26	71	52	Etomidate
67	–	–	–	74	42	73	19	24	45	Levallorphan
67	–	–	–	76	51	75	11	27	33	Haloperidol
67	56	62	63	76	84	87	5	57	66	Benzocaine
67	56	40	73	78	79	83	41	74	62	Medazepam
67	37	51	55	79	78	–	63	–	–	Sulfadoxine
67	–	–	–	83	31	55	64	47	66	Profenamine/Ethopropazine
67	9	8	41	83	45	51	53	60	51	Flurazepam Hy/DCFB
67	–	–	–	88	63	77	36	64	59	Piminodine
68	–	–	–	–	–	–	1	35	65	Brocresine
68	–	–	–	–	–	–	7	45	67	Dimethocaine
68	–	–	–	–	–	–	16	75	58	Dimoxyline
68	–	–	–	–	–	–	59	50	38	Norpipanone
68	–	–	–	–	52	–	67	64	54	Dicycloverine/Dicyclomine
68	–	–	–	8	88	–	53	5	67	Azapropazone
68	5	4	37	28	76	76	1	35	13	Brotizolam M (6-Hydroxy-)
68	–	–	–	44	82	87	0	37	42	Dipyridamole
68	–	–	–	54	57	65	10	54	19	Hydroxyzine
68	35	46	53	64	84	86	0	36	55	Nitrazepam
68	–	–	–	69	83	79	28	67	55	Crotethamide
68	–	–	–	71	79	–	1	65	68	Nifedipine
68	–	–	–	74	50	81	16	39	49	Phenazocine
68	–	–	–	75	42	68	11	39	58	Aconitine
68	–	–	–	78	74	–	4	65	58	Oxypertine
68	–	–	–	79	70	72	13	70	61	Etonitazene
68	–	–	–	81	45	–	58	54	55	Isoaminile
68	–	–	–	82	50	63	58	55	54	Propoxyphene/Dextropropoxyphene/Levopropoxyphene
68	–	–	–	83	43	75	66	61	59	Trihexyphenidyl/Benzhexol

Table 8.8. (Continued)

TLC System 7	1	2	3	4	5	6	8	9	10	[System 7: MeOH–NH4OH (100+1.5) / Silica-KOH]
68	–	–	–	84	43	85	55	68	56	Prenylamine
68	11	29	46	86	–	–	–	–	–	Butalamine
69	–	–	–	–	–	–	0	56	68	Clefamide
69	–	–	–	–	–	–	26	45	67	Amprotropine
69	–	–	–	–	–	–	37	60	29	Furethidine
69	–	–	–	–	–	–	60	66	56	Proadifen
69	29	51	43	9	89	–	–	–	–	Sulfaphenazole
69	11	31	34	36	86	90	0	31	48	Chloramphenicol
69	3	3	33	40	68	60	0	32	42	Chlordiazepoxide M (Nor-)
69	–	–	–	42	70	69	1	23	17	Iproniazid
69	9	2	51	57	68	66	6	55	22	Adinazolam
69	–	–	–	66	51	53	18	54	46	Tetramisole
69	–	–	–	71	63	78	4	59	49	Fluspirilene
69	–	–	–	76	80	86	8	52	59	Fenyramidol
69	–	–	–	77	76	80	2	74	63	Reserpine
69	–	–	–	78	51	80	18	39	47	Phenomorphan
69	–	–	–	79	80	–	41	78	67	Tetrabenazine
69	–	–	–	80	69	79	42	73	65	Bupivacaine
69	–	–	–	81	48	56	59	57	47	Butetamate
69	–	–	–	82	54	52	55	81	55	Piperidolate
70	–	–	–	–	–	–	0	62	64	Norbormide
70	–	–	–	–	–	–	9	60	67	Levophenacylmorphan
70	–	–	–	–	–	–	15	58	66	Diperodon
70	–	–	–	–	–	–	39	73	66	Tolycaine
70	–	–	–	–	–	–	48	69	63	Mebenazine
70	–	–	–	–	–	–	51	65	64	Aminoxytriphene
70	–	–	–	–	–	–	61	43	43	Diethylthiambutene
70	–	–	–	–	62	60	43	55	41	Pramoxine
70	–	–	–	–	68	64	2	25	4	Nialamide
70	17	42	28	4	87	–	–	–	–	Sulfacetamid
70	5	24	15	5	–	–	–	–	–	Aminosalicylic Acid,p-
70	–	–	–	45	26	51	1	24	11	Clomipramine M (2-Hydroxy-)
70	7	23	38	49	84	–	–	–	–	Methocarbamol
70	3	4	41	53	73	72	3	52	8	Midazolam M (α-Hydroxy-)
70	–	–	–	57	72	76	2	40	33	Harman
70	–	–	–	60	77	–	–	–	–	Pargyline
70	–	–	–	61	73	74	1	45	42	Piritramide
70	50	53	59	66	84	–	5	–	–	Ethosuximide
70	–	–	–	70	70	77	11	63	36	Lidoflazine
70	–	–	–	74	25	83	29	69	57	Cropropamide
70	–	–	–	74	52	81	9	32	22	Loperamide
70	–	–	–	78	14	–	57	67	56	Azapetine
70	–	–	–	78	34	30	42	60	55	Cinnamoylcocaine
70	–	–	–	78	70	77	43	74	58	Fentanyl
70	–	–	–	78	79	84	36	80	56	Methaqualone
70	–	–	–	79	30	91	48	40	20	Hexetidine

Table 8.8. (Continued)

TLC System 7	1	2	3	4	5	6	8	9	10	[System 7: MeOH–NH4OH (100+1.5) / Silica-KOH]
70	–	–	–	80	72	69	35	71	63	Lidocaine
70	–	–	–	86	33	86	57	59	68	Clofazimine
71	–	–	–	–	–	–	48	76	62	Carbifene
71	–	–	–	–	–	–	51	72	65	Diampromide
71	4	3	39	42	75	74	4	49	21	Triazolam M (α-Hydroxy-)
71	7	12	24	44	80	83	0	20	49	Nitrazepam M (7-Acetamido-)
71	7	5	44	50	71	60	2	53	25	Estazolam
71	10	20	30	54	77	79	0	30	46	Nitrazepam M (7-Amino-)
71	38	37	64	57	81	80	1	59	52	Flunitrazepam M (3-Hydroxy-)
71	30	14	61	61	76	72	3	63	44	Flumazenil/RO 15-1788
71	–	–	–	62	78	81	4	55	52	Carnidazole
71	–	–	–	65	92	96	41	79	70	Bezitramide
71	42	52	58	69	88	90	2	55	59	Quazepam M (3-Hydroxy-2-oxo-)
71	–	–	–	70	51	67	21	48	44	Dexetimide
71	–	–	–	71	73	82	3	60	40	Pimozide
71	55	55	70	72	85	88	12	–	52	Prazepam M (3-Hydroxy-)
71	–	–	–	74	77	81	23	65	48	Metomidate
71	61	49	65	74	81	–	32	–	–	Propyphenazone
71	–	–	–	75	84	86	20	74	61	Isocarboxazid
71	–	–	–	76	70	82	26	64	58	Phenoperidine
71	–	–	–	78	78	81	21	78	69	Disulfiram
71	–	–	–	83	44	76	7	30	64	Butacaine
72	–	–	–	–	–	–	4	77	64	Methoserpidine
72	–	–	–	–	–	–	5	59	61	Cyprenorphine
72	38	43	49	10	87	88	0	–	–	Chlorpropamide
72	4	4	37	37	76	76	1	44	18	Alprazolam M (α-Hydroxy-)
72	63	47	68	42	75	75	1	–	–	Carbimazole
72	7	6	41	45	78	78	2	46	31	Brotizolam M (α-Hydroxy-)
72	15	5	53	52	72	71	5	52	27	Brotizolam
72	11	19	35	54	78	75	0	34	47	Flunitrazepam M (7-Aminonor-)
72	13	5	53	60	69	70	6	60	19	Midazolam
72	–	–	–	60	72	74	7	64	63	Pentetrazol
72	35	45	56	67	85	87	0	53	61	Clonazepam
72	–	–	–	73	84	88	2	69	61	Bromocriptine
72	–	–	–	81	73	–	3	77	66	Deserpidine
72	37	41	54	81	80	79	5	57	57	Nitrazepam M Hy/DAB
72	–	–	–	82	54	64	62	68	55	Amiodarone
72	–	–	–	86	88	–	47	78	67	Fenbutrazate
72	–	–	–	87	90	–	37	74	67	Famprofazone
73	–	–	–	–	–	–	–	–	–	Diethyl-m-toluamide,N,N-/Deet
73	–	–	–	–	–	–	7	61	63	Etorphine
73	–	–	–	–	–	–	60	67	63	Amylocaine
73	7	11	30	42	79	82	0	22	45	Clonazepam M (7-Acetamido-)
73	–	–	–	43	9	74	23	77	69	Methoxyamphetamine,4-/PMA
73	19	11	41	51	78	76	0	40	49	Flunitrazepam M (7-Acetamido-)
73	11	20	40	52	76	77	0	30	47	Clonazepam M (7-Amino-)

Table 8.8. (Continued)

TLC System 7	1	2	3	4	5	6	8	9	10	[System 7: MeOH–NH4OH (100+1.5) / Silica-KOH]
73	9	22	30	67	66	63	4	51	47	Medazepam M (Nor-)
73	35	41	57	72	82	86	5	58	56	Delorazepam
73	39	45	63	75	78	84	0	66	59	Cloxazolam
73	–	–	–	76	76	88	5	68	61	Ambucetamide
73	2	5	26	77	55	76	13	–	–	Trifluperidol
73	–	–	–	79	60	66	12	56	51	Anileridine
73	–	–	–	79	72	78	42	71	60	Dextromoramide/Racemoramide/Levomoramide
73	–	7	56	80	77	80	11	67	37	Miconazole
73	–	–	–	81	77	79	1	75	64	Rescinnamine
73	–	23	–	82	67	75	39	68	60	Fluanisone
73	66	66	80	83	76	86	31	77	68	Nimetazepam Hy/MNB
73	–	–	–	87	60	–	67	70	70	Benzphetamine
73	–	–	–	87	84	97	63	76	68	Phenoxybenzamine
74	–	–	–	–	–	–	58	45	43	Isomethadone
74	–	–	–	–	–	–	62	80	73	Bialamicol
74	5	5	37	49	76	77	1	43	15	Midazolam M (4-Hydroxy-)
74	–	–	–	61	51	72	4	49	39	Spiperone
74	19	28	54	61	81	82	2	46	55	Flurazepam M (N(1)-Hydroxyethyl-)
74	–	–	–	62	33	83	3	14	50	Buphenine
74	21	21	52	63	77	74	1	55	52	Flunitrazepam M (7-Amino-)
74	46	50	65	75	80	90	11	66	62	Haloxazolam
74	53	46	71	77	81	81	12	70	55	Nimetazepam
74	–	–	–	78	78	–	52	79	71	Broxaldine
74	–	–	–	80	82	87	17	76	66	Bisacodyl
74	41	46	55	82	80	80	5	59	61	Clonazepam Hy/DCB
74	78	71	78	83	87	96	27	75	76	Quazepam
74	–	–	–	87	90	92	42	84	70	Diphenoxylate
75	–	–	–	–	–	–	44	72	57	Benzethidine
75	–	–	–	–	–	–	61	83	72	Buclizine
75	–	–	–	–	83	90	6	63	70	Butyl Aminobenzoate
75	0	0	7	8	82	72	0	2	0	Cinchophen
75	–	9	–	11	74	66	1	30	11	Theophylline
75	2	5	33	38	78	81	4	25	14	Midazolam M (α,4-Dihydroxy-)
75	4	21	15	40	–	–	–	–	–	Quinethazone
75	9	36	32	56	63	87	0	32	58	Meprobamate
75	–	–	–	64	64	67	0	47	29	Ergosine
75	32	43	54	64	84	90	0	52	61	Clobazam M (Nor-)
75	–	–	–	70	74	76	13	58	51	Bromolysergide,2-
75	34	45	60	72	83	88	2	56	58	Flurazepam M (N(1)-Dealkyl-)/Quazepam M (N-Dealkyl-2-oxo-)
75	47	55	60	73	82	–	–	–	–	Phenprobamate
75	–	–	–	76	66	80	36	64	56	Articaine
75	58	49	72	76	82	85	27	73	59	Diazepam/Ketazolam M/Medazepam M
75	71	59	72	77	–	–	–	–	–	Phensuximide
75	–	–	–	78	59	54	44	63	55	Ethylpiperidyl Benzilate/JB-318

Table 8.8. (Continued)

TLC System 7	1	2	3	4	5	6	8	9	10	[System 7: MeOH–NH4OH (100+1.5) / Silica-KOH]
75	63	62	70	80	86	89	31	–	–	Glutethimide
75	65	61	73	81	85	92	29	72	70	Pinazepam
75	71	69	81	83	77	90	48	77	69	Fludiazepam Hy/CFMB
75	81	69	81	87	82	93	49	81	7	Lormetazepam Hy/MDB
75	–	–	–	88	84	–	69	79	71	Bromhexine
76	–	–	–	–	–	–	6	77	69	Thiambutosine
76	–	–	14	–	81	–	0	1	8	Sulfacarbamid
76	–	–	–	5	83	–	0	65	29	Glymidine
76	51	55	62	12	88	88	–	–	–	Tolbutamide
76	–	0	–	25	38	81	0	1	4	Fenoterol
76	–	–	–	41	69	66	0	41	8	Domperidone
76	43	65	51	43	–	–	–	–	–	Ethoxzolamide
76	31	51	47	60	79	–	–	–	–	Clorexolone
76	35	46	56	68	84	85	0	58	59	Flunitrazepam M (Nor-)
76	38	53	59	71	–	–	–	–	–	Pheneturide
76	33	25	65	73	81	84	15	68	58	Metaclazepam M (N-Nor-)
76	53	58	62	74	85	93	0	62	67	Ethyl Loflazepate
76	–	–	–	75	65	–	14	54	61	Butanilicaine
76	56	50	67	75	78	86	24	69	63	Fludiazepam
76	55	32	69	75	82	83	12	73	65	Camazepam
76	49	59	58	76	86	87	5	–	–	Ethinamate
76	–	–	–	80	80	–	9	68	69	Buprenorphine
76	–	–	–	84	72	89	17	60	60	Penfluridol
76	–	–	–	85	55	56	62	63	64	Amfepramone/Diethylpropion
76	–	–	–	86	79	87	54	78	65	Cinnarizine
76	–	–	–	87	80	88	61	79	70	Meclozine
76	–	–	–	87	90	–	–	–	–	Mesuximide
77	–	–	–	–	–	–	53	68	56	Pipoxolan
77	52	62	57	9	90	–	0	–	–	Oxyphenbutazone/Phenylbutazone M (Hydroxy-)
77	–	–	–	57	74	77	0	59	50	Ergocristine
77	18	60	26	59	–	–	–	–	–	Cyclothiazide
77	26	42	52	63	89	93	0	52	56	Budesonide
77	35	68	–	65	–	–	–	–	–	Tybamate
77	–	–	–	67	46	–	26	42	22	Meclofenoxate
77	53	54	65	71	78	95	15	68	66	Oxazolam
77	35	45	56	73	85	92	5	58	60	Tetrazepam M (Nor-)
77	–	–	–	75	62	79	29	64	60	Prilocaine
77	47	35	71	79	82	84	40	73	62	Metaclazepam
77	48	50	57	82	87	91	7	60	66	Flurazepam Hy/HCFB
77	–	–	–	83	29	82	37	12	63	Phenelzine
77	64	61	66	85	86	88	9	68	69	Bromazepam Hy/ABP
77	71	65	68	85	90	91	5	69	69	Nitrazepam Hy/ANB
77	71	66	68	86	87	91	4	69	70	Flunitrazepam M Hy/ANFB
77	72	66	70	86	87	93	4	69	70	Clonazepam Hy/ANCB
77	7	63	80	86	89	90	20	79	71	Flunitrazepam M Hy/MNFB

Table 8.8. (Continued)

TLC System 7	1	2	3	4	5	6	8	9	10	[System 7: MeOH–NH4OH (100+1.5) / Silica-KOH]
78	–	–	–	59	74	77	0	58	50	Ergocryptine
78	–	3	–	60	62	69	3	51	32	Benperidol
78	–	–	–	62	62	81	3	32	53	Isoxsuprine
78	–	–	–	78	76	73	31	69	52	Clemizole
78	57	49	67	79	84	89	32	69	66	Tetrazepam
78	55	48	66	80	84	87	31	69	66	Clotiazepam
78	59	57	70	80	87	90	16	89	71	Quazepam M (2-Oxo-)
78	59	59	70	81	89	91	15	6	71	Halazepam
78	76	68	74	86	89	93	11	75	70	Flurazepam Hy/ACFB
78	76	69	76	86	90	91	12	76	71	Oxazepam Hy/ACB
78	77	70	76	87	89	93	11	76	70	Lorazepam Hy/ADB
79	–	–	–	–	–	–	1	37	50	Orthocaine
79	19	38	39	55	–	–	–	–	–	Clopamide
79	78	68	76	65	87	–	–	–	–	Phenylbutazone
79	–	–	–	82	82	89	11	79	71	Phenothiazine
79	–	–	–	83	88	92	64	84	74	Phenanthrene
79	74	73	80	85	77	93	44	77	73	Quazepam Hy/CFTB
79	82	72	82	88	89	93	53	81	71	Diazepam Hy/MACB
80	–	–	–	–	–	–	7	78	72	Tiocarlide
80	–	–	–	–	–	–	46	74	62	Dioxaphetyl Butyrate
80	–	–	–	–	80	83	58	77	69	Phenadoxone
80	15	60	23	14	88	–	–	–	–	Trichlormethiazide
80	–	–	–	58	83	85	0	55	6	Flubendazole
80	–	–	–	75	78	80	9	61	–	Econazole
80	83	74	82	88	92	95	58	82	74	Prazepam Hy/CCB
81	–	–	–	–	–	–	56	67	65	Phenampromide
82	–	–	–	–	–	–	1	28	41	Diazoxide
82	14	64	23	33	–	–	–	–	–	Diclofenamide
82	11	34	32	52	87	–	0	–	–	Chlorphenesin Carbamate
82	–	–	–	88	83	90	45	75	64	Flunarizine
82	81	74	81	89	94	95	49	81	75	Halazepam Hy/TCB
83	–	–	–	58	54	–	0	21	27	Apomorphine
83	70	69	74	83	81	93	11	71	69	Haloxazolam Hy/ABFB
83	–	–	–	83	84	–	–	–	–	Crotamiton
83	–	–	–	86	85	90	53	77	66	Clocinizine
83	75	73	80	86	86	91	48	77	72	Pinazepam Hy/CPB
83	–	–	–	86	93	97	44	78	75	Proclonol
84	–	–	–	–	–	–	1	20	25	Salinazid
84	0	0	2	2	85	59	0	1	2	Metamizol/Dipyrone
84	34	46	57	68	83	87	3	56	60	Clorazepic Acid
84	91	93	94	99	81	87	40	99	99	Azinphos-methyl
85	4	31	18	5	84	–	–	–	–	Acetazolamide
85	54	61	56	33	88	90	0	51	47	Chlorzoxazone
86	91	92	95	99	80	88	55	99	99	Methidathion
87	–	–	–	–	–	–	0	41	5	Glipizide
88	–	–	–	2	87	46	0	10	10	Dimenhydrinate

Table 8.8. (Continued)

TLC System 7	1	2	3	4	5	6	8	9	10	[System 7: MeOH–NH4OH (100+1.5) / Silica-KOH]
90	–	–	–	47	90	–	2	67	70	Buclosamide
91	–	–	–	23	91	–	–	–	–	Niclosamide
95	–	–	–	26	96	98	0	10	90	Closantel

8.9 hR$_f$ Data in TLC System 8 in Ascending Numerical Order vs. Substance Names

TLC System 8	1	2	3	4	5	6	7	9	10	[System 8: Cych-Tol-DEA (75+15+10) / Silica-KOH]
0	–	–	–	–	–	–	0	0	0	Diminazene Aceturate
0	–	–	–	–	–	–	1	1	3	Benserazide
0	–	–	–	–	–	–	1	0	0	Betanidine
0	–	–	–	–	–	–	1	0	0	Debrisoquine
0	–	–	–	–	–	–	1	0	0	Hydroxystilbamidine
0	–	–	–	–	–	–	1	1	0	Pentapiperide
0	–	–	–	–	–	–	1	0	1	Stilbamidine
0	–	–	–	–	–	–	2	0	0	Aminophylline
0	–	–	–	–	–	–	2	0	0	Buformin
0	–	–	–	–	–	–	2	0	0	Clamoxyquine
0	–	–	–	–	–	–	3	1	1	Chlorproguanil
0	–	–	–	–	–	–	3	0	0	Guanoclor
0	–	–	–	–	–	–	3	0	0	Phenformin
0	–	–	–	–	–	–	13	5	0	Codeine M (Nor-)
0	–	–	–	–	–	–	43	40	40	Furaltadone
0	–	–	–	–	–	–	44	47	59	Furazolidone
0	–	–	–	–	–	–	45	1	4	Practolol
0	–	–	–	–	–	–	46	0	0	Pseudomorphine
0	–	–	–	–	–	–	51	6	3	Broxyquinoline
0	–	–	–	–	–	–	51	3	0	Minoxidil
0	–	–	–	–	–	–	56	11	12	Metopimazine
0	–	–	–	–	–	–	59	21	17	Rolicyprine
0	–	–	–	–	–	–	64	30	33	Norharman
0	–	–	–	–	–	–	65	36	55	Ethionamide
0	–	–	–	–	–	–	65	56	67	Ethoxazene
0	–	–	–	–	–	–	66	53	23	Naftazone
0	–	–	–	–	–	–	69	56	68	Clefamide
0	–	–	–	–	–	–	70	62	64	Norbormide
0	–	–	–	–	–	–	87	41	5	Glipizide
0	–	–	–	–	65	68	63	22	28	Harmine
0	–	–	0	–	69	–	5	1	1	Cytarabine
0	–	–	–	–	73	–	59	0	0	Isoetarine
0	–	–	14	–	81	–	76	1	8	Sulfacarbamid
0	–	–	–	0	0	–	–	–	–	Atropine Methonitrate
0	–	–	–	0	0	0	0	0	0	Diquat Dibromide
0	–	–	–	0	0	0	0	0	0	Paraquat
0	–	–	–	0	0	0	1	0	0	Suxamethonium Chloride
0	–	–	–	0	0	1	0	0	0	Hexamethonium
0	–	–	–	0	0	1	0	0	0	Pentolinium
0	–	–	–	0	0	2	0	0	0	Decamethonium
0	–	–	–	0	0	2	0	0	0	Gallamine
0	–	–	–	0	0	11	0	0	0	Tubocurarine
0	–	–	–	0	0	12	0	0	0	Homatropine Methylbromide
0	–	–	–	0	0	15	1	0	0	Pancuronium Bromide

Table 8.9. (Continued)

TLC System 8	1	2	3	4	5	6	7	9	10	[System 8: Cych-Tol-DEA (75+15+10) / Silica-KOH]
0	–	–	–	0	0	29	0	1	0	Cetrimide
0	–	–	–	0	0	29	0	1	0	Cetylpyridinium
0	–	–	–	0	1	–	15	1	2	Amidefrine
0	–	–	–	0	1	25	2	0	0	Thiazinamium Metilsulfate
0	–	–	–	0	3	76	1	0	0	Guanoxan
0	–	–	–	0	4	23	0	0	0	Carbachol
0	–	–	–	0	16	11	–	0	–	Betaine
0	–	–	–	0	16	11	17	0	0	Cocaine M/Ecgonine
0	–	0	–	0	27	–	–	–	–	Lisinopril
0	–	–	–	0	30	2	–	0	0	Codeine M (-6-glucuronide)
0	–	–	–	0	34	14	52	3	0	Narceine
0	–	0	–	0	44	–	–	–	–	Flumazenil M/RO 15-3890
0	–	–	–	0	53	8	1	0	0	Obidoxime Chloride
0	0	0	0	0	70	16	58	0	0	Lysergic Acid
0	–	–	–	0	80	1	5	0	0	Psilocybine
0	–	–	–	0	81	–	–	–	–	Strophanthin
0	–	–	–	1	1	13	2	–	–	Neostigmine
0	–	–	–	1	2	18	1	0	0	Thiamine Hydrochloride
0	–	–	–	1	2	36	3	1	0	Oxyphenonium Bromide
0	–	–	–	1	3	18	2	0	0	Methscopolamine Bromide
0	–	–	–	1	3	30	1	2	0	Guanethidine
0	–	–	–	2	3	–	5	–	–	Emepronium Bromide
0	–	–	–	2	3	41	8	0	0	Butylscopolammonium Bromide,N-
0	–	–	–	2	4	66	2	4	0	Benzalkonium Chloride
0	–	–	–	2	22	14	21	1	3	Cocaine M/N-Benzoylecgonine
0	0	0	2	2	85	59	84	1	2	Metamizol/Dipyrone
0	–	0	–	2	87	–	–	–	–	Clenbuterol M #2
0	–	–	–	2	87	46	88	10	10	Dimenhydrinate
0	–	–	–	3	3	41	5	5	0	Isopropamide Iodide
0	–	–	–	3	43	–	0	0	0	Levarterenol/Norepinephrine
0	–	–	–	3	44	–	–	–	–	Pirenzepine M (N-Desmethyl-)
0	–	–	–	4	3	31	4	4	0	Propantheline Bromide
0	–	–	–	5	4	39	2	4	0	Fenpiverinium
0	–	–	–	5	83	–	76	65	29	Glymidine
0	–	–	–	5	85	–	–	–	–	Glisoxepide
0	1	7	7	6	86	–	–	–	–	Furosemide
0	–	19	–	6	87	–	–	–	–	Clenbuterol M #3
0	–	–	–	7	0	3	13	0	0	Histamine
0	–	–	–	7	5	44	17	0	0	Morphine M (Nor-)
0	52	50	66	7	86	–	–	–	–	Tolazamide
0	–	–	–	8	19	–	–	–	–	Clozapine M (N-Oxyde)
0	0	0	7	8	82	72	75	2	0	Cinchophen
0	–	–	–	9	84	–	–	–	–	Gliclazide
0	52	62	57	9	90	–	77	–	–	Oxyphenbutazone/Phenylbutazone M (Hydroxy-)
0	38	43	49	10	87	88	72	–	–	Chlorpropamide

Table 8.9. (Continued)

TLC System 8	1	2	3	4	5	6	7	9	10	[System 8: Cych-Tol-DEA (75+15+10) / Silica-KOH]
0	30	30	57	11	90	–	–	–	–	Glibenclamide/Glyburide
0	–	–	–	13	3	–	0	1	0	Epinephrine
0	38	64	36	13	93	–	–	–	–	Xipamide
0	–	–	–	14	3	–	10	3	1	Azacyclonol
0	–	–	–	15	75	67	59	8	5	Pyridoxine Hydrochloride
0	–	–	–	16	27	20	41	9	0	Clomipramine M (N-Oxid)
0	52	48	60	16	92	92	–	–	–	Acenocoumarol
0	–	–	–	17	11	–	–	–	–	Octopamine
0	–	50	–	17	93	–	–	–	–	Chlorphenprocoumon,p-
0	–	–	–	18	7	79	3	1	1	Proguanil/Chlorguanide
0	–	–	–	18	23	16	37	9	0	Mianserin M (N-Oxide)
0	–	–	–	20	18	23	37	9	1	Morphine
0	–	–	–	21	14	69	40	1	3	Isoprenaline
0	–	–	–	22	14	–	45	2	2	Atenolol
0	–	–	0	24	6	74	24	1	2	Amiloride
0	3	1	29	24	75	71	63	29	8	Triazolam M (4-Hydroxy-)
0	–	–	–	25	5	7	16	17	1	Brucine
0	–	0	–	25	38	81	76	1	4	Fenoterol
0	–	–	–	25	70	59	48	12	12	Diprophylline/Propyphylline
0	–	–	–	26	96	98	95	10	90	Closantel
0	–	–	–	27	53	55	54	9	12	Cimetidine
0	–	–	–	28	81	–	–	–	–	Clindamycin
0	–	1	–	29	32	–	–	–	–	Labetalol
0	–	–	70	30	40	88	56	5	9	Clioquinol
0	32	34	44	30	94	98	–	–	–	Hexachlorophene
0	1	1	25	31	62	50	64	–	–	Bromazepam M (N(py)-Oxide)
0	–	–	–	33	10	34	35	1	1	Bufotenine
0	–	–	–	33	13	71	47	3	6	Acebutolol
0	–	–	–	33	62	60	57	12	8	Ergometrine
0	–	–	–	33	69	63	55	37	12	Colchicine
0	54	61	56	33	88	90	85	51	47	Chlorzoxazone
0	–	–	–	34	17	–	38	–	–	Sulpiride
0	–	–	–	35	56	–	53	41	11	Demecolcine
0	–	–	–	36	57	51	60	19	7	Lysergamide
0	–	–	–	36	58	–	–	–	–	Nifenazone
0	–	11	–	36	81	81	60	23	40	Pemoline
0	11	31	34	36	86	90	69	31	48	Chloramphenicol
0	–	–	–	38	20	–	–	–	–	Dibenzepin M (Dinor-,5-desmethyl-)
0	–	6	–	38	66	–	–	–	–	Etofylline
0	–	–	–	38	69	–	–	–	–	Benzquinamide M (Desacetyl-,N-deethyl)
0	–	–	0	38	70	–	55	4	8	Ambazone
0	–	–	–	39	18	–	–	–	–	Propafenone M (5-OH)
0	0	6	26	39	73	61	61	24	28	Bromazepam M (3-Hydroxy-)
0	–	–	–	40	60	–	–	–	–	Dihydroergotamine M (8′-OH)
0	3	3	33	40	68	60	69	32	42	Chlordiazepoxide M (Nor-)
0	–	–	–	40	68	66	54	21	27	Nicotinamide

Table 8.9. (Continued)

TLC System 8	1	2	3	4	5	6	7	9	10	[System 8: Cych-Tol-DEA (75+15+10) / Silica-KOH]
0	–	–	–	41	69	–	62	14	12	Methylergometrine
0	–	–	–	41	69	66	76	41	8	Domperidone
0	15	24	42	41	81	83	63	35	51	Demoxepam/Chlordiazepoxide M
0	–	27	–	41	87	–	–	–	–	Methylprednisolon
0	7	11	30	42	79	82	73	22	45	Clonazepam M (7-Acetamido-)
0	–	–	–	43	14	59	18	0	0	Dopamine
0	–	–	–	43	41	52	65	17	14	Pipamazine
0	–	–	–	44	52	45	53	32	12	Pilocarpine
0	7	12	24	44	80	83	71	20	49	Nitrazepam M (7-Acetamido-)
0	–	–	–	44	82	87	68	37	42	Dipyridamole
0	–	–	20	45	45	59	55	22	12	Trimethoprim
0	15	32	26	45	77	–	–	–	–	Paracetamol
0	22	35	42	45	82	82	56	40	51	Oxazepam/M of Chlordiazepoxide, Clorazepic Acid, Camazepam, Diazepam, Halazepam, Ketazolam, Medazepam, Oxazolam, Prazepam
0	–	28	–	45	86	–	–	–	–	Hydrocortisone
0	–	9	–	46	81	–	–	–	–	Metaclazepam M (Dinor-)
0	–	–	–	49	63	–	–	–	–	Ketanserin M/Ketanserinol
0	–	–	–	49	87	–	52	1	3	Dobutamine
0	15	28	35	49	88	89	67	30	38	Quazepam M (3-Hydroxy-N-dealkyl-2-oxo-)
0	38	55	43	50	83	84	–	–	–	Salicylamide
0	–	–	–	51	54	–	–	–	–	Astemizole M (Nor-)
0	19	11	41	51	78	76	73	40	49	Flunitrazepam M (7-Acetamido-)
0	11	20	40	52	76	77	73	30	47	Clonazepam M (7-Amino-)
0	11	34	32	52	87	–	82	–	–	Chlorphenesin Carbamate
0	10	20	30	54	77	79	71	30	46	Nitrazepam M (7-Amino-)
0	11	19	35	54	78	75	72	34	47	Flunitrazepam M (7-Aminonor-)
0	9	36	32	56	63	87	75	32	58	Meprobamate
0	–	–	–	57	74	77	77	59	50	Ergocristine
0	–	–	–	58	54	–	83	21	27	Apomorphine
0	–	–	–	58	83	85	80	55	6	Flubendazole
0	–	–	–	59	71	–	65	54	34	Bamifylline
0	–	–	–	59	74	77	78	58	50	Ergocryptine
0	–	–	–	60	80	84	65	59	49	Mebendazole
0	10	35	35	60	82	85	–	33	56	Mebutamate
0	–	–	–	62	66	–	–	–	–	Cisapride
0	26	42	52	63	89	93	77	52	56	Budesonide
0	–	–	–	64	64	67	75	47	29	Ergosine
0	35	46	53	64	84	86	68	36	55	Nitrazepam
0	32	43	54	64	84	90	75	52	61	Clobazam M (Nor-)
0	35	45	56	67	85	87	72	53	61	Clonazepam
0	35	46	56	68	84	85	76	58	59	Flunitrazepam M (Nor-)
0	–	–	–	70	86	–	67	51	62	Benorilate
0	53	58	62	74	85	93	76	62	67	Ethyl Loflazepate
0	39	45	63	75	78	84	73	66	59	Cloxazolam

Table 8.9. (Continued)

TLC System 8	1	2	3	4	5	6	7	9	10	[System 8: Cych-Tol-DEA (75+15+10) / Silica-KOH]
0	–	–	–	76	3	–	2	–	–	Methanthelinium Bromide
0	–	–	0	80	93	48	1	0	0	Metformin
0	–	–	–	81	71	–	–	–	–	Doxazosin
0	81	74	82	86	87	–	–	–	–	Ethchlorvynol
1	–	–	–	–	–	–	1	1	1	Picloxydine
1	–	–	–	–	–	–	1	1	1	Propamidine
1	–	–	–	–	–	–	4	1	1	Acefylline Piperazine
1	–	–	–	–	–	–	18	3	0	Codeine M (N-Oxide)
1	–	–	–	–	–	–	25	0	10	Nordefrin Hydrochloride
1	–	–	–	–	–	–	33	7	24	Tuaminoheptane
1	–	–	–	–	–	–	40	10	2	Cytisine
1	–	–	–	–	–	–	42	2	24	Ethylnorepinephrine
1	–	–	–	–	–	–	44	1	2	Cycloserine
1	–	–	–	–	–	–	53	19	8	Cephaeline
1	–	–	–	–	–	–	54	44	53	Niridazole
1	–	–	–	–	–	–	57	43	16	Inositol Nicotinate
1	–	–	–	–	–	–	65	21	15	Methysergide
1	–	–	–	–	–	–	65	3	6	Protokylol
1	–	–	–	–	–	–	66	62	48	Ergotoxine
1	–	–	–	–	–	–	68	35	65	Brocresine
1	–	–	–	–	–	–	79	37	50	Orthocaine
1	–	–	–	–	–	–	82	28	41	Diazoxide
1	–	–	–	–	–	–	84	20	25	Salinazid
1	–	–	0	–	0	–	1	0	0	Pentamidine
1	–	–	–	–	1	4	5	1	0	Piperazine
1	–	–	–	–	77	–	66	38	57	Protionamide
1	–	–	–	2	60	75	49	1	1	Methyldopa
1	–	–	–	3	44	–	36	3	3	Penicillamine
1	–	–	–	6	2	18	2	3	0	Oxyphencyclimine
1	–	9	–	11	74	66	75	30	11	Theophylline
1	–	–	–	12	8	67	33	1	0	Phenylephrine
1	–	–	–	18	13	76	42	1	24	Metaraminol
1	–	–	–	18	21	77	48	3	6	Orciprenaline
1	–	–	–	20	14	–	42	1	1	Nadolol
1	–	0	–	20	16	74	46	1	4	Salbutamol/Albuterol
1	–	–	–	21	18	77	47	1	5	Terbutaline
1	–	–	–	22	14	–	23	2	5	Heptaminol
1	–	–	–	26	6	–	23	2	11	Tryptamine
1	3	2	26	26	73	69	67	25	7	Alprazolam M (4-Hydroxy-)
1	5	4	37	28	76	76	68	35	13	Brotizolam M (6-Hydroxy-)
1	–	–	18	29	55	49	47	11	20	Isoniazid
1	–	–	–	30	19	–	53	3	5	Sotalol
1	–	–	13	30	50	65	51	8	4	Triamterene
1	–	–	–	32	8	75	33	4	1	Clomipramine M (8-Hydroxy-nor-)
1	–	–	–	32	57	59	59	23	29	Nalorphine
1	–	–	–	33	6	–	32	3	2	Phentolamine

Table 8.9. (Continued)

TLC System 8	1	2	3	4	5	6	7	9	10	[System 8: Cych-Tol-DEA (75+15+10) / Silica-KOH]
1	–	–	–	34	12	–	–	–	–	Propafenone M (N-Desmethyl)
1	–	4	–	34	59	54	53	31	21	Theobromine
1	–	–	–	34	80	–	9	1	1	Oxymetazoline
1	49	64	56	36	86	–	–	–	–	Reposal
1	–	–	–	37	7	–	45	2	3	Hydroxychloroquine
1	4	4	37	37	76	76	72	44	18	Alprazolam M (α-Hydroxy-)
1	–	–	–	38	20	–	–	–	–	Dibenzepin M (Nor-,5-desmethyl-)
1	–	–	–	39	17	33	49	5	9	Procainamide
1	–	–	–	40	16	–	–	–	–	Tiapride
1	3	1	36	40	26	15	40	48	5	Loprazolam
1	1	0	28	40	49	39	63	36	2	Adinazolam M (Nor-)
1	–	–	–	41	19	–	–	–	–	Clozapine M (Nor-)
1	–	–	–	41	80	–	62	52	59	Thiamazole
1	–	–	–	42	58	–	60	28	16	Dihydroergotamine
1	–	–	–	42	70	69	69	23	17	Iproniazid
1	63	47	68	42	75	75	72	–	–	Carbimazole
1	–	–	–	43	30	–	–	–	–	Clenbuterol M #1 (Hydroxy-)
1	–	–	–	43	33	61	56	12	8	Pipamperone
1	23	41	42	43	82	82	52	36	28	Lorazepam/Lormetazepam M
1	–	–	–	44	68	64	63	34	22	Ergotamine
1	5	2	41	44	68	65	60	40	16	Triazolam
1	54	67	57	44	87	–	–	–	–	Butalbital
1	–	–	–	45	26	51	70	24	11	Clomipramine M (2-Hydroxy-)
1	–	3	–	46	81	78	67	38	40	Glafenine
1	–	0	–	47	21	–	–	–	–	Carazolol
1	7	2	40	47	67	66	67	57	14	Alprazolam
1	5	5	37	49	76	77	74	43	15	Midazolam M (4-Hydroxy-)
1	–	–	–	51	17	–	47	7	13	Metoclopramide
1	13	39	30	53	–	–	62	20	53	Styramate
1	–	–	–	56	68	–	–	–	–	Bupivacaine M1 (4-OH)
1	–	–	–	57	46	–	60	60	29	Vinblastine
1	38	37	64	57	81	80	71	59	52	Flunitrazepam M (3-Hydroxy-)
1	–	–	–	58	69	–	66	48	38	Ergoloid
1	–	–	–	59	68	74	60	47	49	Prazosin
1	–	–	–	61	73	74	70	45	42	Piritramide
1	–	–	–	62	85	–	–	–	–	Valnoctamine
1	21	21	52	63	77	74	74	55	52	Flunitrazepam M (7-Amino-)
1	–	–	–	64	56	–	–	–	–	Astemizole
1	–	–	–	68	84	80	66	63	57	Chlormezanone
1	–	–	–	70	80	–	59	50	53	Phenazopyridine
1	–	–	–	71	12	–	3	0	1	Hexoprenaline
1	–	–	–	71	79	–	68	65	68	Nifedipine
1	–	–	–	78	87	–	–	–	–	Nimodipine
1	–	–	–	80	82	–	–	–	–	Nicardipine
1	–	–	–	81	77	79	73	75	64	Rescinnamine
2	–	–	–	–	–	–	9	1	1	Dextrorphan M/Levorphanol M (Nor-)

Table 8.9. (Continued)

TLC System										[System 8:
8	1	2	3	4	5	6	7	9	10	Cych-Tol-DEA (75+15+10) / Silica-KOH]
2	–	–	–	–	–	–	35	2	11	Hydroxyamphetamine/Amphetamine M
2	–	–	–	–	–	–	57	54	34	Tozalinone
2	–	–	–	–	–	–	66	58	41	Pipobroman
2	–	–	–	–	68	64	70	25	4	Nialamide
2	23	49	33	6	87	–	–	–	–	Sulfasomizole
2	–	–	–	18	12	–	25	3	1	Dihydromorphine
2	–	–	–	22	15	74	41	2	3	Etilefrine
2	–	–	–	25	3	55	13	2	2	Tolazoline
2	–	–	–	28	6	37	27	8	1	Dothiepin M (Nor-,Sulfoxide)
2	–	–	–	37	26	–	47	9	6	Ketobemidone
2	11	17	40	39	81	–	–	–	–	Guaifenesin
2	0	0	14	40	20	60	58	18	29	Flurazepam M (Dideethyl-)
2	–	–	–	40	21	69	44	7	3	Hydroquinine
2	–	–	–	43	18	78	49	5	8	Pindolol
2	–	–	–	44	42	74	60	23	–	Tocainide
2	–	–	–	45	26	65	51	11	4	Quinine
2	7	6	41	45	78	78	72	46	31	Brotizolam M (α-Hydroxy-)
2	–	–	–	46	65	14	52	52	21	Endralazine Mesilate
2	–	–	32	46	75	70	58	36	40	Metronidazole
2	–	–	–	47	24	–	–	–	–	Trimethobenzamide
2	–	–	–	47	68	–	–	–	–	Benzquinamide M (Nor-)
2	–	–	–	47	90	–	90	67	70	Buclosamide
2	–	–	–	49	71	–	58	33	29	Proxyphylline
2	7	5	44	50	71	60	71	53	25	Estazolam
2	10	10	53	52	76	77	62	50	22	Chlordiazepoxide
2	–	–	–	55	87	89	64	43	57	Mephenesin
2	–	–	–	56	79	75	60	56	47	Carbamazepine
2	–	–	–	57	72	76	70	40	33	Harman
2	–	–	–	58	66	–	61	31	21	Pyrimethamine
2	–	–	–	58	71	73	67	48	36	Droperidol
2	19	28	54	61	81	82	74	46	55	Flurazepam M (N(1)-Hydroxyethyl-)
2	–	–	–	62	–	–	61	33	57	Amiphenazole
2	42	52	58	69	88	90	71	55	59	Quazepam M (3-Hydroxy-2-oxo-)
2	34	45	60	72	83	88	75	56	58	Flurazepam M (N(1)-Dealkyl-)/Quazepam M (N-Dealkyl-2-oxo-)
2	–	–	–	73	84	88	72	69	61	Bromocriptine
2	–	–	–	77	76	80	69	74	63	Reserpine
2	–	–	–	80	87	–	–	–	–	Nitrendipine
2	–	–	68	80	88	–	49	10	4	Procarbazine
3	–	–	–	–	–	–	58	33	28	Propoxycaine
3	–	–	–	–	–	–	65	68	69	Metisazone
3	–	–	–	–	33	32	53	25	3	Acetophenazine
3	–	–	71	6	80	–	53	8	4	Azathioprine
3	–	–	–	11	8	–	25	1	18	Synephrine/Oxedrine
3	–	–	–	18	12	14	23	9	2	Hydromorphone
3	–	–	–	18	99	–	–	–	–	Lorcainide M

Table 8.9. (Continued)

TLC System 8	1	2	3	4	5	6	7	9	10	[System 8: Cych-Tol-DEA (75+15+10) / Silica-KOH]
3	–	–	–	24	6	63	20	10	12	Mescaline
3	–	–	–	25	15	–	36	18	2	Pholcodine
3	–	–	–	27	3	52	14	6	3	Naphazoline
3	–	–	–	27	9	–	29	3	3	Pholedrine
3	–	–	–	30	11	–	38	6	1	Mesoridazine
3	–	–	–	38	23	–	55	6	0	Bamethan
3	–	–	–	39	21	–	–	–	–	Dibenzepin M (Dinor-)
3	–	–	–	40	18	–	42	34	6	Nicomorphine
3	–	–	–	43	20	–	–	–	–	Mepindolol
3	–	–	–	43	20	70	45	8	5	Hydroquinidine
3	–	–	–	48	17	–	–	–	–	Propafenone
3	–	–	–	48	25	–	54	16	15	Fenpipramide
3	15	10	55	52	59	55	52	58	25	Caffeine
3	–	–	–	53	40	59	66	32	21	Pipotiazine
3	3	4	41	53	73	72	70	52	8	Midazolam M (α-Hydroxy-)
3	–	–	–	54	44	–	55	45	14	Fenetylline
3	–	–	44	54	71	70	63	42	46	Pyrazinamide
3	–	–	–	54	86	–	49	45	17	Phanquinone
3	3	–	–	56	60	59	60	39	18	Lysergide/LSD
3	–	–	–	58	60	–	57	44	33	Nimorazole
3	21	40	30	59	75	–	–	–	–	Paracetamol M/p-Aminophenol
3	–	3	–	60	62	69	78	51	32	Benperidol
3	30	14	61	61	76	72	71	63	44	Flumazenil/RO 15-1788
3	–	–	–	62	33	83	74	14	50	Buphenine
3	–	–	–	62	62	81	78	32	53	Isoxsuprine
3	–	–	–	65	73	–	–	–	–	Ketanserin
3	34	46	57	68	83	87	84	56	60	Clorazepic Acid
3	34	45	57	69	82	83	62	55	60	Nordazepam/M of Chlordiazepoxide, Clorazepic Acid, Diazepam, Halazepam, Ketazolam, Medazepam, Oxazolam, Prazepam
3	–	–	–	71	73	82	71	60	40	Pimozide
3	–	–	0	76	12	–	30	2	2	Ethambutol
3	–	–	–	76	87	–	64	59	55	Ethenzamide
3	–	–	–	81	73	–	72	77	66	Deserpidine
4	–	–	–	–	–	–	72	77	64	Methoserpidine
4	–	–	–	–	4	–	6	2	2	Tramazoline
4	–	–	–	–	12	12	30	13	3	Methenamine
4	–	–	–	8	1	33	–	–	–	Tropine
4	–	0	–	30	12	75	44	4	33	Ephedrine M/Phenylpropanolamine
4	–	–	–	33	11	13	25	20	4	Hydrocodone
4	2	5	33	38	78	81	75	25	14	Midazolam M (α,4-Dihydroxy-)
4	–	–	–	41	53	–	–	–	–	Cotinine
4	4	3	39	42	75	74	71	49	21	Triazolam M (α-Hydroxy-)
4	18	14	50	45	66	66	65	–	–	Phenazone/Antipyrine
4	–	0	–	47	42	–	–	–	–	Zopiclone
4	–	–	–	48	74	69	56	17	22	Nicotinyl Alcohol

Table 8.9. (Continued)

TLC System 8	1	2	3	4	5	6	7	9	10	[System 8: Cych-Tol-DEA (75+15+10) / Silica-KOH]
4	–	–	–	49	30	63	51	12	6	Quinidine
4	–	–	–	51	26	51	51	17	10	Clomipramine M (8-Hydroxy-)
4	–	–	–	51	46	61	58	16	18	Periciazine
4	–	–	–	52	39	–	–	–	–	Dibenzepin M (5-Desmethyl-)
4	–	–	–	54	46	52	60	37	18	Mianserin M (8-Hydroxy-)
4	4	5	29	55	42	–	57	38	17	Clozapine
4	–	–	–	61	51	72	74	49	39	Spiperone
4	–	–	–	62	74	–	–	–	–	Oxatomide
4	–	–	–	62	78	81	71	55	52	Carnidazole
4	9	22	30	67	66	63	73	51	47	Medazepam M (Nor-)
4	–	–	–	70	81	–	–	–	–	Pirenzepine M (Desamide)
4	–	–	–	71	17	–	59	35	17	Veratrine
4	–	–	–	71	63	78	69	59	49	Fluspirilene
4	–	–	–	71	86	–	–	–	–	Ethinylestradiol
4	36	53	59	75	85	79	–	–	–	Carisoprodol
4	–	–	–	78	74	–	68	65	58	Oxypertine
4	71	66	68	86	87	91	77	69	70	Flunitrazepam M Hy/ANFB
4	72	66	70	86	87	93	77	69	70	Clonazepam Hy/ANCB
5	–	–	–	–	–	–	22	4	1	Trimetazidine
5	–	–	–	–	–	–	35	12	4	Neopine
5	–	–	–	–	–	–	40	6	5	Hordenine
5	–	–	–	–	–	–	43	4	10	Tacrine
5	–	–	–	–	–	–	52	24	21	Amopyroquine
5	–	–	–	–	–	–	59	23	37	Chloroprocaine
5	–	–	–	–	–	–	72	59	61	Cyprenorphine
5	–	–	–	–	45	55	63	26	31	Butethamine
5	–	–	–	23	7	27	1	1	1	Homatropine
5	–	–	–	24	5	28	18	3	1	Atropine/Hyoscyamine
5	–	–	–	25	10	64	30	5	1	Ephedrine
5	–	–	–	27	11	14	–	13	3	Metopon
5	–	–	–	31	6	61	26	11	14	Dimethoxyphenethylamine,3,4-
5	–	–	–	39	39	–	54	27	7	Carfenazine
5	–	–	–	45	45	49	63	23	10	Fluphenazine
5	–	–	–	47	14	48	39	9	9	Psilocin
5	15	5	53	52	72	71	72	52	27	Brotizolam
5	50	53	59	66	84	–	70	–	–	Ethosuximide
5	–	–	–	71	36	42	54	31	30	Procaine
5	35	41	57	72	82	86	73	58	56	Delorazepam
5	35	45	56	73	85	92	77	58	60	Tetrazepam M (Nor-)
5	–	–	–	76	76	88	73	68	61	Ambucetamide
5	56	62	63	76	84	87	67	57	66	Benzocaine
5	49	59	58	76	86	87	76	–	–	Ethinamate
5	37	41	54	81	80	79	72	57	57	Nitrazepam M Hy/DAB
5	41	46	55	82	80	80	74	59	61	Clonazepam Hy/DCB
5	71	65	68	85	90	91	77	69	69	Nitrazepam Hy/ANB
6	–	–	–	–	–	–	43	21	2	Dehydroemetine

Table 8.9. (Continued)

TLC System 8	1	2	3	4	5	6	7	9	10	[System 8: Cych-Tol-DEA (75+15+10) / Silica-KOH]
6	–	–	–	–	–	–	56	19	17	Piperacetazine
6	–	–	–	–	–	–	76	77	69	Thiambutosine
6	–	–	–	–	83	90	75	63	70	Butyl Aminobenzoate
6	–	–	–	24	82	21	–	28	7	Morphine-3-acetate
6	–	–	–	27	3	–	10	6	0	Mequitazine
6	–	–	–	29	7	70	33	11	12	Trimethoxyamphetamine,3,4,5-
6	–	–	–	35	21	22	33	18	3	Codeine
6	–	–	–	36	25	–	42	23	6	Viloxazine
6	–	–	–	38	35	39	54	22	7	Opipramol
6	–	–	–	39	13	–	–	–	–	Homofenazine
6	–	–	–	40	19	–	–	–	–	Dibenzepin M (Nor-)
6	–	–	–	41	20	23	41	23	8	Benzylmorphine
6	–	–	–	44	19	61	49	12	5	Cinchonine
6	–	–	–	44	24	55	49	8	6	Cinchonidine
6	–	–	–	44	25	27	46	26	8	Morphine-6-acetate/Diamorphine M
6	–	–	–	46	50	–	62	33	–	Flupentixol
6	–	–	–	47	5	66	31	7	3	Antazoline
6	–	–	–	48	49	47	55	37	18	Scopolamine
6	–	–	–	49	21	79	50	10	7	Propranolol
6	–	–	–	49	28	–	–	–	–	Flecainide
6	–	–	–	50	20	75	52	11	9	Timolol
6	–	–	–	57	52	–	–	–	–	Haloperidol M
6	9	2	51	57	68	66	69	55	22	Adinazolam
6	46	45	60	59	82	82	52	61	50	Lormetazepam
6	13	5	53	60	69	70	72	60	19	Midazolam
6	13	18	47	63	73	69	61	41	53	Bromazepam
6	–	–	–	64	66	70	63	38	52	Yohimbine
6	–	–	–	64	78	–	66	45	68	Leucinocaine
6	–	–	–	66	72	–	55	66	46	Pentifylline
6	–	–	–	67	59	–	–	–	–	Bromhexine M/Ambroxol
6	35	51	52	68	84	–	–	–	–	Bromisoval
7	–	–	–	–	–	–	65	69	36	Benzquinamide
7	–	–	–	–	–	–	67	64	69	Isobutyl 4-Aminobenzoate
7	–	–	–	–	–	–	68	45	67	Dimethocaine
7	–	–	–	–	–	–	73	61	63	Etorphine
7	–	–	–	–	–	–	80	78	72	Tiocarlide
7	–	–	–	–	75	–	67	54	53	Tiabendazole
7	–	–	–	26	5	60	13	2	2	Tetryzoline
7	–	–	–	30	5	64	13	5	3	Xylometazoline
7	–	–	–	32	8	63	–	8	2	Pethidine M (Nor-)
7	–	–	–	36	21	26	40	22	6	Ethylmorphine
7	–	–	–	42	40	40	55	29	9	Perphenazine
7	–	–	–	43	22	22	46	34	6	Thioproperazine
7	–	–	–	44	45	–	56	32	11	Clopenthixol/Zuclopenthixol/cis-Ordinol
7	–	–	–	53	46	65	63	13	13	Mazindol
7	–	–	–	60	9	7	45	8	13	Disopyramide

Table 8.9. (Continued)

TLC System 8	1	2	3	4	5	6	7	9	10	[System 8: Cych-Tol-DEA (75+15+10) / Silica-KOH]
7	–	–	–	60	72	74	72	64	63	Pentetrazol
7	48	50	57	82	87	91	77	60	66	Flurazepam Hy/HCFB
7	–	–	–	83	44	76	71	30	64	Butacaine
8	–	–	–	–	–	–	42	6	8	Profadol
8	–	–	–	–	–	–	54	49	32	Morinamide
8	–	–	–	–	–	–	62	40	37	Amodiaquine
8	–	–	–	–	–	–	66	69	60	Clonitazene
8	–	–	–	29	11	19	26	13	2	Dihydrocodeine
8	–	–	–	32	8	11	26	19	2	Strychnine
8	–	0	–	48	22	–	–	–	–	Bisoprolol
8	0	0	15	50	28	62	58	19	11	Flurazepam M (Monodeethyl-)
8	–	–	–	58	61	–	58	46	31	Morazone
8	51	47	65	62	82	82	53	59	53	Temazepam/M of Camazepam, Diazepam, Ketazolam, Medazepam
8	–	–	–	69	74	74	61	65	47	Papaverine
8	–	–	–	70	44	76	62	31	53	Clonidine
8	53	47	70	75	84	85	62	70	62	Clobazam
8	–	–	–	76	80	86	69	52	59	Fenyramidol
8	–	–	–	76	90	–	65	65	64	Beclamide
9	–	–	–	–	–	–	37	9	3	Diphenazoline
9	–	–	–	–	–	–	60	42	18	Etoxeridine
9	–	–	–	–	–	–	70	60	67	Levophenacylmorphan
9	–	–	–	47	74	–	65	66	63	Naloxone
9	–	–	–	50	14	39	40	9	6	Dimethyltryptamine,N,N-/DMT
9	–	–	–	64	53	52	56	29	31	Nomifensine
9	–	–	–	65	15	67	–	15	25	Psilocin-(eth)/CZ-74
9	–	–	–	74	52	81	70	32	22	Loperamide
9	–	–	–	75	78	80	80	61	–	Econazole
9	57	65	60	78	86	–	–	–	–	Hexapropymate
9	–	–	–	80	80	–	76	68	69	Buprenorphine
9	64	61	66	85	86	88	77	68	69	Bromazepam Hy/ABP
10	–	–	–	–	–	–	31	5	4	Metazocine
10	–	–	–	16	12	–	47	44	6	Hexobendine
10	–	–	–	33	27	36	48	37	30	Oxymorphone
10	–	–	–	38	61	–	52	7	15	Oxetacaine
10	–	–	–	44	20	74	49	8	9	Metoprolol
10	–	–	–	44	26	24	49	40	7	Tiotixene
10	–	–	–	54	57	65	68	54	19	Hydroxyzine
10	–	–	–	66	64	61	63	58	37	Trazodone
10	54	47	72	74	80	82	63	72	63	Flunitrazepam
11	–	–	–	–	5	42	26	0	24	Betazole
11	–	–	–	45	20	78	48	11	13	Oxprenolol
11	–	–	–	46	20	22	45	28	5	Dothiepin M (Sulfoxide)
11	–	–	–	46	24	–	–	–	–	Erythromycin
11	–	–	–	49	22	–	52	12	11	Alprenolol
11	–	0	–	51	22	–	–	–	–	Betaxolol

Table 8.9. (Continued)

TLC System 8	1	2	3	4	5	6	7	9	10	[System 8: Cych-Tol-DEA (75+15+10) / Silica-KOH]
11	–	–	–	51	25	60	46	31	6	Mianserin M (Nor-)
11	–	–	–	55	19	30	51	20	6	Triprolidine
11	–	1	–	56	28	–	–	–	–	Gallopamil M (Nor-)
11	–	–	–	68	80	–	61	72	52	Trimetozine
11	–	–	–	70	70	77	70	63	36	Lidoflazine
11	–	–	–	71	77	78	–	73	63	Meconin
11	–	–	–	75	42	68	68	39	58	Aconitine
11	46	50	65	75	80	90	74	66	62	Haloxazolam
11	–	–	–	76	51	75	67	27	33	Haloperidol
11	–	7	56	80	77	80	73	67	37	Miconazole
11	–	–	–	82	82	89	79	79	71	Phenothiazine
11	70	69	74	83	81	93	83	71	69	Haloxazolam Hy/ABFB
11	76	68	74	86	89	93	78	75	70	Flurazepam Hy/ACFB
11	77	70	76	87	89	93	78	76	70	Lorazepam Hy/ADB
12	–	–	–	–	–	–	44	15	6	Methyldesorphine
12	–	–	–	–	–	–	54	39	17	Etamiphyllin
12	–	0	–	44	21	–	–	–	–	Toliprolol
12	–	–	–	46	18	–	–	–	–	Fluvoxamine
12	–	–	–	47	19	–	43	42	10	Nicocodine
12	–	0	–	49	20	–	–	–	–	Bupranolol
12	–	–	–	49	34	–	–	–	–	Bupivacaine M/Mepivacaine M
12	–	–	–	53	62	–	–	–	–	Benzquinamide M (Desacetyl-)
12	–	–	–	55	41	38	55	36	18	Physostigmine
12	55	55	70	72	85	88	71	–	52	Prazepam M (3-Hydroxy-)
12	55	32	69	75	82	83	76	73	65	Camazepam
12	53	55	64	75	85	87	–	–	–	Carbromal
12	53	46	71	77	81	81	74	70	55	Nimetazepam
12	–	55	–	78	86	–	–	–	–	Mexazolam
12	–	–	–	79	60	66	73	56	51	Anileridine
12	76	69	76	86	90	91	78	76	71	Oxazepam Hy/ACB
13	–	–	–	–	–	–	19	5	15	Primaquine
13	–	–	–	–	–	–	38	11	17	Bromo-STP/Bromo-2,5-dimethoxy-4-methyl-amphetamine
13	–	1	–	43	17	–	–	–	–	Tertatolol
13	–	0	–	50	22	80	50	15	8	Penbutolol
13	–	–	–	53	19	–	54	34	12	Emetine
13	–	–	–	58	22	–	–	–	–	Clenbuterol
13	–	–	–	61	49	43	–	59	25	Orientalidine
13	–	–	–	66	43	–	56	48	28	Dimetiotazine
13	–	–	–	70	74	76	75	58	51	Bromolysergide,2-
13	–	3	–	73	50	–	–	–	–	Bromperidol
13	–	–	–	74	45	–	–	–	–	Terfenadine
13	2	5	26	77	55	76	73	–	–	Trifluperidol
13	–	–	–	79	70	72	68	70	61	Etonitazene
14	–	–	–	–	–	–	34	9	2	Racemorphan
14	–	–	–	42	10	49	35	4	3	Dextrorphan/Levorphanol

Table 8.9. (Continued)

TLC System 8	1	2	3	4	5	6	7	9	10	[System 8: Cych-Tol-DEA (75+15+10) / Silica-KOH]
14	–	–	0	46	4	14	38	4	2	Chloroquine
14	–	–	–	46	34	45	50	27	14	Phenmetrazine/Phendimetrazine M (Nor-)
14	–	–	–	61	49	43	55	57	26	Isothebaine
14	–	–	–	63	14	56	46	10	11	Diethyltryptamine,N,N-/DET
14	45	45	62	74	83	80	66	64	66	Ketazolam
14	–	–	–	75	65	–	76	54	61	Butanilicaine
15	–	–	–	–	–	–	43	5	9	Mepacrine
15	–	–	–	–	–	–	59	48	24	Dimefline
15	–	–	–	–	–	–	70	58	66	Diperodon
15	–	–	–	41	9	76	51	17	16	Dimethoxy-4-methylamphetamine,2,5-/STP/DOM
15	–	–	–	43	10	76	36	14	17	Bromo-2,5-dimethoxyamphetamine,4-/DOB
15	–	–	–	49	26	33	47	38	4	Diamorphine
15	–	–	–	54	25	42	49	19	9	Clomipramine M (10-Hydroxy-)
15	–	–	–	59	71	67	59	56	29	Nikethamide
15	–	–	–	64	43	39	57	32	16	Tetracaine/Amethocaine
15	–	–	–	65	24	74	53	13	25	Cyclazocine
15	–	–	–	71	59	57	61	64	52	Hydrastine
15	53	54	65	71	78	95	77	68	66	Oxazolam
15	33	25	65	73	81	84	76	68	58	Metaclazepam M (N-Nor-)
15	59	59	70	81	89	91	78	6	71	Halazepam
16	–	–	–	–	–	–	66	74	67	Diloxanide
16	–	–	–	–	–	–	68	75	58	Dimoxyline
16	–	–	–	64	66	–	58	58	41	Metyrapone
16	–	–	–	70	34	72	61	12	28	Pentazocine
16	–	–	–	74	50	81	68	39	49	Phenazocine
16	59	57	70	80	87	90	78	89	71	Quazepam M (2-Oxo-)
17	–	–	–	–	–	–	40	8	8	Phenglutarimide
17	–	–	–	–	–	–	52	26	5	Diethylcarbamazine
17	–	–	–	35	7	77	23	7	4	Amantadine
17	–	–	–	43	54	–	–	–	–	Maprotiline M (Nor-)
17	–	–	–	44	19	22	45	28	5	Dothiepin M (Nor-)
17	–	–	–	48	12	–	47	13	6	Pipazetate
17	–	–	–	55	25	78	40	4	9	Mexiletine
17	–	–	–	62	18	–	48	24	12	Carbocromen
17	–	–	–	75	23	–	61	35	29	Lobeline
17	–	–	–	80	82	87	74	76	66	Bisacodyl
17	–	–	–	84	72	89	76	60	60	Penfluridol
18	–	–	–	–	–	–	22	7	2	Apoatropine
18	–	–	–	–	–	–	33	17	8	Desomorphine
18	–	–	–	–	–	–	46	13	12	Eucatropine
18	–	–	–	–	–	–	57	15	20	Clorprenaline
18	–	–	–	–	–	–	67	77	68	Octaverine
18	–	–	–	–	21	38	39	24	6	Dimethoxanate
18	–	–	–	36	6	71	15	5	2	Maprotiline
18	–	–	–	38	6	69	19	7	2	Protriptyline

Table 8.9. (Continued)

TLC System										[System 8:
8	1	2	3	4	5	6	7	9	10	Cych-Tol-DEA (75+15+10) / Silica-KOH]
18	–	–	–	42	10	76	39	12	17	Methylenedioxyamphetamine,3,4-/MDA
18	–	–	–	48	14	77	44	17	8	Chlorphentermine
18	–	–	–	59	24	33	48	36	25	Oxomemazine
18	–	–	–	65	53	52	62	48	42	Levamisole
18	–	–	–	66	51	53	69	54	46	Tetramisole
18	–	–	–	78	51	80	69	39	47	Phenomorphan
19	–	–	–	40	7	71	26	11	3	Desipramine/Imipramine M
19	–	–	–	45	8	75	36	10	17	Dimethoxyamphetamine,2,5-
19	–	–	–	74	42	73	67	24	45	Levallorphan
20	–	–	–	43	12	75	43	9	18	Amphetamine/Amphetaminil M artifact
20	–	–	–	49	24	25	45	34	11	Thebacon
20	–	–	–	72	69	–	64	70	54	Doxapram
20	–	–	–	72	80	–	66	70	55	Propanidid
20	–	–	–	75	84	86	71	74	61	Isocarboxazid
20	7	63	80	86	89	90	77	79	71	Flunitrazepam M Hy/MNFB
21	–	–	–	59	39	50	–	48	12	Cocaine M (Nor-)
21	25	10	58	62	70	68	66	–	–	Aminophenazone/Amidopyrine
21	–	7	–	69	56	–	–	–	–	Gallopamil M (D517)
21	–	–	–	70	51	67	71	48	44	Dexetimide
21	–	–	–	78	72	75	64	74	64	Noscapine
21	–	–	–	78	78	81	71	78	69	Disulfiram
22	–	–	–	55	38	22	54	35	14	Dibenzepin
22	–	–	–	73	14	50	–	46	43	Cryptopine
22	61	60	64	76	84	96	–	64	69	Cholesterol
23	–	–	–	43	9	74	35	10	18	Methoxy-4,5-methylenedioxyamphetamine,2-/MMDA-2
23	–	–	–	43	9	74	73	77	69	Methoxyamphetamine,4-/PMA
23	–	–	–	54	26	25	44	43	12	Acetylcodeine
23	–	–	–	73	43	61	59	70	42	Verapamil
23	–	–	–	74	77	81	71	65	48	Metomidate
23	–	–	–	83	54	–	62	41	36	Oxybuprocaine
24	–	–	–	–	–	–	13	4	1	Deptropine
24	–	–	–	11	12	73	55	4	38	Methoxamine
24	–	20	–	39	89	–	–	–	–	Methylenedioxymethamphetamine,3,4-/MDMA/XTC
24	–	–	–	43	8	75	36	10	14	Dimethoxy-4-ethylamphetamine,2,5-/DOET
24	–	–	–	45	8	75	34	13	3	Clomipramine M (Nor-)
24	–	–	–	45	23	32	45	37	5	Thebaine
24	–	–	–	52	24	–	–	–	–	Ketotifen
24	–	4	–	71	46	–	–	–	–	Gallopamil
24	56	50	67	75	78	86	76	69	63	Fludiazepam
25	–	–	–	29	12	–	42	5	46	Pseudoephedrine M (Nor-)/Cathine
25	–	–	–	47	21	23	48	37	3	Perazine
25	–	–	–	62	30	33	50	51	39	Oxycodone
25	–	–	–	67	42	–	63	34	35	Cinchocaine/Dibucaine

Table 8.9. (Continued)

TLC System 8	1	2	3	4	5	6	7	9	10	[System 8: Cych-Tol-DEA (75+15+10) / Silica-KOH]
26	–	–	–	–	–	–	49	14	14	Dimethylphenylthiazolanimin (Dimethyl-5-phenyl-2-thiazolidinimine,3,4-
26	–	–	–	–	–	–	62	41	35	Proxymetacaine
26	–	–	–	–	–	–	69	45	67	Amprotropine
26	–	–	–	32	7	–	23	4	2	Methoxyphenamine
26	–	–	–	36	6	–	13	6	2	Benzatropine
26	–	–	–	48	11	78	46	24	12	Phentermine
26	–	–	–	48	14	–	45	21	7	Tofenacin
26	–	–	–	50	13	16	48	19	4	Carbinoxamine
26	–	–	–	63	28	–	48	24	12	Acepromazine
26	–	–	–	67	46	–	77	42	22	Meclofenoxate
26	–	–	–	73	35	46	66	55	38	Pitofenone
26	–	–	–	73	78	85	67	71	52	Etomidate
26	–	–	–	76	70	82	71	64	58	Phenoperidine
26	–	–	–	82	83	–	–	–	–	Nifedipine M
27	–	–	–	–	–	–	57	39	26	Cyclopentolate
27	–	–	–	–	–	–	62	45	11	Morpheridine
27	–	–	–	48	20	–	47	25	–	Zimeldine
27	–	–	–	65	36	46	57	45	20	Mebhydrolin
27	58	49	72	76	82	85	75	73	59	Diazepam/Ketazolam M/Medazepam M
27	78	71	78	83	87	96	74	75	76	Quazepam
28	–	–	–	–	–	–	22	11	25	Octamylamine
28	–	–	–	–	–	–	48	11	18	Butaxamine
28	–	–	–	42	9	63	31	13	5	Deoxyephedrine,(+/-)-
28	–	–	–	42	9	63	31	13	5	Methamfetamine
28	–	–	–	46	87	71	34	16	4	Amitriptyline M/Nortriptyline
28	–	–	–	52	26	–	53	37	5	Butaperazine
28	–	–	–	54	16	–	–	–	–	Encainide
28	–	–	–	54	44	–	49	28	39	Phenethylamine
28	–	–	–	69	83	79	68	67	55	Crotethamide
28	–	–	–	86	39	72	65	50	62	Ibogaine/Rotenone
29	–	–	–	–	–	–	50	25	11	Betaprodine
29	–	–	–	74	25	83	70	69	57	Cropropamide
29	–	–	–	75	62	79	77	64	60	Prilocaine
29	65	61	73	81	85	92	75	72	70	Pinazepam
30	–	–	–	–	–	–	43	22	12	Methopromazine
30	–	–	–	–	–	–	50	35	11	Alphaprodine
30	–	–	–	–	–	–	50	53	16	Myrophine
30	–	–	–	–	–	–	52	30	17	Dimazole/Diamthazole
30	–	–	–	53	27	–	51	41	8	Thiethylperazine
30	–	2	–	61	29	–	–	–	–	Melperone M (FG 5155)
30	3	3	41	71	52	45	62	48	40	Flurazepam
30	–	5	–	71	64	–	–	–	–	Buspirone
30	–	–	–	73	58	50	60	58	33	Methylpiperidyl Benzilate/JB-336
31	–	–	–	–	–	–	52	44	19	Moxisylyte
31	–	–	–	–	–	–	60	63	50	Thurfyl Nicotinate

Table 8.9. (Continued)

TLC System 8	1	2	3	4	5	6	7	9	10	[System 8: Cych-Tol-DEA (75+15+10) / Silica-KOH]
31	–	–	–	66	63	60	65	62	48	Mepivacaine
31	–	–	–	78	76	73	78	69	52	Clemizole
31	55	48	66	80	84	87	78	69	66	Clotiazepam
31	63	62	70	80	86	89	75	–	–	Glutethimide
31	66	66	80	83	76	86	73	77	68	Nimetazepam Hy/MNB
32	–	–	–	–	–	–	29	15	6	Methdilazine
32	–	–	–	66	6	68	20	10	2	Cyclopentamine
32	61	49	65	74	81	–	71	–	–	Propyphenazone
32	–	–	–	75	78	–	57	27	29	Octacaine
32	57	49	67	79	84	89	78	69	66	Tetrazepam
33	–	–	–	–	–	–	1	40	16	Alphameprodine
33	–	–	–	49	12	–	45	16	6	Brompheniramine
33	–	–	–	55	30	29	53	30	8	Trifluoperazine
33	–	–	–	59	30	–	50	32	17	Nefopam
34	–	–	–	–	–	–	25	8	2	Mephentermine
34	–	–	–	–	–	–	26	–	–	Propylhexedrine
34	–	–	–	–	–	–	33	5	3	Pentaquine
34	–	–	–	–	–	–	58	57	37	Betameprodine
34	–	–	–	–	–	–	60	47	36	Ethomoxane
34	–	–	–	55	26	26	49	37	7	Prochlorperazine
34	–	–	–	68	30	52	55	42	26	Propiomazine
34	–	–	–	82	86	–	–	–	–	Nitrendipine M
35	–	–	–	–	–	–	58	49	24	Allylprodine
35	–	–	–	46	12	21	45	18	2	Chlorphenamine/Chlorpheniramine
35	–	–	–	46	14	26	45	13	3	Pheniramine
35	–	–	–	47	10	–	42	13	6	Dimetindene
35	–	–	–	50	13	28	42	22	6	Tropacocaine
35	–	–	–	56	14	–	44	9	15	Etafedrine
35	–	–	–	66	40	70	57	41	23	Methylphenidate
35	–	–	–	74	62	59	61	53	42	Thiopropazate
35	–	–	–	80	72	69	70	71	63	Lidocaine
36	–	–	–	15	–	–	55	2	1	Dihydralazine
36	–	–	–	50	16	–	44	22	9	Benzydamine
36	–	–	–	54	49	–	–	–	–	Loxapine
36	–	–	–	62	49	41	57	51	24	Phendimetrazine
36	–	–	–	65	29	73	55	38	13	Noracymethadol
36	–	–	–	65	30	44	50	35	17	Promethazine
36	–	–	–	76	66	80	75	64	56	Articaine
36	–	–	–	77	48	–	62	67	45	Flavoxate
36	–	–	–	78	79	84	70	80	56	Methaqualone
36	64	55	72	81	84	89	65	74	63	Prazepam
36	–	–	–	88	63	77	67	64	59	Piminodine
37	–	–	–	–	–	–	59	40	42	Aletamine
37	–	–	–	–	–	–	69	60	29	Furethidine
37	–	–	–	60	34	40	52	34	11	Pethidine/Meperidine
37	–	–	–	68	23	49	46	28	8	Diphenylpyraline

Table 8.9. (Continued)

TLC System 8	1	2	3	4	5	6	7	9	10	[System 8: Cych-Tol-DEA (75+15+10) / Silica-KOH]
37	–	–	–	72	74	77	61	66	59	Methyl Nicotinate
37	–	–	–	79	68	72	63	63	64	Ketamine
37	74	77	73	80	87	95	–	–	–	Cyclandelate
37	–	–	–	83	29	82	77	12	63	Phenelzine
37	–	–	–	87	90	–	72	74	67	Famprofazone
38	–	–	–	–	–	–	51	11	64	Hydralazine
38	–	–	–	8	4	56	7	0	0	Berberine
38	–	–	–	26	1	22	2	1	0	Cotarnine
38	–	–	–	52	20	–	50	44	12	Thenalidine
38	–	–	–	62	18	35	44	30	11	Promazine
38	–	–	–	65	22	31	55	28	14	Thonzylamine
38	–	–	–	67	32	–	53	48	15	Phenyltoloxamine
38	–	–	–	83	70	78	–	73	59	Fluorofentanyl,p-
39	–	–	–	–	–	–	42	21	7	Tigloidine
39	–	–	–	–	–	–	56	44	16	Properidine
39	–	–	–	–	–	–	70	73	66	Tolycaine
39	–	–	–	37	5	70	26	13	5	Coniine
39	–	–	–	58	22	33	51	25	14	Mepyramine/Pyrilamine
39	–	–	–	58	41	67	54	33	48	Tranylcypromine
39	–	–	–	61	39	22	54	35	13	Nicotine
39	–	–	–	68	48	50	58	58	23	Mianserin
39	–	23	–	82	67	75	73	68	60	Fluanisone
40	–	–	–	59	24	–	49	43	13	Bamipine
40	–	–	–	63	33	–	56	34	18	Methadone M (Nor-)
40	47	35	71	79	82	84	77	73	62	Metaclazepam
40	–	–	–	86	32	–	63	53	49	Mebeverine
40	91	93	94	99	81	87	84	99	99	Azinphos-methyl
41	–	–	–	–	–	–	58	41	17	Trimeperidine
41	–	–	–	–	–	–	59	59	23	Clotiapine
41	–	–	–	–	32	–	52	37	29	Clofedanol
41	–	–	–	60	12	–	48	10	9	Doxylamine
41	–	–	–	60	19	–	48	16	11	Fenfluramine
41	–	–	0	63	22	–	52	28	17	Chloropyramine
41	–	–	–	64	22	35	52	30	14	Isothipendyl
41	–	–	–	65	92	96	71	79	70	Bezitramide
41	–	–	–	66	21	24	52	26	13	Methapyrilene
41	–	–	–	68	35	–	56	35	20	Nicametate
41	–	–	–	70	23	79	–	33	29	Phenylcyclohexylamine,1-
41	–	–	–	71	11	28	54	21	9	Diamocaine
41	56	40	73	78	79	83	67	74	62	Medazepam
41	–	–	–	79	80	–	69	78	67	Tetrabenazine
42	–	–	–	–	–	–	61	70	61	Nicofuranose
42	–	–	–	–	–	–	63	71	60	Benzyl Nicotinate
42	–	–	–	–	–	–	67	70	59	Clorgiline
42	–	–	–	–	19	36	53	25	12	Thenyldiamine

Table 8.9. (Continued)

TLC System 8	1	2	3	4	5	6	7	9	10	[System 8: Cych-Tol-DEA (75+15+10) / Silica-KOH]
42	–	–	–	47	10	42	33	18	6	Dextromethorphan/Racemethorphan/Levomethorphan
42	–	–	–	55	12	–	–	–	–	Aprindine M (Desethyl-)
42	–	–	–	60	12	–	50	19	8	Aminopromazine
42	–	–	–	66	27	74	54	27	17	Eucaine,β-
42	–	–	–	67	20	55	48	30	13	Thioridazine
42	–	–	–	67	35	52	57	46	14	Chlorcyclizine
42	–	–	–	69	27	48	54	43	13	Bromazine/Bromdiphenhydramine
42	–	–	–	78	34	30	70	60	55	Cinnamoylcocaine
42	–	–	–	79	72	78	73	71	60	Dextromoramide/Racemoramide/Levomoramide
42	–	–	–	80	69	79	69	73	65	Bupivacaine
42	–	–	–	87	90	92	74	84	70	Diphenoxylate
43	–	–	–	–	62	60	70	55	41	Pramoxine
43	–	–	–	59	15	29	47	23	9	Prothipendyl
43	–	–	–	66	29	–	53	35	18	Noxiptiline
43	–	–	–	75	38	–	–	–	–	Mefenorex
43	–	–	–	77	52	57	66	53	53	Benactyzine
43	–	–	–	78	70	77	70	74	58	Fentanyl
44	–	–	–	–	–	–	75	72	57	Benzethidine
44	–	–	–	65	27	48	55	33	15	Diphenhydramine
44	–	–	–	68	22	34	55	27	15	Tripelennamine
44	–	–	–	76	80	85	64	69	58	Clomethiazole
44	–	–	–	78	59	54	75	63	55	Ethylpiperidyl Benzilate/JB-318
44	74	73	80	85	77	93	79	77	73	Quazepam Hy/CFTB
44	–	–	–	86	93	97	83	78	75	Proclonol
45	–	–	–	–	–	–	46	30	10	Proheptazine
45	–	–	–	–	–	–	63	69	62	Butoxyethyl Nicotinate,2-
45	–	–	–	55	12	41	40	19	4	Ethoheptazine
45	–	–	–	61	21	–	50	25	12	Metixene
45	–	–	–	64	28	–	48	–	–	Pizotifen
45	–	–	–	64	30	50	51	44	13	Cyproheptadine
45	–	–	–	68	28	48	55	33	15	Dimenhydrinate
45	–	–	–	68	41	49	63	57	21	Phenindamine
45	–	–	–	70	25	45	49	35	17	Chlorpromazine
45	–	–	–	77	35	30	65	47	54	Cocaine
45	–	–	–	84	64	77	–	75	64	Alphamethylfentanyl
45	–	–	–	88	83	90	82	75	64	Flunarizine
46	–	–	–	–	–	–	54	42	21	Methaphenilene
46	–	–	–	–	–	–	80	74	62	Dioxaphetyl Butyrate
46	–	–	–	37	1	15	0	0	0	Hydrastinine
47	–	–	–	–	–	–	63	55	43	Methadone Intermediate/Pre-Methadone
47	–	–	–	59	16	74	47	19	8	Etilamfetamine
47	–	1	–	65	27	–	53	44	16	Pecazine/Mepazine
47	–	–	–	68	16	49	48	22	14	Pentoxyverine/Carbetapentane
47	–	–	–	70	29	–	53	36	17	Chlorphenoxamine

Table 8.9. (Continued)

TLC System 8	1	2	3	4	5	6	7	9	10	[System 8: Cych-Tol-DEA (75+15+10) / Silica-KOH]
47	–	–	–	75	32	49	54	35	22	Triflupromazine
47	–	–	–	76	32	49	57	38	46	Levomepromazine/Methotrimeprazine
47	–	–	–	86	88	–	72	78	67	Fenbutrazate
48	–	–	–	–	–	–	64	59	43	Benethamine
48	–	–	–	–	–	–	70	69	63	Mebenazine
48	–	–	–	–	–	–	71	76	62	Carbifene
48	–	–	–	63	24	45	51	37	13	Doxepin
48	–	–	–	67	21	47	48	23	13	Imipramine
48	–	–	–	68	25	49	55	33	16	Orphenadrine
48	–	–	–	79	30	91	70	40	20	Hexetidine
48	–	–	–	80	41	–	–	–	–	Lorcainide
48	71	69	81	83	77	90	75	77	69	Fludiazepam Hy/CFMB
48	75	73	80	86	86	91	83	77	72	Pinazepam Hy/CPB
49	–	–	–	–	–	–	28	3	3	Conessine
49	–	–	–	–	–	–	47	34	16	Iprindole/Glycophene
49	–	–	–	–	–	–	60	40	25	Dyclonine
49	–	–	–	58	18	49	46	25	9	Clemastine
49	–	–	–	68	40	52	57	41	16	Cyclizine
49	–	–	–	70	27	41	51	42	16	Dothiepin
49	–	–	–	77	47	–	66	–	–	Captodiame
49	–	–	–	82	61	–	–	–	–	Prenoxdiazine/Tibexin
49	81	69	81	87	82	93	75	81	7	Lormetazepam Hy/MDB
49	81	74	81	89	94	95	82	81	75	Halazepam Hy/TCB
50	–	–	–	–	–	–	37	11	35	Rimantadine
50	–	–	–	–	–	–	54	32	17	Embramine
50	–	–	–	–	–	–	62	41	21	Dimethylthiambutene
50	–	–	–	69	27	51	51	32	15	Amitriptyline
50	–	–	–	73	25	–	–	–	–	Melperone
51	–	–	–	–	–	–	16	2	4	Mecamylamine
51	–	–	–	–	–	–	70	65	64	Aminoxytriphene
51	–	–	–	–	–	–	71	72	65	Diampromide
51	–	–	–	67	19	–	50	22	19	Oxeladine
51	–	–	–	74	34	51	56	51	25	Chlorprothixene
52	–	–	–	–	–	–	67	68	62	Tiletamine
52	–	–	–	65	27	–	–	–	–	Melitracene
52	–	–	–	68	26	–	51	32	15	Tolpropamine
52	–	–	–	78	43	–	64	41	35	Naftidrofuryl
52	–	–	–	78	78	–	74	79	71	Broxaldine
53	–	–	–	–	–	–	77	68	56	Pipoxolan
53	–	–	–	8	88	–	68	5	67	Azapropazone
53	–	–	–	72	26	54	51	34	18	Clomipramine
53	–	–	–	76	21	56	55	37	27	Piperocaine
53	9	8	41	83	45	51	67	60	51	Flurazepam Hy/DCFB
53	–	–	–	86	85	90	83	77	66	Clocinizine
53	82	72	82	88	89	93	79	81	71	Diazepam Hy/MACB
54	–	–	–	17	99	–	33	4	63	Pseudoephedrine

Table 8.9. (Continued)

TLC System 8	1	2	3	4	5	6	7	9	10	[System 8: Cych-Tol-DEA (75+15+10) / Silica-KOH]
54	–	–	–	71	25	66	54	37	18	Pyrrobutamine
54	–	–	–	77	32	46	58	39	31	Trimeprazine/Alimemazine
54	–	–	–	82	55	57	–	60	62	Eucaine,α-
54	–	–	–	86	79	87	76	78	65	Cinnarizine
55	–	–	–	–	–	–	58	36	25	Cyclomethycaine
55	–	–	–	–	–	–	61	46	28	Quinisocaine
55	–	–	–	–	–	–	64	38	36	Betacetylmethadol
55	–	–	–	82	54	52	69	81	55	Piperidolate
55	–	–	–	83	49	–	64	60	51	Adiphenine
55	–	–	–	84	43	85	68	68	56	Prenylamine
55	91	92	95	99	80	88	86	99	99	Methidathion
56	–	–	–	–	–	–	58	43	38	Alphamethadol
56	–	–	–	–	–	–	61	45	51	Difenidol
56	–	–	–	–	–	–	81	67	65	Phenampromide
56	–	–	–	77	38	–	59	52	43	Benzoctamine
56	–	–	–	82	30	–	60	52	35	Clomifene
57	–	–	–	–	–	–	59	42	28	Ethylmethylthiambutene
57	–	–	–	–	–	–	66	67	64	Amolanone
57	–	–	–	59	8	–	41	8	6	Perhexiline
57	–	–	–	77	33	54	58	51	39	Diethazine
57	–	–	–	78	14	–	70	67	56	Azapetine
57	–	–	–	86	33	86	70	59	68	Clofazimine
58	–	–	–	–	–	–	67	62	49	Diethylaminoethyl Diphenylpropionate
58	–	–	–	–	–	–	74	45	43	Isomethadone
58	–	–	–	–	80	83	80	77	69	Phenadoxone
58	–	24	–	75	25	–	–	–	–	Bornaprine
58	–	–	–	81	45	–	68	54	55	Isoaminile
58	–	–	–	82	50	63	68	55	54	Propoxyphene/Dextropropoxyphene/Levopropoxyphene
58	83	74	82	88	92	95	80	82	74	Prazepam Hy/CCB
59	–	–	–	–	–	–	63	67	61	Piperoxan
59	–	–	–	–	–	–	68	50	38	Norpipanone
59	–	–	–	77	16	60	48	20	27	Methadone/Levomethadone
59	–	–	–	81	19	79	54	38	39	Pipradrol
59	–	–	–	81	48	56	69	57	47	Butetamate
60	–	–	–	–	–	–	61	52	42	Etymemazine
60	–	–	–	–	–	–	65	58	52	Trifluomeprazine
60	–	–	–	–	–	–	69	66	56	Proadifen
60	–	–	–	–	–	–	73	67	63	Amylocaine
61	–	–	–	–	–	–	59	48	38	Butriptyline
61	–	–	–	–	–	–	70	43	43	Diethylthiambutene
61	–	–	–	–	–	–	75	83	72	Buclizine
61	–	–	–	87	80	88	76	79	70	Meclozine
62	–	–	–	–	–	–	74	80	73	Bialamicol
62	–	–	–	10	80	88	5	0	0	Phencyclidine interm./1-Piperidino-1-cyclohexanecarbonitrile/PCC

Table 8.9. (Continued)

TLC System 8	1	2	3	4	5	6	7	9	10	[System 8: Cych-Tol-DEA (75+15+10) / Silica-KOH]
62	–	–	–	74	20	68	48	31	23	Procyclidine
62	–	–	–	77	21	–	54	34	30	Fencamfamine
62	–	–	–	80	18	–	52	19	26	Terodiline
62	–	–	–	80	36	56	59	54	37	Trimipramine
62	–	–	–	82	54	64	72	68	55	Amiodarone
62	–	–	–	84	38	62	57	45	50	Alphacetylmethadol
62	–	–	–	85	55	56	76	63	64	Amfepramone/Diethylpropion
62	–	–	–	87	72	73	–	74	64	Phenylcyclohexyl)morpholine,4-(1-/PCM
63	–	–	–	76	20	–	–	–	–	Aprindine
63	–	–	–	78	30	73	56	37	26	Volazocine
63	37	51	55	79	78	–	67	–	–	Sulfadoxine
63	–	–	–	87	84	97	73	76	68	Phenoxybenzamine
64	–	–	–	83	31	55	67	47	66	Profenamine/Ethopropazine
64	–	–	–	83	88	92	79	84	74	Phenanthrene
65	–	–	–	–	–	–	66	39	38	Alverine
65	–	–	–	79	21	78	–	27	37	Eticyclidine/PCE/N-Ethyl-1-phenylcyclohexyl-amine
66	–	–	–	79	15	65	–	25	26	Rolicyclidine/PHP/1-(1-phenylcyclo-hexyl)pyrrolidine
66	–	–	–	82	27	77	55	44	36	Prozapine
66	–	–	–	83	43	75	68	61	59	Trihexyphenidyl/Benzhexol
67	–	–	–	–	–	–	66	61	60	Cycrimine
67	–	–	–	–	52	–	68	64	54	Dicycloverine/Dicyclomine
67	–	–	–	40	4	10	5	3	5	Sparteine
67	–	–	–	79	22	–	50	32	25	Prolintane
67	–	–	–	87	27	72	66	33	70	Dipipanone
67	–	–	–	87	60	–	73	70	70	Benzphetamine
68	–	–	–	–	–	–	24	3	10	Pempidine
68	–	–	–	83	45	–	64	64	64	Biperiden
68	–	–	–	86	23	75	55	36	52	Diisopromine
69	–	–	–	88	84	–	75	79	71	Bromhexine
72	–	–	–	87	31	78	–	47	68	Methyl-1-phenylcyclohexyl)piperidine,1-(4-
73	–	–	–	84	23	69	59	35	66	Phencyclidine
73	–	–	–	86	38	70	–	54	68	Tenocyclidine/TCP/1-[1-(2-Thienyl)cyclohexyl] piperidine

8.10 hR_f Data in TLC System 9 in Ascending Numerical Order vs. Substance Names

TLC System 9	1	2	3	4	5	6	7	8	10	[System 9: Chl–MeOH (90+10) / Silica-KOH]
0	–	–	–	–	–	–	0	0	0	Diminazene Aceturate
0	–	–	–	–	–	–	1	0	0	Betanidine
0	–	–	–	–	–	–	1	0	0	Debrisoquine
0	–	–	–	–	–	–	1	0	0	Hydroxystilbamidine
0	–	–	–	–	–	–	1	0	1	Stilbamidine
0	–	–	–	–	–	–	2	0	0	Aminophylline
0	–	–	–	–	–	–	2	0	0	Buformin
0	–	–	–	–	–	–	2	0	0	Clamoxyquine
0	–	–	–	–	–	–	3	0	0	Guanoclor
0	–	–	–	–	–	–	3	0	0	Phenformin
0	–	–	–	–	–	–	25	1	10	Nordefrin Hydrochloride
0	–	–	–	–	–	–	46	0	0	Pseudomorphine
0	–	–	0	–	0	–	1	1	0	Pentamidine
0	–	–	–	–	5	42	26	11	24	Betazole
0	–	–	–	–	73	–	59	0	0	Isoetarine
0	–	–	–	0	0	0	0	0	0	Diquat Dibromide
0	–	–	–	0	0	0	0	0	0	Paraquat
0	–	–	–	0	0	0	1	0	0	Suxamethonium Chloride
0	–	–	–	0	0	1	0	0	0	Hexamethonium
0	–	–	–	0	0	1	0	0	0	Pentolinium
0	–	–	–	0	0	2	0	0	0	Decamethonium
0	–	–	–	0	0	2	0	0	0	Gallamine
0	–	–	–	0	0	11	0	0	0	Tubocurarine
0	–	–	–	0	0	12	0	0	0	Homatropine Methylbromide
0	–	–	–	0	0	15	1	0	0	Pancuronium Bromide
0	–	–	–	0	1	25	2	0	0	Thiazinamium Metilsulfate
0	–	–	–	0	3	76	1	0	0	Guanoxan
0	–	–	–	0	4	23	0	0	0	Carbachol
0	–	–	–	0	16	11	–	0	–	Betaine
0	–	–	–	0	16	11	17	0	0	Cocaine M/Ecgonine
0	–	–	–	0	30	2	–	0	0	Codeine M (-6-glucuronide)
0	–	–	–	0	53	8	1	0	0	Obidoxime Chloride
0	0	0	0	0	70	16	58	0	0	Lysergic Acid
0	–	–	–	0	80	1	5	0	0	Psilocybine
0	–	–	–	1	2	18	1	0	0	Thiamine Hydrochloride
0	–	–	–	1	3	18	2	0	0	Methscopolamine Bromide
0	–	–	–	2	3	41	8	0	0	Butylscopolammonium Bromide,N-
0	–	–	–	3	43	–	0	0	0	Levarterenol/Norepinephrine
0	–	–	–	7	0	3	13	0	0	Histamine
0	–	–	–	7	5	44	17	0	0	Morphine M (Nor-)
0	–	–	–	8	4	56	7	38	0	Berberine
0	–	–	–	10	80	88	5	62	0	Phencyclidine interm./1-Piperidino-1-cyclohexanecarbonitrile/PCC
0	–	–	–	37	1	15	0	46	0	Hydrastinine

Table 8.10. (Continued)

TLC System 9	1	2	3	4	5	6	7	8	10	[System 9: Chl–MeOH (90+10) / Silica-KOH]
0	–	–	–	43	14	59	18	0	0	Dopamine
0	–	–	–	71	12	–	3	1	1	Hexoprenaline
0	–	–	0	80	93	48	1	0	0	Metformin
1	–	–	–	–	–	–	1	0	3	Benserazide
1	–	–	–	–	–	–	1	0	0	Pentapiperide
1	–	–	–	–	–	–	1	1	1	Picloxydine
1	–	–	–	–	–	–	1	1	1	Propamidine
1	–	–	–	–	–	–	3	0	1	Chlorproguanil
1	–	–	–	–	–	–	4	1	1	Acefylline Piperazine
1	–	–	–	–	–	–	9	2	1	Dextrorphan M/Levorphanol M (Nor-)
1	–	–	–	–	–	–	44	1	2	Cycloserine
1	–	–	–	–	–	–	45	0	4	Practolol
1	–	–	–	–	1	4	5	1	0	Piperazine
1	–	–	0	–	69	–	5	0	1	Cytarabine
1	–	–	14	–	81	–	76	0	8	Sulfacarbamid
1	–	–	–	0	0	29	0	0	0	Cetrimide
1	–	–	–	0	0	29	0	0	0	Cetylpyridinium
1	–	–	–	0	1	–	15	0	2	Amidefrine
1	–	–	–	1	2	36	3	0	0	Oxyphenonium Bromide
1	–	–	–	2	22	14	21	0	3	Cocaine M/N-Benzoylecgonine
1	–	–	–	2	60	75	49	1	1	Methyldopa
1	0	0	2	2	85	59	84	0	2	Metamizol/Dipyrone
1	–	–	–	11	8	–	25	3	18	Synephrine/Oxedrine
1	–	–	–	12	8	67	33	1	0	Phenylephrine
1	–	–	–	13	3	–	0	0	0	Epinephrine
1	–	–	–	18	7	79	3	0	1	Proguanil/Chlorguanide
1	–	–	–	18	13	76	42	1	24	Metaraminol
1	–	–	–	20	14	–	42	1	1	Nadolol
1	–	0	–	20	16	74	46	1	4	Salbutamol/Albuterol
1	–	–	–	21	14	69	40	0	3	Isoprenaline
1	–	–	–	21	18	77	47	1	5	Terbutaline
1	–	–	–	23	7	27	1	5	1	Homatropine
1	–	–	0	24	6	74	24	0	2	Amiloride
1	–	0	–	25	38	81	76	0	4	Fenoterol
1	–	–	–	26	1	22	2	38	0	Cotarnine
1	–	–	–	33	10	34	35	0	1	Bufotenine
1	–	–	–	34	80	–	9	1	1	Oxymetazoline
1	–	–	–	49	87	–	52	0	3	Dobutamine
2	–	–	–	–	–	–	16	51	4	Mecamylamine
2	–	–	–	–	–	–	35	2	11	Hydroxyamphetamine/Amphetamine M
2	–	–	–	–	–	–	42	1	24	Ethylnorepinephrine
2	–	–	–	–	4	–	6	4	2	Tramazoline
2	–	–	–	1	3	30	1	0	0	Guanethidine
2	0	0	7	8	82	72	75	0	0	Cinchophen
2	–	–	–	15	–	–	55	36	1	Dihydralazine
2	–	–	–	22	14	–	23	1	5	Heptaminol

Table 8.10. (Continued)

TLC System 9	1	2	3	4	5	6	7	8	10	[System 9: Chl–MeOH (90+10) / Silica-KOH]
2	–	–	–	22	14	–	45	0	2	Atenolol
2	–	–	–	22	15	74	41	2	3	Etilefrine
2	–	–	–	25	3	55	13	2	2	Tolazoline
2	–	–	–	26	5	60	13	7	2	Tetryzoline
2	–	–	–	26	6	–	23	1	11	Tryptamine
2	–	–	–	37	7	–	45	1	3	Hydroxychloroquine
2	–	–	0	76	12	–	30	3	2	Ethambutol
3	–	–	–	–	–	–	18	1	0	Codeine M (N-Oxide)
3	–	–	–	–	–	–	24	68	10	Pempidine
3	–	–	–	–	–	–	28	49	3	Conessine
3	–	–	–	–	–	–	51	0	0	Minoxidil
3	–	–	–	–	–	–	65	1	6	Protokylol
3	–	–	–	0	34	14	52	0	0	Narceine
3	–	–	–	3	44	–	36	1	3	Penicillamine
3	–	–	–	6	2	18	2	1	0	Oxyphencyclimine
3	–	–	–	14	3	–	10	0	1	Azacyclonol
3	–	–	–	18	12	–	25	2	1	Dihydromorphine
3	–	–	–	18	21	77	48	1	6	Orciprenaline
3	–	–	–	24	5	28	18	5	1	Atropine/Hyoscyamine
3	–	–	–	27	9	–	29	3	3	Pholedrine
3	–	–	–	30	19	–	53	1	5	Sotalol
3	–	–	–	33	6	–	32	1	2	Phentolamine
3	–	–	–	33	13	71	47	0	6	Acebutolol
3	–	–	–	40	4	10	5	67	5	Sparteine
4	–	–	–	–	–	–	13	24	1	Deptropine
4	–	–	–	–	–	–	22	5	1	Trimetazidine
4	–	–	–	–	–	–	43	5	10	Tacrine
4	–	–	–	2	4	66	2	0	0	Benzalkonium Chloride
4	–	–	–	4	3	31	4	0	0	Propantheline Bromide
4	–	–	–	5	4	39	2	0	0	Fenpiverinium
4	–	–	–	11	12	73	55	24	38	Methoxamine
4	–	–	–	17	99	–	33	54	63	Pseudoephedrine
4	–	0	–	30	12	75	44	4	33	Ephedrine M/Phenylpropanolamine
4	–	–	–	32	7	–	23	26	2	Methoxyphenamine
4	–	–	–	32	8	75	33	1	1	Clomipramine M (8-Hydroxy-nor-)
4	–	–	0	38	70	–	55	0	8	Ambazone
4	–	–	–	42	10	49	35	14	3	Dextrorphan/Levorphanol
4	–	–	0	46	4	14	38	14	2	Chloroquine
4	–	–	–	55	25	78	40	17	9	Mexiletine
5	–	–	–	–	–	–	13	0	0	Codeine M (Nor-)
5	–	–	–	–	–	–	19	13	15	Primaquine
5	–	–	–	–	–	–	31	10	4	Metazocine
5	–	–	–	–	–	–	33	34	3	Pentaquine
5	–	–	–	–	–	–	43	15	9	Mepacrine
5	–	–	–	3	3	41	5	0	0	Isopropamide Iodide
5	–	–	–	8	88	–	68	53	67	Azapropazone

Table 8.10. (Continued)

TLC System 9	1	2	3	4	5	6	7	8	10	[System 9: Chl–MeOH (90+10) / Silica-KOH]
5	–	–	–	25	10	64	30	5	1	Ephedrine
5	–	–	–	29	12	–	42	25	46	Pseudoephedrine M (Nor-)/Cathine
5	–	–	–	30	5	64	13	7	3	Xylometazoline
5	–	–	70	30	40	88	56	0	9	Clioquinol
5	–	–	–	36	6	71	15	18	2	Maprotiline
5	–	–	–	39	17	33	49	1	9	Procainamide
5	–	–	–	43	18	78	49	2	8	Pindolol
6	–	–	–	–	–	–	40	5	5	Hordenine
6	–	–	–	–	–	–	42	8	8	Profadol
6	–	–	–	–	–	–	51	0	3	Broxyquinoline
6	–	–	–	27	3	–	10	6	0	Mequitazine
6	–	–	–	27	3	52	14	3	3	Naphazoline
6	–	–	–	30	11	–	38	3	1	Mesoridazine
6	–	–	–	36	6	–	13	26	2	Benzatropine
6	–	–	–	38	23	–	55	3	0	Bamethan
6	59	59	70	81	89	91	78	15	71	Halazepam
7	–	–	–	–	–	–	22	18	2	Apoatropine
7	–	–	–	–	–	–	33	1	24	Tuaminoheptane
7	–	–	–	35	7	77	23	17	4	Amantadine
7	–	–	–	38	6	69	19	18	2	Protriptyline
7	–	–	–	38	61	–	52	10	15	Oxetacaine
7	–	–	–	40	21	69	44	2	3	Hydroquinine
7	–	–	–	47	5	66	31	6	3	Antazoline
7	–	–	–	51	17	–	47	1	13	Metoclopramide
8	–	–	–	–	–	–	25	34	2	Mephentermine
8	–	–	–	–	–	–	40	17	8	Phenglutarimide
8	–	–	71	6	80	–	53	3	4	Azathioprine
8	–	–	–	15	75	67	59	0	5	Pyridoxine Hydrochloride
8	–	–	–	28	6	37	27	2	1	Dothiepin M (Nor-,Sulfoxide)
8	–	–	13	30	50	65	51	1	4	Triamterene
8	–	–	–	32	8	63	–	7	2	Pethidine M (Nor-)
8	–	–	–	43	20	70	45	3	5	Hydroquinidine
8	–	–	–	44	20	74	49	10	9	Metoprolol
8	–	–	–	44	24	55	49	6	6	Cinchonidine
8	–	–	–	59	8	–	41	57	6	Perhexiline
8	–	–	–	60	9	7	45	7	13	Disopyramide
9	–	–	–	–	–	–	34	14	2	Racemorphan
9	–	–	–	–	–	–	37	9	3	Diphenazoline
9	–	–	–	16	27	20	41	0	0	Clomipramine M (N-Oxid)
9	–	–	–	18	12	14	23	3	2	Hydromorphone
9	–	–	–	18	23	16	37	0	0	Mianserin M (N-Oxide)
9	–	–	–	20	18	23	37	0	1	Morphine
9	–	–	–	27	53	55	54	0	12	Cimetidine
9	–	–	–	37	26	–	47	2	6	Ketobemidone
9	–	–	–	43	12	75	43	20	18	Amphetamine/Amphetaminil M artifact
9	–	–	–	47	14	48	39	5	9	Psilocin

Table 8.10. (Continued)

TLC System 9	1	2	3	4	5	6	7	8	10	[System 9: Chl–MeOH (90+10) / Silica-KOH]
9	–	–	–	50	14	39	40	9	6	Dimethyltryptamine,N,N-/DMT
9	–	–	–	56	14	–	44	35	15	Etafedrine
10	–	–	–	–	–	–	40	1	2	Cytisine
10	–	–	–	2	87	46	88	0	10	Dimenhydrinate
10	–	–	–	24	6	63	20	3	12	Mescaline
10	–	–	–	26	96	98	95	0	90	Closantel
10	–	–	–	43	8	75	36	24	14	Dimethoxy-4-ethylamphetamine,2,5-/DOET
10	–	–	–	43	9	74	35	23	18	Methoxy-4,5-methylenedioxyamphetamine,2-/MMDA-2
10	–	–	–	45	8	75	36	19	17	Dimethoxyamphetamine,2,5-
10	–	–	–	49	21	79	50	6	7	Propranolol
10	–	–	–	60	12	–	48	41	9	Doxylamine
10	–	–	–	63	14	56	46	14	11	Diethyltryptamine,N,N-/DET
10	–	–	–	66	6	68	20	32	2	Cyclopentamine
10	–	–	68	80	88	–	49	2	4	Procarbazine
11	–	–	–	–	–	–	22	28	25	Octamylamine
11	–	–	–	–	–	–	37	50	35	Rimantadine
11	–	–	–	–	–	–	38	13	17	Bromo-STP/Bromo-2,5-dimethoxy-4-methyl-amphetamine
11	–	–	–	–	–	–	48	28	18	Butaxamine
11	–	–	–	–	–	–	51	38	64	Hydralazine
11	–	–	–	–	–	–	56	0	12	Metopimazine
11	–	–	–	29	7	70	33	6	12	Trimethoxyamphetamine,3,4,5-
11	–	–	18	29	55	49	47	1	20	Isoniazid
11	–	–	–	31	6	61	26	5	14	Dimethoxyphenethylamine,3,4-
11	–	–	–	40	7	71	26	19	3	Desipramine/Imipramine M
11	–	–	–	45	20	78	48	11	13	Oxprenolol
11	–	–	–	45	26	65	51	2	4	Quinine
11	–	–	–	50	20	75	52	6	9	Timolol
12	–	–	–	–	–	–	35	5	4	Neopine
12	–	–	–	25	70	59	48	0	12	Diprophylline/Propyphylline
12	–	–	–	33	62	60	57	0	8	Ergometrine
12	–	–	–	42	10	76	39	18	17	Methylenedioxyamphetamine,3,4-/MDA
12	–	–	–	43	33	61	56	1	8	Pipamperone
12	–	–	–	44	19	61	49	6	5	Cinchonine
12	–	–	–	49	22	–	52	11	11	Alprenolol
12	–	–	–	49	30	63	51	4	6	Quinidine
12	–	–	–	70	34	72	61	16	28	Pentazocine
12	–	–	–	83	29	82	77	37	63	Phenelzine
13	–	–	–	–	–	–	46	18	12	Eucatropine
13	–	–	–	–	12	12	30	4	3	Methenamine
13	–	–	–	27	11	14	–	5	3	Metopon
13	–	–	–	29	11	19	26	8	2	Dihydrocodeine
13	–	–	–	37	5	70	26	39	5	Coniine
13	–	–	–	42	9	63	31	28	5	Deoxyephedrine,(+/-)-
13	–	–	–	42	9	63	31	28	5	Methamfetamine

Table 8.10. (Continued)

TLC System 9	1	2	3	4	5	6	7	8	10	[System 9: Chl–MeOH (90+10) / Silica-KOH]
13	–	–	–	45	8	75	34	24	3	Clomipramine M (Nor-)
13	–	–	–	46	14	26	45	35	3	Pheniramine
13	–	–	–	47	10	–	42	35	6	Dimetindene
13	–	–	–	48	12	–	47	17	6	Pipazetate
13	–	–	–	53	46	65	63	7	13	Mazindol
13	–	–	–	65	24	74	53	15	25	Cyclazocine
14	–	–	–	–	–	–	49	26	14	Dimethylphenylthiazolanimin (Dimethyl-5-phenyl-2-thiazolidinimine,3,4-
14	–	–	–	41	69	–	62	0	12	Methylergometrine
14	–	–	–	43	10	76	36	15	17	Bromo-2,5-dimethoxyamphetamine,4-/DOB
14	–	–	–	62	33	83	74	3	50	Buphenine
15	–	–	–	–	–	–	29	32	6	Methdilazine
15	–	–	–	–	–	–	44	12	6	Methyldesorphine
15	–	–	–	–	–	–	57	18	20	Clorprenaline
15	–	0	–	50	22	80	50	13	8	Penbutolol
15	–	–	–	65	15	67	–	9	25	Psilocin-(eth)/CZ-74
16	–	–	–	46	87	71	34	28	4	Amitriptyline M/Nortriptyline
16	–	–	–	48	25	–	54	3	15	Fenpipramide
16	–	–	–	49	12	–	45	33	6	Brompheniramine
16	–	–	–	51	46	61	58	4	18	Periciazine
16	–	–	–	60	19	–	48	41	11	Fenfluramine
17	–	–	–	–	–	–	33	18	8	Desomorphine
17	–	–	–	25	5	7	16	0	1	Brucine
17	–	–	–	41	9	76	51	15	16	Dimethoxy-4-methylamphetamine,2,5-/STP/DOM
17	–	–	–	43	41	52	65	0	14	Pipamazine
17	–	–	–	48	14	77	44	18	8	Chlorphentermine
17	–	–	–	48	74	69	56	4	22	Nicotinyl Alcohol
17	–	–	–	51	26	51	51	4	10	Clomipramine M (8-Hydroxy-)
18	–	–	–	25	15	–	36	3	2	Pholcodine
18	–	–	–	35	21	22	33	6	3	Codeine
18	0	0	14	40	20	60	58	2	29	Flurazepam M (Dideethyl-)
18	–	–	–	46	12	21	45	35	2	Chlorphenamine/Chlorpheniramine
18	–	–	–	47	10	42	33	42	6	Dextromethorphan/Racemethorphan/Levomethorphan
19	–	–	–	–	–	–	53	1	8	Cephaeline
19	–	–	–	–	–	–	56	6	17	Piperacetazine
19	–	–	–	32	8	11	26	8	2	Strychnine
19	–	–	–	36	57	51	60	0	7	Lysergamide
19	–	–	–	50	13	16	48	26	4	Carbinoxamine
19	0	0	15	50	28	62	58	8	11	Flurazepam M (Monodeethyl-)
19	–	–	–	54	25	42	49	15	9	Clomipramine M (10-Hydroxy-)
19	–	–	–	55	12	41	40	45	4	Ethoheptazine
19	–	–	–	59	16	74	47	47	8	Etilamfetamine
19	–	–	–	60	12	–	50	42	8	Aminopromazine
19	–	–	–	80	18	–	52	62	26	Terodiline

Table 8.10. (Continued)

TLC System 9	1	2	3	4	5	6	7	8	10	[System 9: Chl–MeOH (90+10) / Silica-KOH]
20	–	–	–	–	–	–	84	1	25	Salinazid
20	–	–	–	33	11	13	25	4	4	Hydrocodone
20	7	12	24	44	80	83	71	0	49	Nitrazepam M (7-Acetamido-)
20	13	39	30	53	–	–	62	1	53	Styramate
20	–	–	–	55	19	30	51	11	6	Triprolidine
20	–	–	–	77	16	60	48	59	27	Methadone/Levomethadone
21	–	–	–	–	–	–	42	39	7	Tigloidine
21	–	–	–	–	–	–	43	6	2	Dehydroemetine
21	–	–	–	–	–	–	59	0	17	Rolicyprine
21	–	–	–	–	–	–	65	1	15	Methysergide
21	–	–	–	40	68	66	54	0	27	Nicotinamide
21	–	–	–	48	14	–	45	26	7	Tofenacin
21	–	–	–	58	54	–	83	0	27	Apomorphine
21	–	–	–	71	11	28	54	41	9	Diamocaine
22	–	–	–	–	–	–	43	30	12	Methopromazine
22	–	–	–	–	65	68	63	0	28	Harmine
22	–	–	–	36	21	26	40	7	6	Ethylmorphine
22	–	–	–	38	35	39	54	6	7	Opipramol
22	7	11	30	42	79	82	73	0	45	Clonazepam M (7-Acetamido-)
22	–	–	20	45	45	59	55	0	12	Trimethoprim
22	–	–	–	50	13	28	42	35	6	Tropacocaine
22	–	–	–	50	16	–	44	36	9	Benzydamine
22	–	–	–	67	19	–	50	51	19	Oxeladine
22	–	–	–	68	16	49	48	47	14	Pentoxyverine/Carbetapentane
23	–	–	–	–	–	–	59	5	37	Chloroprocaine
23	–	–	–	32	57	59	59	1	29	Nalorphine
23	–	–	–	36	25	–	42	6	6	Viloxazine
23	–	11	–	36	81	81	60	0	40	Pemoline
23	–	–	–	41	20	23	41	6	8	Benzylmorphine
23	–	–	–	42	70	69	69	1	17	Iproniazid
23	–	–	–	44	42	74	60	2	–	Tocainide
23	–	–	–	45	45	49	63	5	10	Fluphenazine
23	–	–	–	59	15	29	47	43	9	Prothipendyl
23	–	–	–	67	21	47	48	48	13	Imipramine
24	–	–	–	–	–	–	52	5	21	Amopyroquine
24	–	–	–	–	21	38	39	18	6	Dimethoxanate
24	0	6	26	39	73	61	61	0	28	Bromazepam M (3-Hydroxy-)
24	–	–	–	45	26	51	70	1	11	Clomipramine M (2-Hydroxy-)
24	–	–	–	48	11	78	46	26	12	Phentermine
24	–	–	–	62	18	–	48	17	12	Carbocromen
24	–	–	–	63	28	–	48	26	12	Acepromazine
24	–	–	–	74	42	73	67	19	45	Levallorphan
25	–	–	–	–	–	–	50	29	11	Betaprodine
25	–	–	–	–	19	36	53	42	12	Thenyldiamine
25	–	–	–	–	33	32	53	3	3	Acetophenazine
25	–	–	–	–	68	64	70	2	4	Nialamide

Table 8.10. (Continued)

TLC System 9	1	2	3	4	5	6	7	8	10	[System 9: Chl–MeOH (90+10) / Silica-KOH]
25	3	2	26	26	73	69	67	1	7	Alprazolam M (4-Hydroxy-)
25	2	5	33	38	78	81	75	4	14	Midazolam M (α,4-Dihydroxy-)
25	–	–	–	48	20	–	47	27	–	Zimeldine
25	–	–	–	58	18	49	46	49	9	Clemastine
25	–	–	–	58	22	33	51	39	14	Mepyramine/Pyrilamine
25	–	–	–	61	21	–	50	45	12	Metixene
25	–	–	–	79	15	65	–	66	26	Rolicyclidine/PHP/1-(1-phenylcyclohex-yl)pyrrolidine
26	–	–	–	–	–	–	52	17	5	Diethylcarbamazine
26	–	–	–	–	45	55	63	5	31	Butethamine
26	–	–	–	44	25	27	46	6	8	Morphine-6-acetate/Diamorphine M
26	–	–	–	66	21	24	52	41	13	Methapyrilene
27	–	–	–	39	39	–	54	5	7	Carfenazine
27	–	–	–	46	34	45	50	14	14	Phenmetrazine/Phendimetrazine M (Nor-)
27	–	–	–	66	27	74	54	42	17	Eucaine,β-
27	–	–	–	68	22	34	55	44	15	Tripelennamine
27	–	–	–	75	78	–	57	32	29	Octacaine
27	–	–	–	76	51	75	67	11	33	Haloperidol
27	–	–	–	79	21	78	–	65	37	Eticyclidine/PCE/N-Ethyl-1-phenylcyclohexyl-amine
28	–	–	–	–	–	–	82	1	41	Diazoxide
28	–	–	–	24	82	21	–	6	7	Morphine-3-acetate
28	–	–	–	42	58	–	60	1	16	Dihydroergotamine
28	–	–	–	44	19	22	45	17	5	Dothiepin M (Nor-)
28	–	–	–	46	20	22	45	11	5	Dothiepin M (Sulfoxide)
28	–	–	–	54	44	–	49	28	39	Phenethylamine
28	–	–	0	63	22	–	52	41	17	Chloropyramine
28	–	–	–	65	22	31	55	38	14	Thonzylamine
28	–	–	–	68	23	49	46	37	8	Diphenylpyraline
29	3	1	29	24	75	71	63	0	8	Triazolam M (4-Hydroxy-)
29	–	–	–	42	40	40	55	7	9	Perphenazine
29	–	–	–	64	53	52	56	9	31	Nomifensine
30	–	–	–	–	–	–	46	45	10	Proheptazine
30	–	–	–	–	–	–	52	30	17	Dimazole/Diamthazole
30	–	–	–	–	–	–	64	0	33	Norharman
30	–	9	–	11	74	66	75	1	11	Theophylline
30	15	28	35	49	88	89	67	0	38	Quazepam M (3-Hydroxy-N-dealkyl-2-oxo-)
30	11	20	40	52	76	77	73	0	47	Clonazepam M (7-Amino-)
30	10	20	30	54	77	79	71	0	46	Nitrazepam M (7-Amino-)
30	–	–	–	55	30	29	53	33	8	Trifluoperazine
30	–	–	–	62	18	35	44	38	11	Promazine
30	–	–	–	64	22	35	52	41	14	Isothipendyl
30	–	–	–	67	20	55	48	42	13	Thioridazine
30	–	–	–	83	44	76	71	7	64	Butacaine
31	–	4	–	34	59	54	53	1	21	Theobromine
31	11	31	34	36	86	90	69	0	48	Chloramphenicol

Table 8.10. (Continued)

TLC System 9	1	2	3	4	5	6	7	8	10	[System 9: Chl–MeOH (90+10) / Silica-KOH]
31	–	–	–	51	25	60	46	11	6	Mianserin M (Nor-)
31	–	–	–	58	66	–	61	2	21	Pyrimethamine
31	–	–	–	70	44	76	62	8	53	Clonidine
31	–	–	–	71	36	42	54	5	30	Procaine
31	–	–	–	74	20	68	48	62	23	Procyclidine
32	–	–	–	–	–	–	54	50	17	Embramine
32	3	3	33	40	68	60	69	0	42	Chlordiazepoxide M (Nor-)
32	–	–	–	44	45	–	56	7	11	Clopenthixol/Zuclopenthixol/cis-Ordinol
32	–	–	–	44	52	45	53	0	12	Pilocarpine
32	–	–	–	53	40	59	66	3	21	Pipotiazine
32	9	36	32	56	63	87	75	0	58	Meprobamate
32	–	–	–	59	30	–	50	33	17	Nefopam
32	–	–	–	62	62	81	78	3	53	Isoxsuprine
32	–	–	–	64	43	39	57	15	16	Tetracaine/Amethocaine
32	–	–	–	68	26	–	51	52	15	Tolpropamine
32	–	–	–	69	27	51	51	50	15	Amitriptyline
32	–	–	–	74	52	81	70	9	22	Loperamide
32	–	–	–	79	22	–	50	67	25	Prolintane
33	–	–	–	–	–	–	58	3	28	Propoxycaine
33	–	–	–	46	50	–	62	6	–	Flupentixol
33	–	–	–	49	71	–	58	2	29	Proxyphylline
33	–	–	–	58	41	67	54	39	48	Tranylcypromine
33	10	35	35	60	82	85	–	0	56	Mebutamate
33	–	–	–	62	–	–	61	2	57	Amiphenazole
33	–	–	–	65	27	48	55	44	15	Diphenhydramine
33	–	–	–	68	25	49	55	48	16	Orphenadrine
33	–	–	–	68	28	48	55	45	15	Dimenhydrinate
33	–	–	–	70	23	79	–	41	29	Phenylcyclohexylamine,1-
33	–	–	–	87	27	72	66	67	70	Dipipanone
34	–	–	–	–	–	–	47	49	16	Iprindole/Glycophene
34	–	–	–	40	18	–	42	3	6	Nicomorphine
34	–	–	–	43	22	22	46	7	6	Thioproperazine
34	–	–	–	44	68	64	63	1	22	Ergotamine
34	–	–	–	49	24	25	45	20	11	Thebacon
34	–	–	–	53	19	–	54	13	12	Emetine
34	11	19	35	54	78	75	72	0	47	Flunitrazepam M (7-Aminonor-)
34	–	–	–	60	34	40	52	37	11	Pethidine/Meperidine
34	–	–	–	63	33	–	56	40	18	Methadone M (Nor-)
34	–	–	–	67	42	–	63	25	35	Cinchocaine/Dibucaine
34	–	–	–	72	26	54	51	53	18	Clomipramine
34	–	–	–	77	21	–	54	62	30	Fencamfamine
35	–	–	–	–	–	–	50	30	11	Alphaprodine
35	–	–	–	–	–	–	68	1	65	Brocresine
35	5	4	37	28	76	76	68	1	13	Brotizolam M (6-Hydroxy-)
35	15	24	42	41	81	83	63	0	51	Demoxepam/Chlordiazepoxide M
35	–	–	–	55	38	22	54	22	14	Dibenzepin

Table 8.10. (Continued)

TLC System										
9	1	2	3	4	5	6	7	8	10	[System 9: Chl–MeOH (90+10) / Silica-KOH]
35	–	–	–	61	39	22	54	39	13	Nicotine
35	–	–	–	65	30	44	50	36	17	Promethazine
35	–	–	–	66	29	–	53	43	18	Noxiptiline
35	–	–	–	68	35	–	56	41	20	Nicametate
35	–	–	–	70	25	45	49	45	17	Chlorpromazine
35	–	–	–	71	17	–	59	4	17	Veratrine
35	–	–	–	75	23	–	61	17	29	Lobeline
35	–	–	–	75	32	49	54	47	22	Triflupromazine
35	–	–	–	84	23	69	59	73	66	Phencyclidine
36	–	–	–	–	–	–	58	55	25	Cyclomethycaine
36	–	–	–	–	–	–	65	0	55	Ethionamide
36	1	0	28	40	49	39	63	1	2	Adinazolam M (Nor-)
36	23	41	42	43	82	82	52	1	28	Lorazepam/Lormetazepam M
36	–	–	32	46	75	70	58	2	40	Metronidazole
36	–	–	–	55	41	38	55	12	18	Physostigmine
36	–	–	–	59	24	33	48	18	25	Oxomemazine
36	35	46	53	64	84	86	68	0	55	Nitrazepam
36	–	–	–	70	29	–	53	47	17	Chlorphenoxamine
36	–	–	–	86	23	75	55	68	52	Diisopromine
37	–	–	–	–	–	–	79	1	50	Orthocaine
37	–	–	–	–	32	–	52	41	29	Clofedanol
37	–	–	–	33	27	36	48	10	30	Oxymorphone
37	–	–	–	33	69	63	55	0	12	Colchicine
37	–	–	–	44	82	87	68	0	42	Dipyridamole
37	–	–	–	45	23	32	45	24	5	Thebaine
37	–	–	–	47	21	23	48	25	3	Perazine
37	–	–	–	48	49	47	55	6	18	Scopolamine
37	–	–	–	52	26	–	53	28	5	Butaperazine
37	–	–	–	54	46	52	60	4	18	Mianserin M (8-Hydroxy-)
37	–	–	–	55	26	26	49	34	7	Prochlorperazine
37	–	–	–	63	24	45	51	48	13	Doxepin
37	–	–	–	71	25	66	54	54	18	Pyrrobutamine
37	–	–	–	76	21	56	55	53	27	Piperocaine
37	–	–	–	78	30	73	56	63	26	Volazocine
38	–	–	–	–	–	–	64	55	36	Betacetylmethadol
38	–	–	–	–	77	–	66	1	57	Protionamide
38	–	3	–	46	81	78	67	1	40	Glafenine
38	–	–	–	49	26	33	47	15	4	Diamorphine
38	4	5	29	55	42	–	57	4	17	Clozapine
38	–	–	–	64	66	70	63	6	52	Yohimbine
38	–	–	–	65	29	73	55	36	13	Noracymethadol
38	–	–	–	76	32	49	57	47	46	Levomepromazine/Methotrimeprazine
38	–	–	–	81	19	79	54	59	39	Pipradrol
39	–	–	–	–	–	–	54	12	17	Etamiphyllin
39	–	–	–	–	–	–	57	27	26	Cyclopentolate
39	–	–	–	–	–	–	66	65	38	Alverine

Table 8.10. (Continued)

TLC System 9	1	2	3	4	5	6	7	8	10	[System 9: Chl–MeOH (90+10) / Silica-KOH]
39	3	–	–	56	60	59	60	3	18	Lysergide/LSD
39	–	–	–	74	50	81	68	16	49	Phenazocine
39	–	–	–	75	42	68	68	11	58	Aconitine
39	–	–	–	77	32	46	58	54	31	Trimeprazine/Alimemazine
39	–	–	–	78	51	80	69	18	47	Phenomorphan
40	–	–	–	–	–	–	1	33	16	Alphameprodine
40	–	–	–	–	–	–	43	0	40	Furaltadone
40	–	–	–	–	–	–	59	37	42	Aletamine
40	–	–	–	–	–	–	60	49	25	Dyclonine
40	–	–	–	–	–	–	62	8	37	Amodiaquine
40	–	–	–	44	26	24	49	10	7	Tiotixene
40	5	2	41	44	68	65	60	1	16	Triazolam
40	22	35	42	45	82	82	56	0	51	Oxazepam/M of Chlordiazepoxide, Clorazepic Acid, Camazepam, Diazepam, Halazepam, Ketazolam, Medazepam, Oxazolam, Prazepam
40	19	11	41	51	78	76	73	0	49	Flunitrazepam M (7-Acetamido-)
40	–	–	–	57	72	76	70	2	33	Harman
40	–	–	–	79	30	91	70	48	20	Hexetidine
41	–	–	–	–	–	–	58	41	17	Trimeperidine
41	–	–	–	–	–	–	62	26	35	Proxymetacaine
41	–	–	–	–	–	–	62	50	21	Dimethylthiambutene
41	–	–	–	–	–	–	87	0	5	Glipizide
41	–	–	–	35	56	–	53	0	11	Demecolcine
41	–	–	–	41	69	66	76	0	8	Domperidone
41	–	–	–	53	27	–	51	30	8	Thiethylperazine
41	13	18	47	63	73	69	61	6	53	Bromazepam
41	–	–	–	66	40	70	57	35	23	Methylphenidate
41	–	–	–	68	40	52	57	49	16	Cyclizine
41	–	–	–	78	43	–	64	52	35	Naftidrofuryl
41	–	–	–	83	54	–	62	23	36	Oxybuprocaine
42	–	–	–	–	–	–	54	46	21	Methaphenilene
42	–	–	–	–	–	–	59	57	28	Ethylmethylthiambutene
42	–	–	–	–	–	–	60	9	18	Etoxeridine
42	–	–	–	47	19	–	43	12	10	Nicocodine
42	–	–	44	54	71	70	63	3	46	Pyrazinamide
42	–	–	–	67	46	–	77	26	22	Meclofenoxate
42	–	–	–	68	30	52	55	34	26	Propiomazine
42	–	–	–	70	27	41	51	49	16	Dothiepin
43	–	–	–	–	–	–	57	1	16	Inositol Nicotinate
43	–	–	–	–	–	–	58	56	38	Alphamethadol
43	–	–	–	–	–	–	70	61	43	Diethylthiambutene
43	5	5	37	49	76	77	74	1	15	Midazolam M (4-Hydroxy-)
43	–	–	–	54	26	25	44	23	12	Acetylcodeine
43	–	–	–	55	87	89	64	2	57	Mephenesin
43	–	–	–	59	24	–	49	40	13	Bamipine

Table 8.10. (Continued)

TLC System 9	1	2	3	4	5	6	7	8	10	[System 9: Chl–MeOH (90+10) / Silica-KOH]
43	–	–	–	69	27	48	54	42	13	Bromazine/Bromdiphenhydramine
44	–	–	–	–	–	–	52	31	19	Moxisylyte
44	–	–	–	–	–	–	54	1	53	Niridazole
44	–	–	–	–	–	–	56	39	16	Properidine
44	–	–	–	16	12	–	47	10	6	Hexobendine
44	4	4	37	37	76	76	72	1	18	Alprazolam M (α-Hydroxy-)
44	–	–	–	52	20	–	50	38	12	Thenalidine
44	–	–	–	58	60	–	57	3	33	Nimorazole
44	–	–	–	64	30	50	51	45	13	Cyproheptadine
44	–	1	–	65	27	–	53	47	16	Pecazine/Mepazine
44	–	–	–	82	27	77	55	66	36	Prozapine
45	–	–	–	–	–	–	61	56	51	Difenidol
45	–	–	–	–	–	–	62	27	11	Morpheridine
45	–	–	–	–	–	–	68	7	67	Dimethocaine
45	–	–	–	–	–	–	69	26	67	Amprotropine
45	–	–	–	–	–	–	74	58	43	Isomethadone
45	–	–	–	54	44	–	55	3	14	Fenetylline
45	–	–	–	54	86	–	49	3	17	Phanquinone
45	–	–	–	61	73	74	70	1	42	Piritramide
45	–	–	–	64	78	–	66	6	68	Leucinocaine
45	–	–	–	65	36	46	57	27	20	Mebhydrolin
45	–	–	–	84	38	62	57	62	50	Alphacetylmethadol
46	–	–	–	–	–	–	61	55	28	Quinisocaine
46	7	6	41	45	78	78	72	2	31	Brotizolam M (α-Hydroxy-)
46	–	–	–	58	61	–	58	8	31	Morazone
46	19	28	54	61	81	82	74	2	55	Flurazepam M (N(1)-Hydroxyethyl-)
46	–	–	–	67	35	52	57	42	14	Chlorcyclizine
46	–	–	–	73	14	50	–	22	43	Cryptopine
47	–	–	–	–	–	–	44	0	59	Furazolidone
47	–	–	–	–	–	–	60	34	36	Ethomoxane
47	–	–	–	59	68	74	60	1	49	Prazosin
47	–	–	–	64	64	67	75	0	29	Ergosine
47	–	–	–	77	35	30	65	45	54	Cocaine
47	–	–	–	83	31	55	67	64	66	Profenamine/Ethopropazine
47	–	–	–	87	31	78	–	72	68	Methyl-1-phenylcyclohexyl)piperidine,1-(4-
48	–	–	–	–	–	–	59	15	24	Dimefline
48	–	–	–	–	–	–	59	61	38	Butriptyline
48	3	1	36	40	26	15	40	1	5	Loprazolam
48	–	–	–	58	69	–	66	1	38	Ergoloid
48	–	–	–	58	71	73	67	2	36	Droperidol
48	–	–	–	59	39	50	–	21	12	Cocaine M (Nor-)
48	–	–	–	65	53	52	62	18	42	Levamisole
48	–	–	–	66	43	–	56	13	28	Dimetiotazine
48	–	–	–	67	32	–	53	38	15	Phenyltoloxamine
48	–	–	–	70	51	67	71	21	44	Dexetimide
48	3	3	41	71	52	45	62	30	40	Flurazepam

Table 8.10. (Continued)

TLC System										
9	1	2	3	4	5	6	7	8	10	[System 9: Chl–MeOH (90+10) / Silica-KOH]
49	–	–	–	–	–	–	54	8	32	Morinamide
49	–	–	–	–	–	–	58	35	24	Allylprodine
49	4	3	39	42	75	74	71	4	21	Triazolam M (α-Hydroxy-)
49	–	–	–	61	51	72	74	4	39	Spiperone
50	–	–	–	–	–	–	68	59	38	Norpipanone
50	10	10	53	52	76	77	62	2	22	Chlordiazepoxide
50	–	–	–	70	80	–	59	1	53	Phenazopyridine
50	–	–	–	86	39	72	65	28	62	Ibogaine/Rotenone
51	54	61	56	33	88	90	85	0	47	Chlorzoxazone
51	–	3	–	60	62	69	78	3	32	Benperidol
51	–	–	–	62	30	33	50	25	39	Oxycodone
51	–	–	–	62	49	41	57	36	24	Phendimetrazine
51	9	22	30	67	66	63	73	4	47	Medazepam M (Nor-)
51	–	–	–	70	86	–	67	0	62	Benorilate
51	–	–	–	74	34	51	56	51	25	Chlorprothixene
51	–	–	–	77	33	54	58	57	39	Diethazine
52	–	–	–	–	–	–	61	60	42	Etymemazine
52	–	–	–	41	80	–	62	1	59	Thiamazole
52	–	–	–	46	65	14	52	2	21	Endralazine Mesilate
52	15	5	53	52	72	71	72	5	27	Brotizolam
52	3	4	41	53	73	72	70	3	8	Midazolam M (α-Hydroxy-)
52	26	42	52	63	89	93	77	0	56	Budesonide
52	32	43	54	64	84	90	75	0	61	Clobazam M (Nor-)
52	–	–	–	76	80	86	69	8	59	Fenyramidol
52	–	–	–	77	38	–	59	56	43	Benzoctamine
52	–	–	–	82	30	–	60	56	35	Clomifene
53	–	–	–	–	–	–	50	30	16	Myrophine
53	–	–	–	–	–	–	66	0	23	Naftazone
53	7	5	44	50	71	60	71	2	25	Estazolam
53	35	45	56	67	85	87	72	0	61	Clonazepam
53	–	–	–	74	62	59	61	35	42	Thiopropazate
53	–	–	–	77	52	57	66	43	53	Benactyzine
53	–	–	–	86	32	–	63	40	49	Mebeverine
54	–	–	–	–	–	–	57	2	34	Tozalinone
54	–	–	–	–	75	–	67	7	53	Tiabendazole
54	–	–	–	54	57	65	68	10	19	Hydroxyzine
54	–	–	–	59	71	–	65	0	34	Bamifylline
54	–	–	–	66	51	53	69	18	46	Tetramisole
54	–	–	–	75	65	–	76	14	61	Butanilicaine
54	–	–	–	80	36	56	59	62	37	Trimipramine
54	–	–	–	81	45	–	68	58	55	Isoaminile
54	–	–	–	86	38	70	–	73	68	Tenocyclidine/TCP/1-[1-(2-Thienyl)cyclohexyl] piperidine
55	–	–	–	–	–	–	63	47	43	Methadone Intermediate/Pre-Methadone
55	–	–	–	–	62	60	70	43	41	Pramoxine
55	9	2	51	57	68	66	69	6	22	Adinazolam

Table 8.10. (Continued)

TLC System 9	1	2	3	4	5	6	7	8	10	[System 9: Chl–MeOH (90+10) / Silica-KOH]
55	–	–	–	58	83	85	80	0	6	Flubendazole
55	–	–	–	62	78	81	71	4	52	Carnidazole
55	21	21	52	63	77	74	74	1	52	Flunitrazepam M (7-Amino-)
55	34	45	57	69	82	83	62	3	60	Nordazepam/M of Chlordiazepoxide, Clorazepic Acid, Diazepam, Halazepam, Ketazolam, Medazepam, Oxazolam, Prazepam
55	42	52	58	69	88	90	71	2	59	Quazepam M (3-Hydroxy-2-oxo-)
55	–	–	–	73	35	46	66	26	38	Pitofenone
55	–	–	–	82	50	63	68	58	54	Propoxyphene/Dextropropoxyphene/Levopropoxyphene
56	–	–	–	–	–	–	65	0	67	Ethoxazene
56	–	–	–	–	–	–	69	0	68	Clefamide
56	–	–	–	56	79	75	60	2	47	Carbamazepine
56	–	–	–	59	71	67	59	15	29	Nikethamide
56	34	46	57	68	83	87	84	3	60	Clorazepic Acid
56	34	45	60	72	83	88	75	2	58	Flurazepam M (N(1)-Dealkyl-)/Quazepam M (N-Dealkyl-2-oxo-)
56	–	–	–	79	60	66	73	12	51	Anileridine
57	–	–	–	–	–	–	58	34	37	Betameprodine
57	7	2	40	47	67	66	67	1	14	Alprazolam
57	–	–	–	61	49	43	55	14	26	Isothebaine
57	–	–	–	68	41	49	63	45	21	Phenindamine
57	56	62	63	76	84	87	67	5	66	Benzocaine
57	–	–	–	81	48	56	69	59	47	Butetamate
57	37	41	54	81	80	79	72	5	57	Nitrazepam M Hy/DAB
58	–	–	–	–	–	–	65	60	52	Trifluomeprazine
58	–	–	–	–	–	–	66	2	41	Pipobroman
58	–	–	–	–	–	–	70	15	66	Diperodon
58	15	10	55	52	59	55	52	3	25	Caffeine
58	–	–	–	59	74	77	78	0	50	Ergocryptine
58	–	–	–	64	66	–	58	16	41	Metyrapone
58	–	–	–	66	64	61	63	10	37	Trazodone
58	–	–	–	68	48	50	58	39	23	Mianserin
58	35	46	56	68	84	85	76	0	59	Flunitrazepam M (Nor-)
58	–	–	–	70	74	76	75	13	51	Bromolysergide,2-
58	35	41	57	72	82	86	73	5	56	Delorazepam
58	–	–	–	73	58	50	60	30	33	Methylpiperidyl Benzilate/JB-336
58	35	45	56	73	85	92	77	5	60	Tetrazepam M (Nor-)
59	–	–	–	–	–	–	59	41	23	Clotiapine
59	–	–	–	–	–	–	64	48	43	Benethamine
59	–	–	–	–	–	–	72	5	61	Cyprenorphine
59	–	–	–	57	74	77	77	0	50	Ergocristine
59	38	37	64	57	81	80	71	1	52	Flunitrazepam M (3-Hydroxy-)
59	–	–	–	60	80	84	65	0	49	Mebendazole
59	–	–	–	61	49	43	–	13	25	Orientalidine

Table 8.10. (Continued)

TLC System 9	1	2	3	4	5	6	7	8	10	[System 9: Chl–MeOH (90+10) / Silica-KOH]
59	51	47	65	62	82	82	53	8	53	Temazepam/M of Camazepam, Diazepam, Ketazolam, Medazepam
59	–	–	–	71	63	78	69	4	49	Fluspirilene
59	–	–	–	76	87	–	64	3	55	Ethenzamide
59	41	46	55	82	80	80	74	5	61	Clonazepam Hy/DCB
59	–	–	–	86	33	86	70	57	68	Clofazimine
60	–	–	–	–	–	–	69	37	29	Furethidine
60	–	–	–	–	–	–	70	9	67	Levophenacylmorphan
60	–	–	–	57	46	–	60	1	29	Vinblastine
60	13	5	53	60	69	70	72	6	19	Midazolam
60	–	–	–	71	73	82	71	3	40	Pimozide
60	–	–	–	78	34	30	70	42	55	Cinnamoylcocaine
60	–	–	–	82	55	57	–	54	62	Eucaine,α-
60	48	50	57	82	87	91	77	7	66	Flurazepam Hy/HCFB
60	9	8	41	83	45	51	67	53	51	Flurazepam Hy/DCFB
60	–	–	–	83	49	–	64	55	51	Adiphenine
60	–	–	–	84	72	89	76	17	60	Penfluridol
61	–	–	–	–	–	–	66	67	60	Cycrimine
61	–	–	–	–	–	–	73	7	63	Etorphine
61	46	45	60	59	82	82	52	6	50	Lormetazepam
61	–	–	–	75	78	80	80	9	–	Econazole
61	–	–	–	83	43	75	68	66	59	Trihexyphenidyl/Benzhexol
62	–	–	–	–	–	–	66	1	48	Ergotoxine
62	–	–	–	–	–	–	67	58	49	Diethylaminoethyl Diphenylpropionate
62	–	–	–	–	–	–	70	0	64	Norbormide
62	–	–	–	66	63	60	65	31	48	Mepivacaine
62	53	58	62	74	85	93	76	0	67	Ethyl Loflazepate
63	–	–	–	–	–	–	60	31	50	Thurfyl Nicotinate
63	–	–	–	–	83	90	75	6	70	Butyl Aminobenzoate
63	30	14	61	61	76	72	71	3	44	Flumazenil/RO 15-1788
63	–	–	–	68	84	80	66	1	57	Chlormezanone
63	–	–	–	70	70	77	70	11	36	Lidoflazine
63	–	–	–	78	59	54	75	44	55	Ethylpiperidyl Benzilate/JB-318
63	–	–	–	79	68	72	63	37	64	Ketamine
63	–	–	–	85	55	56	76	62	64	Amfepramone/Diethylpropion
64	–	–	–	–	–	–	67	7	69	Isobutyl 4-Aminobenzoate
64	–	–	–	–	52	–	68	67	54	Dicycloverine/Dicyclomine
64	–	–	–	60	72	74	72	7	63	Pentetrazol
64	–	–	–	71	59	57	61	15	52	Hydrastine
64	45	45	62	74	83	80	66	14	66	Ketazolam
64	–	–	–	75	62	79	77	29	60	Prilocaine
64	–	–	–	76	66	80	75	36	56	Articaine
64	–	–	–	76	70	82	71	26	58	Phenoperidine
64	61	60	64	76	84	96	–	22	69	Cholesterol
64	–	–	–	83	45	–	64	68	64	Biperiden
64	–	–	–	88	63	77	67	36	59	Piminodine

Table 8.10. (Continued)

TLC System 9	1	2	3	4	5	6	7	8	10	[System 9: Chl–MeOH (90+10) / Silica-KOH]
65	–	–	–	–	–	–	70	51	64	Aminoxytriphene
65	–	–	–	5	83	–	76	0	29	Glymidine
65	–	–	–	69	74	74	61	8	47	Papaverine
65	–	–	–	71	79	–	68	1	68	Nifedipine
65	–	–	–	74	77	81	71	23	48	Metomidate
65	–	–	–	76	90	–	65	8	64	Beclamide
65	–	–	–	78	74	–	68	4	58	Oxypertine
66	–	–	–	–	–	–	69	60	56	Proadifen
66	–	–	–	47	74	–	65	9	63	Naloxone
66	–	–	–	66	72	–	55	6	46	Pentifylline
66	–	–	–	72	74	77	61	37	59	Methyl Nicotinate
66	39	45	63	75	78	84	73	0	59	Cloxazolam
66	46	50	65	75	80	90	74	11	62	Haloxazolam
67	–	–	–	–	–	–	63	59	61	Piperoxan
67	–	–	–	–	–	–	66	57	64	Amolanone
67	–	–	–	–	–	–	73	60	63	Amylocaine
67	–	–	–	–	–	–	81	56	65	Phenampromide
67	–	–	–	47	90	–	90	2	70	Buclosamide
67	–	–	–	69	83	79	68	28	55	Crotethamide
67	–	–	–	77	48	–	62	36	45	Flavoxate
67	–	–	–	78	14	–	70	57	56	Azapetine
67	–	7	56	80	77	80	73	11	37	Miconazole
68	–	–	–	–	–	–	65	3	69	Metisazone
68	–	–	–	–	–	–	67	52	62	Tiletamine
68	–	–	–	–	–	–	77	53	56	Pipoxolan
68	53	54	65	71	78	95	77	15	66	Oxazolam
68	33	25	65	73	81	84	76	15	58	Metaclazepam M (N-Nor-)
68	–	–	–	76	76	88	73	5	61	Ambucetamide
68	–	–	–	80	80	–	76	9	69	Buprenorphine
68	–	–	–	82	54	64	72	62	55	Amiodarone
68	–	23	–	82	67	75	73	39	60	Fluanisone
68	–	–	–	84	43	85	68	55	56	Prenylamine
68	61	61	66	85	86	88	77	9	69	Bromazepam Hy/ABP
69	–	–	–	–	–	–	63	45	62	Butoxyethyl Nicotinate,2-
69	–	–	–	–	–	–	65	7	36	Benzquinamide
69	–	–	–	–	–	–	66	8	60	Clonitazene
69	–	–	–	–	–	–	70	48	63	Mebenazine
69	–	–	–	73	84	88	72	2	61	Bromocriptine
69	–	–	–	74	25	83	70	29	57	Cropropamide
69	56	50	67	75	78	86	76	24	63	Fludiazepam
69	–	–	–	76	80	85	64	44	58	Clomethiazole
69	–	–	–	78	76	73	78	31	52	Clemizole
69	57	49	67	79	84	89	78	32	66	Tetrazepam
69	55	48	66	80	84	87	78	31	66	Clotiazepam
69	71	65	68	85	90	91	77	5	69	Nitrazepam Hy/ANB
69	71	66	68	86	87	91	77	4	70	Flunitrazepam M Hy/ANFB

Table 8.10. (Continued)

TLC System 9	1	2	3	4	5	6	7	8	10	[System 9: Chl–MeOH (90+10) / Silica-KOH]
69	72	66	70	86	87	93	77	4	70	Clonazepam Hy/ANCB
70	–	–	–	–	–	–	61	42	61	Nicofuranose
70	–	–	–	–	–	–	67	42	59	Clorgiline
70	–	–	–	72	69	–	64	20	54	Doxapram
70	–	–	–	72	80	–	66	20	55	Propanidid
70	–	–	–	73	43	61	59	23	42	Verapamil
70	53	47	70	75	84	85	62	8	62	Clobazam
70	53	46	71	77	81	81	74	12	55	Nimetazepam
70	–	–	–	79	70	72	68	13	61	Etonitazene
70	–	–	–	87	60	–	73	67	70	Benzphetamine
71	–	–	–	–	–	–	63	42	60	Benzyl Nicotinate
71	–	–	–	73	78	85	67	26	52	Etomidate
71	–	–	–	79	72	78	73	42	60	Dextromoramide/Racemoramide/Levomoramide
71	–	–	–	80	72	69	70	35	63	Lidocaine
71	70	69	74	83	81	93	83	11	69	Haloxazolam Hy/ABFB
72	–	–	–	–	–	–	71	51	65	Diampromide
72	–	–	–	–	–	–	75	44	57	Benzethidine
72	–	–	–	68	80	–	61	11	52	Trimetozine
72	54	47	72	74	80	82	63	10	63	Flunitrazepam
72	65	61	73	81	85	92	75	29	70	Pinazepam
73	–	–	–	–	–	–	70	39	66	Tolycaine
73	–	–	–	71	77	78	–	11	63	Meconin
73	55	32	69	75	82	83	76	12	65	Camazepam
73	58	49	72	76	82	85	75	27	59	Diazepam/Ketazolam M/Medazepam M
73	47	35	71	79	82	84	77	40	62	Metaclazepam
73	–	–	–	80	69	79	69	42	65	Bupivacaine
73	–	–	–	83	70	78	–	38	59	Fluorofentanyl,p-
74	–	–	–	–	–	–	66	16	67	Diloxanide
74	–	–	–	–	–	–	80	46	62	Dioxaphetyl Butyrate
74	–	–	–	75	84	86	71	20	61	Isocarboxazid
74	–	–	–	77	76	80	69	2	63	Reserpine
74	–	–	–	78	70	77	70	43	58	Fentanyl
74	–	–	–	78	72	75	64	21	64	Noscapine
74	56	40	73	78	79	83	67	41	62	Medazepam
74	64	55	72	81	84	89	65	36	63	Prazepam
74	–	–	–	87	72	73	–	62	64	Phenylcyclohexyl)morpholine,4-(1-/PCM
74	–	–	–	87	90	–	72	37	67	Famprofazone
75	–	–	–	–	–	–	68	16	58	Dimoxyline
75	–	–	–	81	77	79	73	1	64	Rescinnamine
75	78	71	78	83	87	96	74	27	76	Quazepam
75	–	–	–	84	64	77	–	45	64	Alphamethylfentanyl
75	76	68	74	86	89	93	78	11	70	Flurazepam Hy/ACFB
75	–	–	–	88	83	90	82	45	64	Flunarizine
76	–	–	–	–	–	–	71	48	62	Carbifene
76	–	–	–	80	82	87	74	17	66	Bisacodyl

Table 8.10. (Continued)

TLC System 9	1	2	3	4	5	6	7	8	10	[System 9: Chl–MeOH (90+10) / Silica-KOH]
76	76	69	76	86	90	91	78	12	71	Oxazepam Hy/ACB
76	–	–	–	87	84	97	73	63	68	Phenoxybenzamine
76	77	70	76	87	89	93	78	11	70	Lorazepam Hy/ADB
77	–	–	–	–	–	–	67	18	68	Octaverine
77	–	–	–	–	–	–	72	4	64	Methoserpidine
77	–	–	–	–	–	–	76	6	69	Thiambutosine
77	–	–	–	–	80	83	80	58	69	Phenadoxone
77	–	–	–	43	9	74	73	23	69	Methoxyamphetamine,4-/PMA
77	–	–	–	81	73	–	72	3	66	Deserpidine
77	66	66	80	83	76	86	73	31	68	Nimetazepam Hy/MNB
77	71	69	81	83	77	90	75	48	69	Fludiazepam Hy/CFMB
77	74	73	80	85	77	93	79	44	73	Quazepam Hy/CFTB
77	–	–	–	86	85	90	83	53	66	Clocinizine
77	75	73	80	86	86	91	83	48	72	Pinazepam Hy/CPB
78	–	–	–	–	–	–	80	7	72	Tiocarlide
78	–	–	–	78	78	81	71	21	69	Disulfiram
78	–	–	–	79	80	–	69	41	67	Tetrabenazine
78	–	–	–	86	79	87	76	54	65	Cinnarizine
78	–	–	–	86	88	–	72	47	67	Fenbutrazate
78	–	–	–	86	93	97	83	44	75	Proclonol
79	–	–	–	65	92	96	71	41	70	Bezitramide
79	–	–	–	78	78	–	74	52	71	Broxaldine
79	–	–	–	82	82	89	79	11	71	Phenothiazine
79	7	63	80	86	89	90	77	20	71	Flunitrazepam M Hy/MNFB
79	–	–	–	87	80	88	76	61	70	Meclozine
79	–	–	–	88	84	–	75	69	71	Bromhexine
80	–	–	–	–	–	–	74	62	73	Bialamicol
80	–	–	–	78	79	84	70	36	56	Methaqualone
81	–	–	–	82	54	52	69	55	55	Piperidolate
81	81	69	81	87	82	93	75	49	7	Lormetazepam Hy/MDB
81	82	72	82	88	89	93	79	53	71	Diazepam Hy/MACB
81	81	74	81	89	94	95	82	49	75	Halazepam Hy/TCB
82	83	74	82	88	92	95	80	58	74	Prazepam Hy/CCB
83	–	–	–	–	–	–	75	61	72	Buclizine
84	–	–	–	83	88	92	79	64	74	Phenanthrene
84	–	–	–	87	90	92	74	42	70	Diphenoxylate
89	59	57	70	80	87	90	78	16	71	Quazepam M (2-Oxo-)
99	91	92	95	99	80	88	86	55	99	Methidathion
99	91	93	94	99	81	87	84	40	99	Azinphos-methyl

8.11 hR_f Data in TLC System 10 in Ascending Numerical Order vs. Substance Names

TLC System 10	1	2	3	4	5	6	7	8	9	[System 10: Ac / Silica-KOH]
0	–	–	–	–	–	–	0	0	0	Diminazene Aceturate
0	–	–	–	–	–	–	1	0	0	Betanidine
0	–	–	–	–	–	–	1	0	0	Debrisoquine
0	–	–	–	–	–	–	1	0	0	Hydroxystilbamidine
0	–	–	–	–	–	–	1	0	1	Pentapiperide
0	–	–	–	–	–	–	2	0	0	Aminophylline
0	–	–	–	–	–	–	2	0	0	Buformin
0	–	–	–	–	–	–	2	0	0	Clamoxyquine
0	–	–	–	–	–	–	3	0	0	Guanoclor
0	–	–	–	–	–	–	3	0	0	Phenformin
0	–	–	–	–	–	–	13	0	5	Codeine M (Nor-)
0	–	–	–	–	–	–	18	1	3	Codeine M (N-Oxide)
0	–	–	–	–	–	–	46	0	0	Pseudomorphine
0	–	–	–	–	–	–	51	0	3	Minoxidil
0	–	–	0	–	0	–	1	1	0	Pentamidine
0	–	–	–	–	1	4	5	1	1	Piperazine
0	–	–	–	–	73	–	59	0	0	Isoetarine
0	–	–	–	0	0	0	0	0	0	Diquat Dibromide
0	–	–	–	0	0	0	0	0	0	Paraquat
0	–	–	–	0	0	0	1	0	0	Suxamethonium Chloride
0	–	–	–	0	0	1	0	0	0	Hexamethonium
0	–	–	–	0	0	1	0	0	0	Pentolinium
0	–	–	–	0	0	2	0	0	0	Decamethonium
0	–	–	–	0	0	2	0	0	0	Gallamine
0	–	–	–	0	0	11	0	0	0	Tubocurarine
0	–	–	–	0	0	12	0	0	0	Homatropine Methylbromide
0	–	–	–	0	0	15	1	0	0	Pancuronium Bromide
0	–	–	–	0	0	29	0	0	1	Cetrimide
0	–	–	–	0	0	29	0	0	1	Cetylpyridinium
0	–	–	–	0	1	25	2	0	0	Thiazinamium Metilsulfate
0	–	–	–	0	3	76	1	0	0	Guanoxan
0	–	–	–	0	4	23	0	0	0	Carbachol
0	–	–	–	0	16	11	17	0	0	Cocaine M/Ecgonine
0	–	–	–	0	20	0	–	–	–	Morphine M (-3-glucuronide)
0	–	–	–	0	30	2	–	0	0	Codeine M (-6-glucuronide)
0	–	–	–	0	34	14	52	0	3	Narceine
0	–	–	–	0	53	8	1	0	0	Obidoxime Chloride
0	0	0	0	0	70	16	58	0	0	Lysergic Acid
0	–	–	–	0	80	1	5	0	0	Psilocybine
0	–	–	–	1	2	18	1	0	0	Thiamine Hydrochloride
0	–	–	–	1	2	36	3	0	1	Oxyphenonium Bromide
0	–	–	–	1	3	18	2	0	0	Methscopolamine Bromide
0	–	–	–	1	3	30	1	0	2	Guanethidine
0	–	–	–	2	3	41	8	0	0	Butylscopolammonium Bromide,N-

Table 8.11. (Continued)

TLC System 10	1	2	3	4	5	6	7	8	9	[System 10: Ac / Silica-KOH]
0	–	–	–	2	4	66	2	0	4	Benzalkonium Chloride
0	–	–	–	3	3	41	5	0	5	Isopropamide Iodide
0	–	–	–	3	43	–	0	0	0	Levarterenol/Norepinephrine
0	–	–	–	4	3	31	4	0	4	Propantheline Bromide
0	–	–	–	5	4	39	2	0	4	Fenpiverinium
0	–	–	–	6	2	18	2	1	3	Oxyphencyclimine
0	–	–	–	7	0	3	13	0	0	Histamine
0	–	–	–	7	5	44	17	0	0	Morphine M (Nor-)
0	–	–	–	8	4	56	7	38	0	Berberine
0	0	0	7	8	82	72	75	0	2	Cinchophen
0	–	–	–	10	80	88	5	62	0	Phencyclidine interm./1-Piperidino-1-cyclo-hexanecarbonitrile/PCC
0	–	–	–	12	8	67	33	1	1	Phenylephrine
0	–	–	–	13	3	–	0	0	1	Epinephrine
0	–	–	–	16	27	20	41	0	9	Clomipramine M (N-Oxid)
0	–	–	–	18	23	16	37	0	9	Mianserin M (N-Oxide)
0	–	–	–	26	1	22	2	38	1	Cotarnine
0	–	–	–	27	3	–	10	6	6	Mequitazine
0	–	–	–	37	1	15	0	46	0	Hydrastinine
0	–	–	–	38	23	–	55	3	6	Bamethan
0	–	–	–	43	14	59	18	0	0	Dopamine
0	–	–	0	80	93	48	1	0	0	Metformin
1	–	–	–	–	–	–	1	0	0	Stilbamidine
1	–	–	–	–	–	–	1	1	1	Picloxydine
1	–	–	–	–	–	–	1	1	1	Propamidine
1	–	–	–	–	–	–	3	0	1	Chlorproguanil
1	–	–	–	–	–	–	4	1	1	Acefylline Piperazine
1	–	–	–	–	–	–	9	2	1	Dextrorphan M/Levorphanol M (Nor-)
1	–	–	–	–	–	–	13	24	4	Deptropine
1	–	–	–	–	–	–	22	5	4	Trimetazidine
1	–	–	0	–	69	–	5	0	1	Cytarabine
1	–	–	–	2	60	75	49	1	1	Methyldopa
1	–	–	–	14	3	–	10	0	3	Azacyclonol
1	–	–	–	15	–	–	55	36	2	Dihydralazine
1	–	–	–	18	7	79	3	0	1	Proguanil/Chlorguanide
1	–	–	–	18	12	–	25	2	3	Dihydromorphine
1	–	–	–	20	14	–	42	1	1	Nadolol
1	–	–	–	20	18	23	37	0	9	Morphine
1	–	–	–	23	7	27	1	5	1	Homatropine
1	–	–	–	24	5	28	18	5	3	Atropine/Hyoscyamine
1	–	–	–	25	5	7	16	0	17	Brucine
1	–	–	–	25	10	64	30	5	5	Ephedrine
1	–	–	–	28	6	37	27	2	8	Dothiepin M (Nor-,Sulfoxide)
1	–	–	–	30	11	–	38	3	6	Mesoridazine
1	–	–	–	32	8	75	33	1	4	Clomipramine M (8-Hydroxy-nor-)
1	–	–	–	33	10	34	35	0	1	Bufotenine

Table 8.11. (Continued)

TLC System 10	1	2	3	4	5	6	7	8	9	[System 10: Ac / Silica-KOH]
1	–	–	–	34	80	–	9	1	1	Oxymetazoline
1	–	–	–	71	12	–	3	1	0	Hexoprenaline
2	–	–	–	–	–	–	22	18	7	Apoatropine
2	–	–	–	–	–	–	25	34	8	Mephentermine
2	–	–	–	–	–	–	34	14	9	Racemorphan
2	–	–	–	–	–	–	40	1	10	Cytisine
2	–	–	–	–	–	–	43	6	21	Dehydroemetine
2	–	–	–	–	–	–	44	1	1	Cycloserine
2	–	–	–	–	4	–	6	4	2	Tramazoline
2	–	–	–	0	1	–	15	0	1	Amidefrine
2	0	0	2	2	85	59	84	0	1	Metamizol/Dipyrone
2	–	–	–	18	12	14	23	3	9	Hydromorphone
2	–	–	–	22	14	–	45	0	2	Atenolol
2	–	–	0	24	6	74	24	0	1	Amiloride
2	–	–	–	25	3	55	13	2	2	Tolazoline
2	–	–	–	25	15	–	36	3	18	Pholcodine
2	–	–	–	26	5	60	13	7	2	Tetryzoline
2	–	–	–	29	11	19	26	8	13	Dihydrocodeine
2	–	–	–	32	7	–	23	26	4	Methoxyphenamine
2	–	–	–	32	8	11	26	8	19	Strychnine
2	–	–	–	32	8	63	–	7	8	Pethidine M (Nor-)
2	–	–	–	33	6	–	32	1	3	Phentolamine
2	–	–	–	36	6	–	13	26	6	Benzatropine
2	–	–	–	36	6	71	15	18	5	Maprotiline
2	–	–	–	38	6	69	19	18	7	Protriptyline
2	1	0	28	40	49	39	63	1	36	Adinazolam M (Nor-)
2	–	–	0	46	4	14	38	14	4	Chloroquine
2	–	–	–	46	12	21	45	35	18	Chlorphenamine/Chlorpheniramine
2	–	–	–	66	6	68	20	32	10	Cyclopentamine
2	–	–	0	76	12	–	30	3	2	Ethambutol
3	–	–	–	–	–	–	1	0	1	Benserazide
3	–	–	–	–	–	–	28	49	3	Conessine
3	–	–	–	–	–	–	33	34	5	Pentaquine
3	–	–	–	–	–	–	37	9	9	Diphenazoline
3	–	–	–	–	–	–	51	0	6	Broxyquinoline
3	–	–	–	–	12	12	30	4	13	Methenamine
3	–	–	–	–	33	32	53	3	25	Acetophenazine
3	–	–	–	2	22	14	21	0	1	Cocaine M/N-Benzoylecgonine
3	–	–	–	3	44	–	36	1	3	Penicillamine
3	–	–	–	21	14	69	40	0	1	Isoprenaline
3	–	–	–	22	15	74	41	2	2	Etilefrine
3	–	–	–	27	3	52	14	3	6	Naphazoline
3	–	–	–	27	9	–	29	3	3	Pholedrine
3	–	–	–	27	11	14	–	5	13	Metopon
3	–	–	–	30	5	64	13	7	5	Xylometazoline
3	–	–	–	35	21	22	33	6	18	Codeine

Table 8.11. (Continued)

TLC System 10	1	2	3	4	5	6	7	8	9	[System 10: Ac / Silica-KOH]
3	–	–	–	37	7	–	45	1	2	Hydroxychloroquine
3	–	–	–	40	7	71	26	19	11	Desipramine/Imipramine M
3	–	–	–	40	21	69	44	2	7	Hydroquinine
3	–	–	–	42	10	49	35	14	4	Dextrorphan/Levorphanol
3	–	–	–	45	8	75	34	24	13	Clomipramine M (Nor-)
3	–	–	–	46	14	26	45	35	13	Pheniramine
3	–	–	–	47	5	66	31	6	7	Antazoline
3	–	–	–	47	21	23	48	25	37	Perazine
3	–	–	–	49	87	–	52	0	1	Dobutamine
4	–	–	–	–	–	–	16	51	2	Mecamylamine
4	–	–	–	–	–	–	31	10	5	Metazocine
4	–	–	–	–	–	–	35	5	12	Neopine
4	–	–	–	–	–	–	45	0	1	Practolol
4	–	–	–	–	68	64	70	2	25	Nialamide
4	–	–	71	6	80	–	53	3	8	Azathioprine
4	–	0	–	20	16	74	46	1	1	Salbutamol/Albuterol
4	–	0	–	25	38	81	76	0	1	Fenoterol
4	–	–	13	30	50	65	51	1	8	Triamterene
4	–	–	–	33	11	13	25	4	20	Hydrocodone
4	–	–	–	35	7	77	23	17	7	Amantadine
4	–	–	–	45	26	65	51	2	11	Quinine
4	–	–	–	46	87	71	34	28	16	Amitriptyline M/Nortriptyline
4	–	–	–	49	26	33	47	15	38	Diamorphine
4	–	–	–	50	13	16	48	26	19	Carbinoxamine
4	–	–	–	55	12	41	40	45	19	Ethoheptazine
4	–	–	68	80	88	–	49	2	10	Procarbazine
5	–	–	–	–	–	–	40	5	6	Hordenine
5	–	–	–	–	–	–	52	17	26	Diethylcarbamazine
5	–	–	–	–	–	–	87	0	41	Glipizide
5	–	–	–	15	75	67	59	0	8	Pyridoxine Hydrochloride
5	–	–	–	21	18	77	47	1	1	Terbutaline
5	–	–	–	22	14	–	23	1	2	Heptaminol
5	–	–	–	30	19	–	53	1	3	Sotalol
5	–	–	–	37	5	70	26	39	13	Coniine
5	–	–	–	40	4	10	5	67	3	Sparteine
5	3	1	36	40	26	15	40	1	48	Loprazolam
5	–	–	–	42	9	63	31	28	13	Deoxyephedrine,(+/-)-
5	–	–	–	42	9	63	31	28	13	Methamfetamine
5	–	–	–	43	20	70	45	3	8	Hydroquinidine
5	–	–	–	44	19	22	45	17	28	Dothiepin M (Nor-)
5	–	–	–	44	19	61	49	6	12	Cinchonine
5	–	–	–	45	23	32	45	24	37	Thebaine
5	–	–	–	46	20	22	45	11	28	Dothiepin M (Sulfoxide)
5	–	–	–	52	26	–	53	28	37	Butaperazine
6	–	–	–	–	–	–	29	32	15	Methdilazine
6	–	–	–	–	–	–	44	12	15	Methyldesorphine

Table 8.11. (Continued)

TLC System 10	1	2	3	4	5	6	7	8	9	[System 10: Ac / Silica-KOH]
6	–	–	–	–	–	–	65	1	3	Protokylol
6	–	–	–	–	21	38	39	18	24	Dimethoxanate
6	–	–	–	16	12	–	47	10	44	Hexobendine
6	–	–	–	18	21	77	48	1	3	Orciprenaline
6	–	–	–	33	13	71	47	0	3	Acebutolol
6	–	–	–	36	21	26	40	7	22	Ethylmorphine
6	–	–	–	36	25	–	42	6	23	Viloxazine
6	–	–	–	37	26	–	47	2	9	Ketobemidone
6	–	–	–	40	18	–	42	3	34	Nicomorphine
6	–	–	–	43	22	22	46	7	34	Thioproperazine
6	–	–	–	44	24	55	49	6	8	Cinchonidine
6	–	–	–	47	10	–	42	35	13	Dimetindene
6	–	–	–	47	10	42	33	42	18	Dextromethorphan/Racemethorphan/Levomethorphan
6	–	–	–	48	12	–	47	17	13	Pipazetate
6	–	–	–	49	12	–	45	33	16	Brompheniramine
6	–	–	–	49	30	63	51	4	12	Quinidine
6	–	–	–	50	13	28	42	35	22	Tropacocaine
6	–	–	–	50	14	39	40	9	9	Dimethyltryptamine,N,N-/DMT
6	–	–	–	51	25	60	46	11	31	Mianserin M (Nor-)
6	–	–	–	55	19	30	51	11	20	Triprolidine
6	–	–	–	58	83	85	80	0	55	Flubendazole
6	–	–	–	59	8	–	41	57	8	Perhexiline
7	–	–	–	–	–	–	42	39	21	Tigloidine
7	–	–	–	24	82	21	–	6	28	Morphine-3-acetate
7	3	2	26	26	73	69	67	1	25	Alprazolam M (4-Hydroxy-)
7	–	–	–	36	57	51	60	0	19	Lysergamide
7	–	–	–	38	35	39	54	6	22	Opipramol
7	–	–	–	39	39	–	54	5	27	Carfenazine
7	–	–	–	44	26	24	49	10	40	Tiotixene
7	–	–	–	48	14	–	45	26	21	Tofenacin
7	–	–	–	49	21	79	50	6	10	Propranolol
7	–	–	–	55	26	26	49	34	37	Prochlorperazine
7	81	69	81	87	82	93	75	49	81	Lormetazepam Hy/MDB
8	–	–	–	–	–	–	33	18	17	Desomorphine
8	–	–	–	–	–	–	40	17	8	Phenglutarimide
8	–	–	–	–	–	–	42	8	6	Profadol
8	–	–	–	–	–	–	53	1	19	Cephaeline
8	–	–	14	–	81	–	76	0	1	Sulfacarbamid
8	3	1	29	24	75	71	63	0	29	Triazolam M (4-Hydroxy-)
8	–	–	–	33	62	60	57	0	12	Ergometrine
8	–	–	0	38	70	–	55	0	4	Ambazone
8	–	–	–	41	20	23	41	6	23	Benzylmorphine
8	–	–	–	41	69	66	76	0	41	Domperidone
8	–	–	–	43	18	78	49	2	5	Pindolol
8	–	–	–	43	33	61	56	1	12	Pipamperone

Table 8.11. (Continued)

TLC System 10	1	2	3	4	5	6	7	8	9	[System 10: Ac / Silica-KOH]
8	–	–	–	44	25	27	46	6	26	Morphine-6-acetate/Diamorphine M
8	–	–	–	48	14	77	44	18	17	Chlorphentermine
8	–	0	–	50	22	80	50	13	15	Penbutolol
8	–	–	–	53	27	–	51	30	41	Thiethylperazine
8	3	4	41	53	73	72	70	3	52	Midazolam M (α-Hydroxy-)
8	–	–	–	55	30	29	53	33	30	Trifluoperazine
8	–	–	–	59	16	74	47	47	19	Etilamfetamine
8	–	–	–	60	12	–	50	42	19	Aminopromazine
8	–	–	–	68	23	49	46	37	28	Diphenylpyraline
9	–	–	–	–	–	–	43	15	5	Mepacrine
9	–	–	70	30	40	88	56	0	5	Clioquinol
9	–	–	–	39	17	33	49	1	5	Procainamide
9	–	–	–	42	40	40	55	7	29	Perphenazine
9	–	–	–	44	20	74	49	10	8	Metoprolol
9	–	–	–	47	14	48	39	5	9	Psilocin
9	–	–	–	50	16	–	44	36	22	Benzydamine
9	–	–	–	50	20	75	52	6	11	Timolol
9	–	–	–	54	25	42	49	15	19	Clomipramine M (10-Hydroxy-)
9	–	–	–	55	25	78	40	17	4	Mexiletine
9	–	–	–	58	18	49	46	49	25	Clemastine
9	–	–	–	59	15	29	47	43	23	Prothipendyl
9	–	–	–	60	12	–	48	41	10	Doxylamine
9	–	–	–	71	11	28	54	41	21	Diamocaine
10	–	–	–	–	–	–	24	68	3	Pempidine
10	–	–	–	–	–	–	25	1	0	Nordefrin Hydrochloride
10	–	–	–	–	–	–	43	5	4	Tacrine
10	–	–	–	–	–	–	46	45	30	Proheptazine
10	–	–	–	2	87	46	88	0	10	Dimenhydrinate
10	–	–	–	45	45	49	63	5	23	Fluphenazine
10	–	–	–	47	19	–	43	12	42	Nicocodine
10	–	–	–	51	26	51	51	4	17	Clomipramine M (8-Hydroxy-)
11	–	–	–	–	–	–	35	2	2	Hydroxyamphetamine/Amphetamine M
11	–	–	–	–	–	–	50	29	25	Betaprodine
11	–	–	–	–	–	–	50	30	35	Alphaprodine
11	–	–	–	–	–	–	62	27	45	Morpheridine
11	–	9	–	11	74	66	75	1	30	Theophylline
11	–	–	–	26	6	–	23	1	2	Tryptamine
11	–	–	–	35	56	–	53	0	41	Demecolcine
11	–	–	–	44	45	–	56	7	32	Clopenthixol/Zuclopenthixol/cis-Ordinol
11	–	–	–	45	26	51	70	1	24	Clomipramine M (2-Hydroxy-)
11	–	–	–	49	22	–	52	11	12	Alprenolol
11	–	–	–	49	24	25	45	20	34	Thebacon
11	0	0	15	50	28	62	58	8	19	Flurazepam M (Monodeethyl-)
11	–	–	–	60	19	–	48	41	16	Fenfluramine
11	–	–	–	60	34	40	52	37	34	Pethidine/Meperidine
11	–	–	–	62	18	35	44	38	30	Promazine

Table 8.11. (Continued)

TLC System 10	1	2	3	4	5	6	7	8	9	[System 10: Ac / Silica-KOH]
11	–	–	–	63	14	56	46	14	10	Diethyltryptamine,N,N-/DET
12	–	–	–	–	–	–	43	30	22	Methopromazine
12	–	–	–	–	–	–	46	18	13	Eucatropine
12	–	–	–	–	–	–	56	0	11	Metopimazine
12	–	–	–	–	19	36	53	42	25	Thenyldiamine
12	–	–	–	24	6	63	20	3	10	Mescaline
12	–	–	–	25	70	59	48	0	12	Diprophylline/Propyphylline
12	–	–	–	27	53	55	54	0	9	Cimetidine
12	–	–	–	29	7	70	33	6	11	Trimethoxyamphetamine,3,4,5-
12	–	–	–	33	69	63	55	0	37	Colchicine
12	–	–	–	41	69	–	62	0	14	Methylergometrine
12	–	–	–	44	52	45	53	0	32	Pilocarpine
12	–	–	20	45	45	59	55	0	22	Trimethoprim
12	–	–	–	48	11	78	46	26	24	Phentermine
12	–	–	–	52	20	–	50	38	44	Thenalidine
12	–	–	–	53	19	–	54	13	34	Emetine
12	–	–	–	54	26	25	44	23	43	Acetylcodeine
12	–	–	–	59	39	50	–	21	48	Cocaine M (Nor-)
12	–	–	–	61	21	–	50	45	25	Metixene
12	–	–	–	62	18	–	48	17	24	Carbocromen
12	–	–	–	63	28	–	48	26	24	Acepromazine
13	5	4	37	28	76	76	68	1	35	Brotizolam M (6-Hydroxy-)
13	–	–	–	45	20	78	48	11	11	Oxprenolol
13	–	–	–	51	17	–	47	1	7	Metoclopramide
13	–	–	–	53	46	65	63	7	13	Mazindol
13	–	–	–	59	24	–	49	40	43	Bamipine
13	–	–	–	60	9	7	45	7	8	Disopyramide
13	–	–	–	61	39	22	54	39	35	Nicotine
13	–	–	–	63	24	45	51	48	37	Doxepin
13	–	–	–	64	30	50	51	45	44	Cyproheptadine
13	–	–	–	65	29	73	55	36	38	Noracymethadol
13	–	–	–	66	21	24	52	41	26	Methapyrilene
13	–	–	–	67	20	55	48	42	30	Thioridazine
13	–	–	–	67	21	47	48	48	23	Imipramine
13	–	–	–	69	27	48	54	42	43	Bromazine/Bromdiphenhydramine
14	–	–	–	–	–	–	49	26	14	Dimethylphenylthiazolanimin (Dimethyl-5-phenyl-2-thiazolidinimine,3,4-
14	–	–	–	31	6	61	26	5	11	Dimethoxyphenethylamine,3,4-
14	2	5	33	38	78	81	75	4	25	Midazolam M (α,4-Dihydroxy-)
14	–	–	–	43	8	75	36	24	10	Dimethoxy-4-ethylamphetamine,2,5-/DOET
14	–	–	–	43	41	52	65	0	17	Pipamazine
14	–	–	–	46	34	45	50	14	27	Phenmetrazine/Phendimetrazine M (Nor-)
14	7	2	40	47	67	66	67	1	57	Alprazolam
14	–	–	–	54	44	–	55	3	45	Fenetylline
14	–	–	–	55	38	22	54	22	35	Dibenzepin
14	–	–	–	58	22	33	51	39	25	Mepyramine/Pyrilamine

Table 8.11. (Continued)

TLC System 10	1	2	3	4	5	6	7	8	9	[System 10: Ac / Silica-KOH]
14	–	–	–	64	22	35	52	41	30	Isothipendyl
14	–	–	–	65	22	31	55	38	28	Thonzylamine
14	–	–	–	67	35	52	57	42	46	Chlorcyclizine
14	–	–	–	68	16	49	48	47	22	Pentoxyverine/Carbetapentane
15	–	–	–	–	–	–	19	13	5	Primaquine
15	–	–	–	–	–	–	65	1	21	Methysergide
15	–	–	–	38	61	–	52	10	7	Oxetacaine
15	–	–	–	48	25	–	54	3	16	Fenpipramide
15	5	5	37	49	76	77	74	1	43	Midazolam M (4-Hydroxy-)
15	–	–	–	56	14	–	44	35	9	Etafedrine
15	–	–	–	65	27	48	55	44	33	Diphenhydramine
15	–	–	–	67	32	–	53	38	48	Phenyltoloxamine
15	–	–	–	68	22	34	55	44	27	Tripelennamine
15	–	–	–	68	26	–	51	52	32	Tolpropamine
15	–	–	–	68	28	48	55	45	33	Dimenhydrinate
15	–	–	–	69	27	51	51	50	32	Amitriptyline
16	–	–	–	–	–	–	1	33	40	Alphameprodine
16	–	–	–	–	–	–	47	49	34	Iprindole/Glycophene
16	–	–	–	–	–	–	50	30	53	Myrophine
16	–	–	–	–	–	–	56	39	44	Properidine
16	–	–	–	–	–	–	57	1	43	Inositol Nicotinate
16	–	–	–	41	9	76	51	15	17	Dimethoxy-4-methylamphetamine,2,5- /STP/DOM
16	–	–	–	42	58	–	60	1	28	Dihydroergotamine
16	5	2	41	44	68	65	60	1	40	Triazolam
16	–	–	–	64	43	39	57	15	32	Tetracaine/Amethocaine
16	–	1	–	65	27	–	53	47	44	Pecazine/Mepazine
16	–	–	–	68	25	49	55	48	33	Orphenadrine
16	–	–	–	68	40	52	57	49	41	Cyclizine
16	–	–	–	70	27	41	51	49	42	Dothiepin
17	–	–	–	–	–	–	38	13	11	Bromo-STP/Bromo-2,5-dimethoxy-4-methyl-amphetamine
17	–	–	–	–	–	–	52	30	30	Dimazole/Diamthazole
17	–	–	–	–	–	–	54	12	39	Etamiphyllin
17	–	–	–	–	–	–	54	50	32	Embramine
17	–	–	–	–	–	–	56	6	19	Piperacetazine
17	–	–	–	–	–	–	58	41	41	Trimeperidine
17	–	–	–	–	–	–	59	0	21	Rolicyprine
17	–	–	–	42	10	76	39	18	12	Methylenedioxyamphetamine,3,4-/MDA
17	–	–	–	42	70	69	69	1	23	Iproniazid
17	–	–	–	43	10	76	36	15	14	Bromo-2,5-dimethoxyamphetamine,4-/DOB
17	–	–	–	45	8	75	36	19	10	Dimethoxyamphetamine,2,5-
17	–	–	–	54	86	–	49	3	45	Phanquinone
17	4	5	29	55	42	–	57	4	38	Clozapine
17	–	–	–	59	30	–	50	33	32	Nefopam
17	–	–	0	63	22	–	52	41	28	Chloropyramine

Table 8.11. (Continued)

TLC System 10	1	2	3	4	5	6	7	8	9	[System 10: Ac / Silica-KOH]
17	–	–	–	65	30	44	50	36	35	Promethazine
17	–	–	–	66	27	74	54	42	27	Eucaine,β-
17	–	–	–	70	25	45	49	45	35	Chlorpromazine
17	–	–	–	70	29	–	53	47	36	Chlorphenoxamine
17	–	–	–	71	17	–	59	4	35	Veratrine
18	–	–	–	–	–	–	48	28	11	Butaxamine
18	–	–	–	–	–	–	60	9	42	Etoxeridine
18	–	–	–	11	8	–	25	3	1	Synephrine/Oxedrine
18	4	4	37	37	76	76	72	1	44	Alprazolam M (α-Hydroxy-)
18	–	–	–	43	9	74	35	23	10	Methoxy-4,5-methylenedioxyamphetamine,2- /MMDA-2
18	–	–	–	43	12	75	43	20	9	Amphetamine/Amphetaminil M artifact
18	–	–	–	48	49	47	55	6	37	Scopolamine
18	–	–	–	51	46	61	58	4	16	Periciazine
18	–	–	–	54	46	52	60	4	37	Mianserin M (8-Hydroxy-)
18	–	–	–	55	41	38	55	12	36	Physostigmine
18	3	–	–	56	60	59	60	3	39	Lysergide/LSD
18	–	–	–	63	33	–	56	40	34	Methadone M (Nor-)
18	–	–	–	66	29	–	53	43	35	Noxiptiline
18	–	–	–	71	25	66	54	54	37	Pyrrobutamine
18	–	–	–	72	26	54	51	53	34	Clomipramine
19	–	–	–	–	–	–	52	31	44	Moxisylyte
19	–	–	–	54	57	65	68	10	54	Hydroxyzine
19	13	5	53	60	69	70	72	6	60	Midazolam
19	–	–	–	67	19	–	50	51	22	Oxeladine
20	–	–	–	–	–	–	57	18	15	Clorprenaline
20	–	–	18	29	55	49	47	1	11	Isoniazid
20	–	–	–	65	36	46	57	27	45	Mebhydrolin
20	–	–	–	68	35	–	56	41	35	Nicametate
20	–	–	–	79	30	91	70	48	40	Hexetidine
21	–	–	–	–	–	–	52	5	24	Amopyroquine
21	–	–	–	–	–	–	54	46	42	Methaphenilene
21	–	–	–	–	–	–	62	50	41	Dimethylthiambutene
21	–	4	–	34	59	54	53	1	31	Theobromine
21	4	3	39	42	75	74	71	4	49	Triazolam M (α-Hydroxy-)
21	–	–	–	46	65	14	52	2	52	Endralazine Mesilate
21	–	–	–	53	40	59	66	3	32	Pipotiazine
21	–	–	–	58	66	–	61	2	31	Pyrimethamine
21	–	–	–	68	41	49	63	45	57	Phenindamine
22	–	–	–	44	68	64	63	1	34	Ergotamine
22	–	–	–	48	74	69	56	4	17	Nicotinyl Alcohol
22	10	10	53	52	76	77	62	2	50	Chlordiazepoxide
22	9	2	51	57	68	66	69	6	55	Adinazolam
22	–	–	–	67	46	–	77	26	42	Meclofenoxate
22	–	–	–	74	52	81	70	9	32	Loperamide
22	–	–	–	75	32	49	54	47	35	Triflupromazine

Table 8.11. (Continued)

TLC System 10	1	2	3	4	5	6	7	8	9	[System 10: Ac / Silica-KOH]
23	–	–	–	–	–	–	59	41	59	Clotiapine
23	–	–	–	–	–	–	66	0	53	Naftazone
23	–	–	–	66	40	70	57	35	41	Methylphenidate
23	–	–	–	68	48	50	58	39	58	Mianserin
23	–	–	–	74	20	68	48	62	31	Procyclidine
24	–	–	–	–	–	–	33	1	7	Tuaminoheptane
24	–	–	–	–	–	–	42	1	2	Ethylnorepinephrine
24	–	–	–	–	–	–	58	35	49	Allylprodine
24	–	–	–	–	–	–	59	15	48	Dimefline
24	–	–	–	–	5	42	26	11	0	Betazole
24	–	–	–	18	13	76	42	1	1	Metaraminol
24	–	–	–	62	49	41	57	36	51	Phendimetrazine
25	–	–	–	–	–	–	22	28	11	Octamylamine
25	–	–	–	–	–	–	58	55	36	Cyclomethycaine
25	–	–	–	–	–	–	60	49	40	Dyclonine
25	–	–	–	–	–	–	84	1	20	Salinazid
25	7	5	44	50	71	60	71	2	53	Estazolam
25	15	10	55	52	59	55	52	3	58	Caffeine
25	–	–	–	59	24	33	48	18	36	Oxomemazine
25	–	–	–	61	49	43	–	13	59	Orientalidine
25	–	–	–	65	15	67	–	9	15	Psilocin-(eth)/CZ-74
25	–	–	–	65	24	74	53	15	13	Cyclazocine
25	–	–	–	74	34	51	56	51	51	Chlorprothixene
25	–	–	–	79	22	–	50	67	32	Prolintane
26	–	–	–	–	–	–	57	27	39	Cyclopentolate
26	–	–	–	61	49	43	55	14	57	Isothebaine
26	–	–	–	68	30	52	55	34	42	Propiomazine
26	–	–	–	78	30	73	56	63	37	Volazocine
26	–	–	–	79	15	65	–	66	25	Rolicyclidine/PHP/1-(1-phenylcyclohexyl)pyrrolidine
26	–	–	–	80	18	–	52	62	19	Terodiline
27	–	–	–	40	68	66	54	0	21	Nicotinamide
27	15	5	53	52	72	71	72	5	52	Brotizolam
27	–	–	–	58	54	–	83	0	21	Apomorphine
27	–	–	–	76	21	56	55	53	37	Piperocaine
27	–	–	–	77	16	60	48	59	20	Methadone/Levomethadone
28	–	–	–	–	–	–	58	3	33	Propoxycaine
28	–	–	–	–	–	–	59	57	42	Ethylmethylthiambutene
28	–	–	–	–	–	–	61	55	46	Quinisocaine
28	–	–	–	–	65	68	63	0	22	Harmine
28	0	6	26	39	73	61	61	0	24	Bromazepam M (3-Hydroxy-)
28	23	41	42	43	82	82	52	1	36	Lorazepam/Lormetazepam M
28	–	–	–	66	43	–	56	13	48	Dimetiotazine
28	–	–	–	70	34	72	61	16	12	Pentazocine
29	–	–	–	–	–	–	69	37	60	Furethidine
29	–	–	–	–	32	–	52	41	37	Clofedanol

Table 8.11. (Continued)

TLC System 10	1	2	3	4	5	6	7	8	9	[System 10: Ac / Silica-KOH]
29	–	–	–	5	83	–	76	0	65	Glymidine
29	–	–	–	32	57	59	59	1	23	Nalorphine
29	0	0	14	40	20	60	58	2	18	Flurazepam M (Dideethyl-)
29	–	–	–	49	71	–	58	2	33	Proxyphylline
29	–	–	–	57	46	–	60	1	60	Vinblastine
29	–	–	–	59	71	67	59	15	56	Nikethamide
29	–	–	–	64	64	67	75	0	47	Ergosine
29	–	–	–	70	23	79	–	41	33	Phenylcyclohexylamine,1-
29	–	–	–	75	23	–	61	17	35	Lobeline
29	–	–	–	75	78	–	57	32	27	Octacaine
30	–	–	–	33	27	36	48	10	37	Oxymorphone
30	–	–	–	71	36	42	54	5	31	Procaine
30	–	–	–	77	21	–	54	62	34	Fencamfamine
31	–	–	–	–	45	55	63	5	26	Butethamine
31	7	6	41	45	78	78	72	2	46	Brotizolam M (α-Hydroxy-)
31	–	–	–	58	61	–	58	8	46	Morazone
31	–	–	–	64	53	52	56	9	29	Nomifensine
31	–	–	–	77	32	46	58	54	39	Trimeprazine/Alimemazine
32	–	–	–	–	–	–	54	8	49	Morinamide
32	–	3	–	60	62	69	78	3	51	Benperidol
33	–	–	–	–	–	–	64	0	30	Norharman
33	–	0	–	30	12	75	44	4	4	Ephedrine M/Phenylpropanolamine
33	–	–	–	57	72	76	70	2	40	Harman
33	–	–	–	58	60	–	57	3	44	Nimorazole
33	–	–	–	73	58	50	60	30	58	Methylpiperidyl Benzilate/JB-336
33	–	–	–	76	51	75	67	11	27	Haloperidol
34	–	–	–	–	–	–	57	2	54	Tozalinone
34	–	–	–	59	71	–	65	0	54	Bamifylline
35	–	–	–	–	–	–	37	50	11	Rimantadine
35	–	–	–	–	–	–	62	26	41	Proxymetacaine
35	–	–	–	67	42	–	63	25	34	Cinchocaine/Dibucaine
35	–	–	–	78	43	–	64	52	41	Naftidrofuryl
35	–	–	–	82	30	–	60	56	52	Clomifene
36	–	–	–	–	–	–	60	34	47	Ethomoxane
36	–	–	–	–	–	–	64	55	38	Betacetylmethadol
36	–	–	–	–	–	–	65	7	69	Benzquinamide
36	–	–	–	58	71	73	67	2	48	Droperidol
36	–	–	–	70	70	77	70	11	63	Lidoflazine
36	–	–	–	82	27	77	55	66	44	Prozapine
36	–	–	–	83	54	–	62	23	41	Oxybuprocaine
37	–	–	–	–	–	–	58	34	57	Betameprodine
37	–	–	–	–	–	–	59	5	23	Chloroprocaine
37	–	–	–	–	–	–	62	8	40	Amodiaquine
37	–	–	–	66	64	61	63	10	58	Trazodone
37	–	–	–	79	21	78	–	65	27	Eticyclidine/PCE/N-Ethyl-1-phenylcyclohexyl-amine

Table 8.11. (Continued)

TLC System 10	1	2	3	4	5	6	7	8	9	[System 10: Ac / Silica-KOH]
37	–	–	–	80	36	56	59	62	54	Trimipramine
37	–	7	56	80	77	80	73	11	67	Miconazole
38	–	–	–	–	–	–	58	56	43	Alphamethadol
38	–	–	–	–	–	–	59	61	48	Butriptyline
38	–	–	–	–	–	–	66	65	39	Alverine
38	–	–	–	–	–	–	68	59	50	Norpipanone
38	–	–	–	11	12	73	55	24	4	Methoxamine
38	15	28	35	49	88	89	67	0	30	Quazepam M (3-Hydroxy-N-dealkyl-2-oxo-)
38	–	–	–	58	69	–	66	1	48	Ergoloid
38	–	–	–	73	35	46	66	26	55	Pitofenone
39	–	–	–	54	44	–	49	28	28	Phenethylamine
39	–	–	–	61	51	72	74	4	49	Spiperone
39	–	–	–	62	30	33	50	25	51	Oxycodone
39	–	–	–	77	33	54	58	57	51	Diethazine
39	–	–	–	81	19	79	54	59	38	Pipradrol
40	–	–	–	–	–	–	43	0	40	Furaltadone
40	–	11	–	36	81	81	60	0	23	Pemoline
40	–	–	32	46	75	70	58	2	36	Metronidazole
40	–	3	–	46	81	78	67	1	38	Glafenine
40	3	3	41	71	52	45	62	30	48	Flurazepam
40	–	–	–	71	73	82	71	3	60	Pimozide
41	–	–	–	–	–	–	66	2	58	Pipobroman
41	–	–	–	–	–	–	82	1	28	Diazoxide
41	–	–	–	–	62	60	70	43	55	Pramoxine
41	–	–	–	64	66	–	58	16	58	Metyrapone
42	–	–	–	–	–	–	59	37	40	Aletamine
42	–	–	–	–	–	–	61	60	52	Etymemazine
42	3	3	33	40	68	60	69	0	32	Chlordiazepoxide M (Nor-)
42	–	–	–	44	82	87	68	0	37	Dipyridamole
42	–	–	–	61	73	74	70	1	45	Piritramide
42	–	–	–	65	53	52	62	18	48	Levamisole
42	–	–	–	73	43	61	59	23	70	Verapamil
42	–	–	–	74	62	59	61	35	53	Thiopropazate
43	–	–	–	–	–	–	63	47	55	Methadone Intermediate/Pre-Methadone
43	–	–	–	–	–	–	64	48	59	Benethamine
43	–	–	–	–	–	–	70	61	43	Diethylthiambutene
43	–	–	–	–	–	–	74	58	45	Isomethadone
43	–	–	–	73	14	50	–	22	46	Cryptopine
43	–	–	–	77	38	–	59	56	52	Benzoctamine
44	30	14	61	61	76	72	71	3	63	Flumazenil/RO 15-1788
44	–	–	–	70	51	67	71	21	48	Dexetimide
45	7	11	30	42	79	82	73	0	22	Clonazepam M (7-Acetamido-)
45	–	–	–	74	42	73	67	19	24	Levallorphan
45	–	–	–	77	48	–	62	36	67	Flavoxate
46	–	–	–	29	12	–	42	25	5	Pseudoephedrine M (Nor-)/Cathine
46	–	–	44	54	71	70	63	3	42	Pyrazinamide

Table 8.11. (Continued)

TLC System 10	1	2	3	4	5	6	7	8	9	[System 10: Ac / Silica-KOH]
46	10	20	30	54	77	79	71	0	30	Nitrazepam M (7-Amino-)
46	–	–	–	66	51	53	69	18	54	Tetramisole
46	–	–	–	66	72	–	55	6	66	Pentifylline
46	–	–	–	76	32	49	57	47	38	Levomepromazine/Methotrimeprazine
47	54	61	56	33	88	90	85	0	51	Chlorzoxazone
47	11	20	40	52	76	77	73	0	30	Clonazepam M (7-Amino-)
47	11	19	35	54	78	75	72	0	34	Flunitrazepam M (7-Aminonor-)
47	–	–	–	56	79	75	60	2	56	Carbamazepine
47	9	22	30	67	66	63	73	4	51	Medazepam M (Nor-)
47	–	–	–	69	74	74	61	8	65	Papaverine
47	–	–	–	78	51	80	69	18	39	Phenomorphan
47	–	–	–	81	48	56	69	59	57	Butetamate
48	–	–	–	–	–	–	66	1	62	Ergotoxine
48	11	31	34	36	86	90	69	0	31	Chloramphenicol
48	–	–	–	58	41	67	54	39	33	Tranylcypromine
48	–	–	–	66	63	60	65	31	62	Mepivacaine
48	–	–	–	74	77	81	71	23	65	Metomidate
49	–	–	–	–	–	–	67	58	62	Diethylaminoethyl Diphenylpropionate
49	7	12	24	44	80	83	71	0	20	Nitrazepam M (7-Acetamido-)
49	19	11	41	51	78	76	73	0	40	Flunitrazepam M (7-Acetamido-)
49	–	–	–	59	68	74	60	1	47	Prazosin
49	–	–	–	60	80	84	65	0	59	Mebendazole
49	–	–	–	71	63	78	69	4	59	Fluspirilene
49	–	–	–	74	50	81	68	16	39	Phenazocine
49	–	–	–	86	32	–	63	40	53	Mebeverine
50	–	–	–	–	–	–	60	31	63	Thurfyl Nicotinate
50	–	–	–	–	–	–	79	1	37	Orthocaine
50	–	–	–	57	74	77	77	0	59	Ergocristine
50	–	–	–	59	74	77	78	0	58	Ergocryptine
50	46	45	60	59	82	82	52	6	61	Lormetazepam
50	–	–	–	62	33	83	74	3	14	Buphenine
50	–	–	–	84	38	62	57	62	45	Alphacetylmethadol
51	–	–	–	–	–	–	61	56	45	Difenidol
51	15	24	42	41	81	83	63	0	35	Demoxepam/Chlordiazepoxide M
51	22	35	42	45	82	82	56	0	40	Oxazepam/M of Chlordiazepoxide, Clorazepic Acid, Camazepam, Diazepam, Halazepam, Ketazolam, Medazepam, Oxazolam, Prazepam
51	–	–	–	70	74	76	75	13	58	Bromolysergide,2-
51	–	–	–	79	60	66	73	12	56	Anileridine
51	9	8	41	83	45	51	67	53	60	Flurazepam Hy/DCFB
51	–	–	–	83	49	–	64	55	60	Adiphenine
52	–	–	–	–	–	–	65	60	58	Trifluomeprazine
52	38	37	64	57	81	80	71	1	59	Flunitrazepam M (3-Hydroxy-)
52	–	–	–	62	78	81	71	4	55	Carnidazole
52	21	21	52	63	77	74	74	1	55	Flunitrazepam M (7-Amino-)

Table 8.11. (Continued)

TLC System										
10	1	2	3	4	5	6	7	8	9	[System 10: Ac / Silica-KOH]
52	–	–	–	64	66	70	63	6	38	Yohimbine
52	–	–	–	68	80	–	61	11	72	Trimetozine
52	–	–	–	71	59	57	61	15	64	Hydrastine
52	55	55	70	72	85	88	71	12	–	Prazepam M (3-Hydroxy-)
52	–	–	–	73	78	85	67	26	71	Etomidate
52	–	–	–	78	76	73	78	31	69	Clemizole
52	–	–	–	86	23	75	55	68	36	Diisopromine
53	–	–	–	–	–	–	54	1	44	Niridazole
53	–	–	–	–	75	–	67	7	54	Tiabendazole
53	13	39	30	53	–	–	62	1	20	Styramate
53	–	–	–	62	62	81	78	3	32	Isoxsuprine
53	51	47	65	62	82	82	53	8	59	Temazepam/M of Camazepam, Diazepam, Ketazolam, Medazepam
53	13	18	47	63	73	69	61	6	41	Bromazepam
53	–	–	–	70	44	76	62	8	31	Clonidine
53	–	–	–	70	80	–	59	1	50	Phenazopyridine
53	–	–	–	77	52	57	66	43	53	Benactyzine
54	–	–	–	–	52	–	68	67	64	Dicycloverine/Dicyclomine
54	–	–	–	72	69	–	64	20	70	Doxapram
54	–	–	–	77	35	30	65	45	47	Cocaine
54	–	–	–	82	50	63	68	58	55	Propoxyphene/Dextropropoxyphene/Levopropoxyphene
55	–	–	–	–	–	–	65	0	36	Ethionamide
55	19	28	54	61	81	82	74	2	46	Flurazepam M (N(1)-Hydroxyethyl-)
55	35	46	53	64	84	86	68	0	36	Nitrazepam
55	–	–	–	69	83	79	68	28	67	Crotethamide
55	–	–	–	72	80	–	66	20	70	Propanidid
55	–	–	–	76	87	–	64	3	59	Ethenzamide
55	53	46	71	77	81	81	74	12	70	Nimetazepam
55	–	–	–	78	34	30	70	42	60	Cinnamoylcocaine
55	–	–	–	78	59	54	75	44	63	Ethylpiperidyl Benzilate/JB-318
55	–	–	–	81	45	–	68	58	54	Isoaminile
55	–	–	–	82	54	52	69	55	81	Piperidolate
55	–	–	–	82	54	64	72	62	68	Amiodarone
56	–	–	–	–	–	–	69	60	66	Proadifen
56	–	–	–	–	–	–	77	53	68	Pipoxolan
56	10	35	35	60	82	85	–	0	33	Mebutamate
56	26	42	52	63	89	93	77	0	52	Budesonide
56	35	41	57	72	82	86	73	5	58	Delorazepam
56	–	–	–	76	66	80	75	36	64	Articaine
56	–	–	–	78	14	–	70	57	67	Azapetine
56	–	–	–	78	79	84	70	36	80	Methaqualone
56	–	–	–	84	43	85	68	55	68	Prenylamine
57	–	–	–	–	–	–	75	44	72	Benzethidine
57	–	–	–	–	77	–	66	1	38	Protionamide
57	–	–	–	55	87	89	64	2	43	Mephenesin

Table 8.11. (Continued)

TLC System										
10	1	2	3	4	5	6	7	8	9	[System 10: Ac / Silica-KOH]
57	–	–	–	62	–	–	61	2	33	Amiphenazole
57	–	–	–	68	84	80	66	1	63	Chlormezanone
57	–	–	–	74	25	83	70	29	69	Cropropamide
57	37	41	54	81	80	79	72	5	57	Nitrazepam M Hy/DAB
58	–	–	–	–	–	–	68	16	75	Dimoxyline
58	9	36	32	56	63	87	75	0	32	Meprobamate
58	34	45	60	72	83	88	75	2	56	Flurazepam M (N(1)-Dealkyl-)/Quazepam M (N-Dealkyl-2-oxo-)
58	33	25	65	73	81	84	76	15	68	Metaclazepam M (N-Nor-)
58	–	–	–	75	42	68	68	11	39	Aconitine
58	–	–	–	76	70	82	71	26	64	Phenoperidine
58	–	–	–	76	80	85	64	44	69	Clomethiazole
58	–	–	–	78	70	77	70	43	74	Fentanyl
58	–	–	–	78	74	–	68	4	65	Oxypertine
59	–	–	–	–	–	–	44	0	47	Furazolidone
59	–	–	–	–	–	–	67	42	70	Clorgiline
59	–	–	–	41	80	–	62	1	52	Thiamazole
59	35	46	56	68	84	85	76	0	58	Flunitrazepam M (Nor-)
59	42	52	58	69	88	90	71	2	55	Quazepam M (3-Hydroxy-2-oxo-)
59	–	–	–	72	74	77	61	37	66	Methyl Nicotinate
59	39	45	63	75	78	84	73	0	66	Cloxazolam
59	–	–	–	76	80	86	69	8	52	Fenyramidol
59	58	49	72	76	82	85	75	27	73	Diazepam/Ketazolam M/Medazepam M
59	–	–	–	83	43	75	68	66	61	Trihexyphenidyl/Benzhexol
59	–	–	–	83	70	78	–	38	73	Fluorofentanyl,p-
59	–	–	–	88	63	77	67	36	64	Piminodine
60	–	–	–	–	–	–	63	42	71	Benzyl Nicotinate
60	–	–	–	–	–	–	66	8	69	Clonitazene
60	–	–	–	–	–	–	66	67	61	Cycrimine
60	34	46	57	68	83	87	84	3	56	Clorazepic Acid
60	34	45	57	69	82	83	62	3	55	Nordazepam/M of Chlordiazepoxide, Clorazepic Acid, Diazepam, Halazepam, Ketazolam, Medazepam, Oxazolam, Prazepam
60	35	45	56	73	85	92	77	5	58	Tetrazepam M (Nor-)
60	–	–	–	75	62	79	77	29	64	Prilocaine
60	–	–	–	79	72	78	73	42	71	Dextromoramide/Racemoramide/Levomoramide
60	–	23	–	82	67	75	73	39	68	Fluanisone
60	–	–	–	84	72	89	76	17	60	Penfluridol
61	–	–	–	–	–	–	61	42	70	Nicofuranose
61	–	–	–	–	–	–	63	59	67	Piperoxan
61	–	–	–	–	–	–	72	5	59	Cyprenorphine
61	32	43	54	64	84	90	75	0	52	Clobazam M (Nor-)
61	35	45	56	67	85	87	72	0	53	Clonazepam
61	–	–	–	73	84	88	72	2	69	Bromocriptine
61	–	–	–	75	65	–	76	14	54	Butanilicaine

Table 8.11. (Continued)

TLC System 10	1	2	3	4	5	6	7	8	9	[System 10: Ac / Silica-KOH]
61	–	–	–	75	84	86	71	20	74	Isocarboxazid
61	–	–	–	76	76	88	73	5	68	Ambucetamide
61	–	–	–	79	70	72	68	13	70	Etonitazene
61	41	46	55	82	80	80	74	5	59	Clonazepam Hy/DCB
62	–	–	–	–	–	–	63	45	69	Butoxyethyl Nicotinate,2-
62	–	–	–	–	–	–	67	52	68	Tiletamine
62	–	–	–	–	–	–	71	48	76	Carbifene
62	–	–	–	–	–	–	80	46	74	Dioxaphetyl Butyrate
62	–	–	–	70	86	–	67	0	51	Benorilate
62	46	50	65	75	80	90	74	11	66	Haloxazolam
62	53	47	70	75	84	85	62	8	70	Clobazam
62	56	40	73	78	79	83	67	41	74	Medazepam
62	47	35	71	79	82	84	77	40	73	Metaclazepam
62	–	–	–	82	55	57	–	54	60	Eucaine,α-
62	–	–	–	86	39	72	65	28	50	Ibogaine/Rotenone
63	–	–	–	–	–	–	70	48	69	Mebenazine
63	–	–	–	–	–	–	73	7	61	Etorphine
63	–	–	–	–	–	–	73	60	67	Amylocaine
63	–	–	–	17	99	–	33	54	4	Pseudoephedrine
63	–	–	–	47	74	–	65	9	66	Naloxone
63	–	–	–	60	72	74	72	7	64	Pentetrazol
63	–	–	–	71	77	78	–	11	73	Meconin
63	54	47	72	74	80	82	63	10	72	Flunitrazepam
63	56	50	67	75	78	86	76	24	69	Fludiazepam
63	–	–	–	77	76	80	69	2	74	Reserpine
63	–	–	–	80	72	69	70	35	71	Lidocaine
63	64	55	72	81	84	89	65	36	74	Prazepam
63	–	–	–	83	29	82	77	37	12	Phenelzine
64	–	–	–	–	–	–	51	38	11	Hydralazine
64	–	–	–	–	–	–	66	57	67	Amolanone
64	–	–	–	–	–	–	70	0	62	Norbormide
64	–	–	–	–	–	–	70	51	65	Aminoxytriphene
64	–	–	–	–	–	–	72	4	77	Methoserpidine
64	–	–	–	76	90	–	65	8	65	Beclamide
64	–	–	–	78	72	75	64	21	74	Noscapine
64	–	–	–	79	68	72	63	37	63	Ketamine
64	–	–	–	81	77	79	73	1	75	Rescinnamine
64	–	–	–	83	44	76	71	7	30	Butacaine
64	–	–	–	83	45	–	64	68	64	Biperiden
64	–	–	–	84	64	77	–	45	75	Alphamethylfentanyl
64	–	–	–	85	55	56	76	62	63	Amfepramone/Diethylpropion
64	–	–	–	87	72	73	–	62	74	Phenylcyclohexyl)morpholine,4-(1-/PCM
64	–	–	–	88	83	90	82	45	75	Flunarizine
65	–	–	–	–	–	–	68	1	35	Brocresine
65	–	–	–	–	–	–	71	51	72	Diampromide
65	–	–	–	–	–	–	81	56	67	Phenampromide

Table 8.11. (Continued)

TLC System 10	1	2	3	4	5	6	7	8	9	[System 10: Ac / Silica-KOH]
65	55	32	69	75	82	83	76	12	73	Camazepam
65	–	–	–	80	69	79	69	42	73	Bupivacaine
65	–	–	–	86	79	87	76	54	78	Cinnarizine
66	–	–	–	–	–	–	70	15	58	Diperodon
66	–	–	–	–	–	–	70	39	73	Tolycaine
66	53	54	65	71	78	95	77	15	68	Oxazolam
66	45	45	62	74	83	80	66	14	64	Ketazolam
66	56	62	63	76	84	87	67	5	57	Benzocaine
66	57	49	67	79	84	89	78	32	69	Tetrazepam
66	–	–	–	80	82	87	74	17	76	Bisacodyl
66	55	48	66	80	84	87	78	31	69	Clotiazepam
66	–	–	–	81	73	–	72	3	77	Deserpidine
66	48	50	57	82	87	91	77	7	60	Flurazepam Hy/HCFB
66	–	–	–	83	31	55	67	64	47	Profenamine/Ethopropazine
66	–	–	–	84	23	69	59	73	35	Phencyclidine
66	–	–	–	86	85	90	83	53	77	Clocinizine
67	–	–	–	–	–	–	65	0	56	Ethoxazene
67	–	–	–	–	–	–	66	16	74	Diloxanide
67	–	–	–	–	–	–	68	7	45	Dimethocaine
67	–	–	–	–	–	–	69	26	45	Amprotropine
67	–	–	–	–	–	–	70	9	60	Levophenacylmorphan
67	–	–	–	8	88	–	68	53	5	Azapropazone
67	53	58	62	74	85	93	76	0	62	Ethyl Loflazepate
67	–	–	–	79	80	–	69	41	78	Tetrabenazine
67	–	–	–	86	88	–	72	47	78	Fenbutrazate
67	–	–	–	87	90	–	72	37	74	Famprofazone
68	–	–	–	–	–	–	67	18	77	Octaverine
68	–	–	–	–	–	–	69	0	56	Clefamide
68	–	–	–	64	78	–	66	6	45	Leucinocaine
68	–	–	–	71	79	–	68	1	65	Nifedipine
68	66	66	80	83	76	86	73	31	77	Nimetazepam Hy/MNB
68	–	–	–	86	33	86	70	57	59	Clofazimine
68	–	–	–	86	38	70	–	73	54	Tenocyclidine/TCP/1-[1-(2-Thienyl)cyclo-hexyl]piperidine
68	–	–	–	87	31	78	–	72	47	Methyl-1-phenylcyclohexyl)piperidine,1-(4-
68	–	–	–	87	84	97	73	63	76	Phenoxybenzamine
69	–	–	–	–	–	–	65	3	68	Metisazone
69	–	–	–	–	–	–	67	7	64	Isobutyl 4-Aminobenzoate
69	–	–	–	–	–	–	76	6	77	Thiambutosine
69	–	–	–	–	80	83	80	58	77	Phenadoxone
69	–	–	–	43	9	74	73	23	77	Methoxyamphetamine,4-/PMA
69	61	60	64	76	84	96	–	22	64	Cholesterol
69	–	–	–	78	78	81	71	21	78	Disulfiram
69	–	–	–	80	80	–	76	9	68	Buprenorphine
69	71	69	81	83	77	90	75	48	77	Fludiazepam Hy/CFMB
69	70	69	74	83	81	93	83	11	71	Haloxazolam Hy/ABFB

Table 8.11. (Continued)

TLC System										
10	1	2	3	4	5	6	7	8	9	[System 10: Ac / Silica-KOH]
69	64	61	66	85	86	88	77	9	68	Bromazepam Hy/ABP
69	71	65	68	85	90	91	77	5	69	Nitrazepam Hy/ANB
70	–	–	–	–	83	90	75	6	63	Butyl Aminobenzoate
70	–	–	–	47	90	–	90	2	67	Buclosamide
70	–	–	–	65	92	96	71	41	79	Bezitramide
70	65	61	73	81	85	92	75	29	72	Pinazepam
70	71	66	68	86	87	91	77	4	69	Flunitrazepam M Hy/ANFB
70	72	66	70	86	87	93	77	4	69	Clonazepam Hy/ANCB
70	76	68	74	86	89	93	78	11	75	Flurazepam Hy/ACFB
70	–	–	–	87	27	72	66	67	33	Dipipanone
70	–	–	–	87	60	–	73	67	70	Benzphetamine
70	–	–	–	87	80	88	76	61	79	Meclozine
70	77	70	76	87	89	93	78	11	76	Lorazepam Hy/ADB
70	–	–	–	87	90	92	74	42	84	Diphenoxylate
71	–	–	–	78	78	–	74	52	79	Broxaldine
71	59	57	70	80	87	90	78	16	89	Quazepam M (2-Oxo-)
71	59	59	70	81	89	91	78	15	6	Halazepam
71	–	–	–	82	82	89	79	11	79	Phenothiazine
71	7	63	80	86	89	90	77	20	79	Flunitrazepam M Hy/MNFB
71	76	69	76	86	90	91	78	12	76	Oxazepam Hy/ACB
71	–	–	–	88	84	–	75	69	79	Bromhexine
71	82	72	82	88	89	93	79	53	81	Diazepam Hy/MACB
72	–	–	–	–	–	–	75	61	83	Buclizine
72	–	–	–	–	–	–	80	7	78	Tiocarlide
72	75	73	80	86	86	91	83	48	77	Pinazepam Hy/CPB
73	–	–	–	–	–	–	74	62	80	Bialamicol
73	74	73	80	85	77	93	79	44	77	Quazepam Hy/CFTB
74	–	–	–	83	88	92	79	64	84	Phenanthrene
74	83	74	82	88	92	95	80	58	82	Prazepam Hy/CCB
75	–	–	–	86	93	97	83	44	78	Proclonol
75	81	74	81	89	94	95	82	49	81	Halazepam Hy/TCB
76	78	71	78	83	87	96	74	27	75	Quazepam
90	–	–	–	26	96	98	95	0	10	Closantel
99	91	92	95	99	80	88	86	55	99	Methidathion
99	91	93	94	99	81	87	84	40	99	Azinphos-methyl

8.12 hR$_f$ Data in TLC System T (Toxi-Lab A) vs. Substance Names

hR$_f$	Color Code 1	2	3	4	Substance Name
1	1	6	9	3	Zomepirac
1	4	0	9	3	Fenoprofen L
1	4	0	9	3	Nadolol L
1	4	2	9	3	Fenoprofen
1	4	2	9	3	Nadolol
1	4	3	6	3	Nadolol H
1	7	3	6	3	Nafcillin
2	5	0	6	0	Meclofenoxate
2	5	0	9	3	Indometacin
3	5	5	9	3	Perazine M #3
3	5	9	9	3	Flurbiprofen
3	9	9	7	3	Acemetacine L
3	9	9	9	3	Acemetacine
3	9	9	9	3	Carprofen
4	3	3	6	3	Isoprenaline
4	7	0	9	3	Naproxen L
4	7	3	9	3	Naproxen
4	7	4	9	3	Promazine Artifact
5	0	0	6	3	Ketoprofen
5	0	0	9	3	Atropine/Hyoscyamine
5	0	0	9	3	Ketoprofen H
5	0	0	9	3	Pirenzepine
5	2	0	9	3	Ranitidine
5	3	3	2	3	Amitriptyline-oxide L
5	3	3	6	3	Ranitidine H
5	4	0	7	3	Ranitidine
5	4	2	6	3	Phenylephrine
5	5	3	9	3	Amitriptyline-oxide
5	5	5	9	3	Phenothiazine M
5	5	6	9	3	Phenothiazine M
5	6	0	7	3	Desipramine M #2
5	6	0	9	3	Atropine/Hyoscyamine H
5	7	4	9	3	Phenothiazine M
5	7	5	9	3	Phenothiazine M
6	0	0	1	3	Opipramol M #5 L
6	0	1	1	3	Opipramol M #5
6	1	0	6	3	Difenoxine
6	1	1	1	3	Opipramol M #5 H
6	4	3	9	3	Bumadizon
6	6	3	9	3	Codeine M (Nor-)
7	0	0	6	3	Lonazolac
7	3	3	7	3	Synephrine/Oxedrine L
7	3	3	9	3	Synephrine/Oxedrine
7	5	5	9	3	Mequitazine H

Table 8.12. (Continued)

hR_f	Color Code 1	2	3	4	Substance Name
7	7	4	9	3	Mequitazine
8	0	0	9	3	Sparteine
8	5	5	9	3	Perazine M #2 H
8	7	4	9	3	Perazine M #2
8	9	3	9	3	Azapropazone
9	0	0	7	3	Ibuprofen L
9	3	3	9	3	Pirprofen L
9	5	0	6	3	Mofebutazone
9	5	3	7	3	Artifact
9	5	3	9	3	Pirprofen
9	6	0	7	3	Ibuprofen
10	0	0	7	3	Tiotixene M
10	0	0	9	3	Disopyramide M
10	0	0	9	3	Polyethyleneglycol
10	4	0	7	3	Tiotixene M H
10	5	0	6	3	Atenolol
10	5	0	9	3	Atenolol H
10	5	3	6	3	Octopamine
10	5	5	7	3	Viloxazine M L
10	5	5	9	3	Phenothiazine M
10	5	5	9	3	Viloxazine M
10	5	6	9	3	Phenothiazine M
10	6	0	6	3	Doxylamine M
10	6	3	9	3	Hydromorphone
10	7	4	9	3	Phenothiazine M
10	7	5	9	3	Phenothiazine M
10	9	3	9	3	Dobutamine
11	1	0	6	3	Benzatropine H
11	1	0	9	0	Benzatropine
11	3	3	3	3	Salbutamol H/Albuterol H
11	3	3	7	3	Orciprenaline L
11	3	3	7	3	Salbutamol/Albuterol
11	3	3	9	3	Orciprenaline
11	7	4	9	3	Promethazine M #2
12	0	0	9	3	Celiprolol
12	3	1	9	3	Dihydrocodeine L
12	5	2	9	3	Strychnine
12	5	4	9	9	Diclofenac
12	6	0	9	0	Diflunisal
12	6	2	9	3	Dihydrocodeine
13	0	0	6	3	Chloroquine
14	0	0	6	3	Acebutolol
14	0	0	9	3	Propylhexedrine
14	1	0	9	3	Pseudoephedrine L
14	1	7	6	3	Pseudoephedrine
14	3	3	4	3	Amitriptyline M/Nortriptyline M #2

Table 8.12. (Continued)

hR$_f$	Color Code 1	2	3	4	Substance Name
14	3	3	7	3	Terbutaline L
14	3	3	9	3	Terbutaline
14	4	0	6	3	Etilefrine L
14	4	2	6	3	Etilefrine
14	7	7	6	3	Pseudoephedrine
15	0	0	9	3	Isometheptene L
15	1	0	6	3	Pethidine M/Meperidine M (Nor-)
15	1	0	7	3	Verapamil M #2
15	1	7	6	3	Pethidine M/Meperidine M (Nor-)
15	2	0	7	3	Desipramine M #1
15	2	6	7	3	Desipramine M #1 H
15	3	0	9	3	Morphine L
15	4	0	9	3	Thymol M L
15	4	2	9	3	Thymol M
15	5	2	6	3	Metoclopramide M
15	5	3	2	3	Maprotiline L
15	5	3	9	3	Maprotiline
15	5	3	9	3	Morphine
15	6	0	9	3	Meclofenamic Acid L
15	6	3	6	3	Isometheptene
15	6	3	6	3	Meclofenamic Acid
15	6	7	9	3	Mefenamic Acid
16	0	0	6	3	Niflumic Acid
16	0	0	7	3	Clomipramine M #2
16	0	0	9	3	Propyphenazone M (Nor-)
16	2	2	9	3	Oxyphenbutazone/Phenylbutazone M
16	5	3	7	3	Procyclidine M
16	5	5	6	3	Flufenamic Acid H
16	5	5	9	3	Flufenamic Acid
17	3	3	4	3	Amitriptyline M/Nortriptyline M #1
17	5	0	6	3	Azatadine
18	0	0	1	3	Opipramol M #4 L
18	0	0	6	3	Sulpiride
18	0	1	1	3	Opipramol M #4
18	1	1	1	3	Opipramol M #4 H
18	2	2	7	3	Spironolactone M #2
18	3	0	9	3	Hydrocodone L
18	5	0	1	3	Tramadol M #3 L
18	5	0	6	3	Sotalol
18	5	0	9	3	Trihexyphenidyl M #2 L
18	5	3	7	3	Trihexyphenidyl M #2
18	5	6	1	3	Tramadol M #3
18	6	1	9	3	Hydrocodone
18	7	4	9	3	Promethazine M #1
19	0	0	9	0	Tolbutamide
19	0	0	9	3	Tolbutamide H

Table 8.12. (Continued)

hR_f	Color Code 1	2	3	4	Substance Name
19	3	3	2	3	Dextrorphan/Levorphanol
19	3	3	4	3	Dextrorphan/Levorphanol H
19	4	0	6	3	Methoxyphenamine
19	4	0	9	3	Methoxyphenamine L
19	4	4	6	3	Carteolol
19	4	4	9	3	Carteolol H
20	0	0	9	3	Amantadine
20	0	0	9	3	Pipamperone
20	0	0	9	3	Procainamide
20	1	7	9	3	Propoxyphene M (Nor-)
20	3	7	6	3	Propoxyphene M (Nor-) H
20	9	7	6	3	Propoxyphene M (Nor-) H
21	0	0	7	3	Chlorphenamine/Chlorpheniramine H
21	0	0	7	3	Dothiepin M L
21	0	0	9	3	Chlorphenamine/Chlorpheniramine
21	0	0	9	3	Tiapride
21	5	5	9	3	Perazine M #1 H
21	6	0	6	3	Piroxicam
21	6	0	7	3	Dothiepin M
21	7	0	7	3	Dothiepin M H
21	7	4	9	3	Perazine M #1
22	0	0	9	3	Cimetidine L
22	1	0	9	3	Methamfetamine L
22	1	0	9	3	Phenethylamine
22	1	7	6	3	Methamfetamine
22	1	7	6	3	Phenethylamine
22	3	7	6	3	Methamfetamine H
22	3	7	6	3	Phenethylamine H
22	5	5	3	3	Thioridazine M
22	6	0	9	3	Cimetidine
22	6	3	9	3	Methylenedioxymethamphetamine,3,4-/MDMA/XTC
23	1	0	9	3	Ephedrine M/Phenylpropanolamine L
23	1	7	6	3	Ephedrine M/Phenylpropanolamine
23	3	2	9	3	Colchicine
23	7	7	6	3	Ephedrine M/Phenylpropanolamine H
24	0	0	6	3	Glymidine
24	0	0	9	3	Brompheniramine
24	0	6	7	3	Desipramine/Imipramine M H
24	0	6	9	3	Desipramine/Imipramine M
24	3	0	9	3	Codeine L
24	6	3	9	3	Codeine
25	0	0	6	3	Emetine L
25	0	0	6	3	Nifenazone
25	0	0	6	3	Quinidine
25	0	0	6	3	Quinine
25	0	0	7	3	Tiotixene M

Table 8.12. (Continued)

hR_f	Color Code 1	2	3	4	Substance Name
25	0	0	9	3	Glibornuride
25	1	3	6	3	Emetine
25	3	0	6	3	Bunitrolol
25	3	2	1	3	Ketobemidone
25	4	0	7	3	Tiotixene M H
25	5	2	9	3	Phenylbutazone M
25	5	3	6	3	Metoprolol
25	5	3	9	3	Metoprolol H
25	6	0	6	3	Emetine
26	0	0	1	3	Opipramol L
26	0	0	6	3	Loprazolam L
26	0	0	9	3	Acetaminophenazone M (Bis-nor-) AC
26	0	0	9	3	Loprazolam
26	0	1	1	3	Opipramol
26	1	1	1	3	Opipramol H
26	4	0	6	3	Gliquidone
26	6	0	6	3	Dextromethorphan/Racemethorphan/Levomethorphan
26	6	0	9	3	Dextromethorphan/Racemethorphan/Levomethorphan
26	6	3	6	3	Dextromethorphan H
27	0	0	9	3	Carbinoxamine
27	0	4	6	3	Trazodone M
27	0	4	9	3	Trazodone M L
27	0	6	7	3	Clomipramine M (Nor-)
27	0	6	9	3	Clomipramine M (Nor-) L
27	5	2	7	3	Tiotixene
27	5	5	7	3	Dimetindene
27	5	5	9	3	Dimetindene H
27	6	4	6	3	Trazodone M H
28	0	0	6	3	Piprinhydrinate M
28	1	0	6	3	Oxeladine M
28	2	1	6	3	Methoxy-4,5-methylenedioxyamphetamine,3-/MMDA
28	3	7	4	3	Oxeladine M H
28	4	0	6	3	Bisoprolol L
28	4	4	3	3	Bisoprolol
29	2	0	6	3	Phenprocoumon L
29	3	3	3	3	Clozapine M
29	4	0	6	0	Phenprocoumon
29	4	0	9	3	Phenprocoumon H
29	5	5	7	3	Viloxazine L
29	5	5	9	3	Viloxazine
30	0	0	9	3	Hydroxyzine M L
30	0	1	6	3	Hydroxyzine M
30	0	2	9	3	Metoclopramide L
30	0	3	9	3	Erythromycin L
30	0	4	9	3	Metoclopramide L
30	1	0	6	0	Diphenhydramine M

Table 8.12. (Continued)

hR_f	Color Code				Substance Name
	1	2	3	4	
30	1	1	9	3	Hydroxyzine M
30	2	2	7	3	Spironolactone M #1
30	2	4	6	3	Metoclopramide
30	3	3	4	3	Amitriptyline M/Nortriptyline
30	3	3	4	3	Erythromycin H
30	3	5	9	3	Erythromycin H
30	5	0	1	3	Tramadol M #2 L
30	5	0	6	3	Fluoxetine
30	5	3	9	3	Morphine-6-Acetate/Diamorphine M
30	5	5	9	3	Perphenazine H
30	5	6	1	3	Tramadol M #2
30	5	6	6	3	Tramadol M #2 H
30	6	0	6	3	Methylenedioxyamphetamine,3,4-/MDA L
30	6	0	9	3	Timolol
30	6	3	6	3	Methylenedioxyamphetamine,3,4-/MDA
30	6	3	9	3	Erythromycin
30	6	3	9	3	Methylenedioxyamphetamine,3,4-/MDA H
30	6	4	9	3	Metoclopramide L
30	7	0	6	3	Fluoxetine L
30	7	0	9	3	Fluoxetine L
30	7	4	9	3	Perphenazine
31	0	0	6	3	Triamterene L
31	0	0	9	9	Etofylline
31	0	1	6	3	Triamterene L
31	1	1	6	3	Triamterene
31	5	5	3	3	Oxprenolol
31	5	5	7	3	Oxprenolol L
31	5	5	9	3	Perazine H
31	7	4	9	3	Perazine
32	0	0	9	3	Disopyramide
32	1	0	9	3	Amphetamine/Amphetaminil M artifact L
32	1	7	6	3	Amphetamine/Amphetaminil M artifact
32	3	7	6	3	Amphetamine/Amphetaminil M artifact H
32	5	0	6	3	Trimethobenzamide
32	5	0	9	3	Trimethobenzamide L
32	5	5	9	3	Thiethylperazine
32	6	0	9	3	Zimeldine
32	7	3	9	3	Pindolol
32	7	9	9	3	Pindolol H
33	0	0	9	3	Cotinine
33	0	0	9	3	Phenazone M/Antipyrine M
33	1	1	7	3	Carbamazepine M #3
33	2	0	7	3	Imipramine M
33	2	6	7	3	Imipramine M H
33	4	0	6	3	Toliprolol L
33	4	4	3	3	Toliprolol

Table 8.12. (Continued)

hR$_f$	Color Code 1	2	3	4	Substance Name
33	5	5	7	3	Psilocin
33	5	9	4	3	Dihydroergotamine L
33	5	9	9	3	Dihydroergotamine
34	1	0	6	3	Labetalol L
34	1	7	6	3	Phentermine
34	2	7	6	3	Labetalol
34	3	7	6	3	Phentermine H
35	0	0	9	3	Fluvoxamine
35	3	4	9	3	Metipranolol
35	5	3	6	3	Betaxolol
35	5	3	9	3	Betaxolol H
35	5	5	9	3	Mepindolol
36	0	0	9	3	Nifenalol
36	0	0	9	3	Perhexiline
36	1	3	6	3	Propafenone L
36	1	7	6	3	Propoxyphene M (Nor-)
36	2	3	3	3	Propafenone
36	3	3	4	3	Amitriptyline M #2
36	3	7	6	3	Propoxyphene M (Nor-) H
36	4	0	1	3	Clopenthixol artifact / Zuclophenthixol artifact
36	4	0	1	3	Flupentixol
36	4	4	6	3	Bupranolol
36	4	4	9	3	Bupranolol H
36	5	5	2	3	Sulforidazine H
36	5	5	9	3	Dixyracine H
36	5	5	9	3	Fluphenazine H
36	7	0	2	3	Propranolol L
36	7	3	9	3	Propranolol
36	7	4	9	3	Dixyracine
36	7	4	9	3	Fluphenazine
36	7	4	9	3	Sulforidazine
36	9	7	6	3	Propoxyphene M (Nor-) H
37	1	0	9	3	Flecainide
38	0	5	6	3	Mianserin M (Nor-)
38	6	0	9	3	Doxylamine L
38	6	3	7	3	Doxylamine
39	0	0	6	3	Alprazolam
39	0	0	7	3	Triazolam
39	0	0	9	3	Phenmetrazine L
39	0	4	4	3	Prajmaline H
39	0	4	6	3	Prajmaline tautomeric form
39	1	0	6	3	Phenmetrazine
39	1	0	7	3	Piprinhydrinate
39	1	0	9	3	Phenmetrazine L
39	2	0	6	3	Tocainide
39	2	0	9	3	Tocainide L

Table 8.12. (Continued)

hR_f	Color Code 1	2	3	4	Substance Name
39	3	3	4	3	Amitriptyline M #1
39	5	0	9	3	Trihexyphenidyl M #1 L
39	5	3	7	3	Trihexyphenidyl M #1
39	5	5	3	3	Alprenolol
39	5	5	7	3	Alprenolol L
39	5	5	9	3	Trifluoperazine
39	5	6	7	3	Carazolol L
39	5	6	9	3	Carazolol
39	6	0	9	3	Amoxapine L
39	6	2	6	3	Amoxapine H
39	6	2	9	3	Amoxapine
39	7	4	9	3	Trifluoperazine L
40	0	0	1	3	Clomipramine M #1
40	0	0	6	3	Nicotine H
40	0	0	9	3	Nicotine
40	0	0	9	3	Propyphenazone M (Hydroxypropyl-)
40	0	0	9	9	Theophylline
40	2	1	1	3	Trimethoprim L
40	2	1	6	3	Trimethoprim H
40	2	1	9	3	Trimethoprim
40	2	5	6	3	Clemastine L
40	2	7	4	7	Clemastine
40	5	5	3	3	Penbutolol
40	5	5	7	3	Penbutolol L
40	5	9	4	3	Ergotamine L
40	5	9	9	3	Ergotamine
41	1	7	6	3	Fenetylline
41	3	3	7	3	Bencyclane
41	3	3	9	3	Bencyclane H
41	3	7	3	3	Fenetylline H
41	5	5	9	3	Promazine H
41	7	4	9	3	Promazine
42	0	0	9	3	Clindamycin L
42	0	0	9	3	Fenfluramine
42	4	0	9	3	Methoxyphenamine M
42	6	0	6	3	Verapamil M #1 H
42	6	0	9	3	Clindamycin
42	6	0	9	3	Verapamil M #1
43	0	0	1	3	Ketotifen
43	0	4	4	3	Ajmaline H
43	0	4	6	3	Ajmaline
43	2	3	6	3	Prothipendyl
43	3	3	2	3	Prothipendyl H
43	5	5	9	3	Periciazine H
43	7	4	9	3	Periciazine
44	3	2	1	3	Metixene

Table 8.12. (Continued)

hR_f	Color Code 1	2	3	4	Substance Name
44	5	5	1	3	Tramadol M #1 L
44	5	5	6	3	Tramadol M #1
45	0	0	6	3	Dibenzepin L
45	0	0	7	3	Flurazepam M
45	0	0	9	3	Dibenzepin
45	1	1	7	3	Flurazepam M H
45	5	6	9	3	Thioridazine
45	6	0	6	3	Scopolamine
45	7	3	6	3	Oxomemazine
46	1	0	6	3	Nefopam
46	4	0	7	3	Chlorprothixene M L
46	4	4	7	3	Chlorprothixene M
48	0	0	6	3	Bamipine
48	0	4	4	3	Prajmaline tautomeric form H
48	0	4	6	3	Prajmaline
48	1	0	6	3	Oxeladine
48	1	9	6	3	Pentoxyverine
48	3	3	2	3	Pentazocine M
48	3	3	9	3	Mexiletine
48	3	7	4	3	Oxeladine H
48	3	9	1	3	Mexiletine L
48	4	1	7	3	Mexiletine
48	7	5	9	3	Pentoxyverine H
49	0	0	9	3	Phenazone/Antipyrine/Phenyldimethylpyrazolone
49	4	0	6	3	Mepyramine
49	4	4	6	3	Mepyramine H
49	5	5	3	3	Astemizole
49	5	5	7	3	Astemizole L
49	5	5	9	3	Aceprometazine
49	6	1	9	3	Phenazone/Antipyrine H
50	0	0	9	3	Pethidine/Meperidine L
50	0	6	7	3	Imipramine H
50	0	6	9	3	Imipramine
50	1	0	6	3	Pethidine/Meperidine
50	1	0	7	3	Haloperidol M
50	1	0	9	3	Diltiazem L
50	1	0	9	3	Pethidine/Meperidine L
50	1	2	9	3	Physostigmine
50	2	0	6	3	Lidocaine M H
50	2	0	9	3	Lidocaine M
50	3	0	6	3	Haloperidol M H
50	3	3	9	3	Clozapine
50	3	7	6	3	Pethidine/Meperidine H
50	5	0	7	3	Diltiazem
50	5	0	7	3	Tolpropamine
50	6	3	9	3	Erythromycin ester

Table 8.12. (Continued)

hRf	Color Code 1	2	3	4	Substance Name
50	7	0	6	3	Diltiazem H
50	7	7	6	3	Pethidine/Meperidine H
51	0	0	6	3	Aminophenazone/Amidopyrine
51	0	0	9	3	Aminophenazone/Amidopyrine L
51	1	7	6	3	Tripelenamine
51	4	4	3	3	Isothipendyl H
51	4	4	6	3	Isothipendyl
52	0	0	6	3	Aminophenazone M (Bis-nor-)
52	0	0	7	3	Benzquinamide L
52	0	0	9	3	Hydroxyzine L
52	0	0	9	9	Proxyphylline
52	0	1	6	3	Hydroxyzine
52	1	0	6	0	Orphenadrine
52	1	0	7	3	Benzquinamide
52	1	0	9	0	Orphenadrine L
52	2	0	6	3	Orphenadrine H
52	4	0	6	3	Methapyrilene
52	4	0	7	3	Doxepin L
52	4	4	9	3	Doxepin
52	5	3	6	3	Methapyrilene H
52	6	0	7	3	Benzquinamide L
53	0	0	9	3	Mazindol
53	0	0	9	3	Metamizol/Dipyrone
53	5	0	9	3	Metamizol/Dipyrone H
53	5	0	9	3	Procyclidine L
53	5	3	6	3	Procyclidine
53	5	9	7	3	Proglumetacin
53	5	9	9	3	Proglumetacin H
54	0	0	6	3	Brotizolam
54	5	3	9	3	Promethazine H
54	5	9	9	3	Mebhydrolin H
54	6	9	6	3	Mebhydrolin
54	7	3	9	3	Promethazine
55	0	0	1	3	Opipramol M #3
55	0	0	7	3	Dothiepin L
55	0	0	9	3	Phendimetrazine L
55	1	0	6	0	Diphenhydramine
55	1	0	6	3	Diphenhydramine H
55	1	0	6	3	Phendimetrazine
55	1	0	9	0	Diphenhydramine L
55	1	0	9	3	Diphenhydramine L
55	1	0	9	3	Phendimetrazine L
55	3	5	6	3	Terpin hydrate M
55	4	0	6	3	Phenyltoloxamine
55	4	0	9	3	Phenyltoloxamine L
55	5	5	9	3	Chlorpromazine

Table 8.12. (Continued)

hR$_f$	Color Code 1	2	3	4	Substance Name
55	6	0	7	3	Dothiepin
55	6	2	9	3	Betanaltrexol,6- H
55	7	0	7	3	Betanaltrexol,6-
55	7	4	9	3	Chlorpromazine L
56	0	0	6	3	Methadone M/Levomethadone M
56	0	0	9	3	Methadone M/Levomethadone M H
56	1	0	6	3	Methadone M/Levomethadone M H
56	3	3	2	3	Cyclobenzaprine
56	3	3	4	3	Cyclobenzaprine
56	5	5	1	3	Pizotifen L
56	5	5	9	3	Pizotifen
57	1	9	4	3	Bornaprine
57	1	9	6	3	Bornaprine L
58	0	0	9	3	Chlordiazepoxide
58	0	1	7	3	Carbamazepine M #2 L
58	1	1	7	3	Carbamazepine M #2
58	3	3	2	3	Amitriptyline L
58	3	3	4	3	Amitriptyline
58	5	1	7	3	Dipyridamole
58	5	3	9	3	Amitriptyline H
58	5	9	1	3	Cyproheptadine
58	5	9	9	3	Cyproheptadine H
59	0	6	7	3	Clomipramine
59	0	6	9	3	Clomipramine L
59	5	0	9	3	Droperidol
59	6	6	5	3	Droperidol H
60	0	0	6	3	Midazolam
60	0	4	6	3	Trazodone L
60	0	4	9	3	Trazodone L
60	5	5	7	3	Guaifenesin
60	5	5	9	3	Guaifenesin
60	6	0	9	3	Molindone L
60	6	3	9	3	Molindone
60	6	4	6	3	Trazodone
60	6	6	1	3	Tramadol L
60	6	6	6	3	Tramadol
61	1	1	1	3	Melitracene
61	2	7	3	7	Chlorphenoxamine H
61	2	7	6	7	Chlorphenoxamine
61	3	3	2	3	Pentazocine
61	4	0	6	3	Benperidol L
61	5	0	6	3	Pyrazobutamine
61	5	0	9	3	Benperidol
61	5	3	7	3	Pyrrobutamine
61	5	9	1	3	Benperidol H
62	0	0	9	3	Melperone

Table 8.12. (Continued)

hRf	Color Code 1	2	3	4	Substance Name
62	3	2	9	3	Oxycodone H
62	5	3	4	3	Aprindine
62	5	5	9	3	Gallopamil
62	6	0	6	3	Verapamil
62	6	0	9	3	Oxycodone
62	6	0	9	3	Verapamil L
63	2	0	1	3	Butinoline
63	7	5	9	3	Triflupromazine
64	0	0	7	3	Flurazepam
64	0	0	9	3	Haloperidol L
64	0	0	9	3	Procaine L
64	0	0	9	9	Caffeine
64	0	6	7	3	Procaine
64	0	9	7	3	Procaine H
64	1	0	6	3	Nomifensine
64	1	0	6	3	Prolintane L
64	1	9	6	3	Prolintane
64	2	3	6	3	Nomifensine H
64	3	9	4	3	Prolintane H
64	6	0	6	3	Haloperidol
65	0	0	4	3	Chlordiazepoxide M H
65	0	0	6	3	Lorcainide
65	0	0	9	3	Chlordiazepoxide M
66	0	5	6	3	Mianserin
66	3	3	3	3	Tranylcypromine
66	3	3	6	3	Tranylcypromine L
66	4	4	4	3	Chlorprothixene H
66	4	4	7	3	Chlorprothixene
66	6	0	6	3	Loxapine
66	6	0	9	3	Loxapine L
66	6	0	9	3	Methadone/Levomethadone
67	6	0	6	3	Loperamide
67	6	0	9	3	Loperamide H
68	5	5	7	3	Methocarbamol H
68	5	5	9	3	Methocarbamol
68	6	0	6	3	Bromperidol
70	0	0	6	3	Carbamazepine
70	0	0	6	3	Prazosin
70	0	0	7	3	Hydrocortisone
70	0	0	9	3	Buspirone
70	0	0	9	3	Hydroxycarisodol
70	1	6	6	3	Menthol M
70	3	3	4	3	Paracetamol H
70	3	3	7	3	Hydrocortisone H
70	3	6	9	3	Menthol M H
70	3	9	9	3	Methylprednisolon

Table 8.12. (Continued)

hR_f	Color Code 1	2	3	4	Substance Name
70	4	0	6	3	Mepivacaine
70	5	0	6	3	Prazosin
70	5	3	6	3	Prazosin
70	6	0	6	3	Trifluperidol
70	6	0	9	3	Paracetamol L
70	6	2	4	3	Paracetamol
70	6	2	9	3	Paracetamol
70	9	5	9	3	Methylprednisolon H
72	0	0	9	9	Pentifylline
72	6	0	9	3	Methyprylon
72	6	3	9	3	Papaverine
73	5	5	9	3	Trimeprazine H/Alimemazine H
73	7	4	9	3	Trimeprazine/Alimemazine
73	7	5	9	3	Trimeprazine/Alimemazine
74	1	1	9	3	Bromazepam
75	1	0	6	3	Mefenorex
75	1	0	7	3	Lorazepam/Lormetazepam M
75	2	2	7	3	Terfenadine L
75	2	2	9	3	Terfenadine
75	2	4	9	3	Pimozide H
75	3	3	9	3	Terfenadine H
75	4	0	6	3	Oxatomide L
75	4	4	6	3	Pimozide L
75	4	4	9	3	Pimozide
75	5	0	9	3	Oxatomide
75	5	5	9	3	Levomepromazine/Methotrimeprazine
75	5	6	1	3	Oxatomide H
75	7	3	4	3	Propranolol M
77	3	5	6	3	Terpin hydrate
78	0	6	7	3	Trimipramine
78	0	6	9	3	Trimipramine L
78	4	2	6	3	Bromhexine M/Ambroxol
78	4	4	9	0	Bromhexine M/Ambroxol H
79	0	0	6	3	Lormetazepam
79	6	2	9	3	Naloxone
79	6	9	7	3	Bromocriptine L
79	6	9	9	3	Bromocriptine
80	0	0	6	3	Clonidine
80	2	0	9	0	Artifact
80	4	0	9	0	Artifact
80	6	0	9	0	Methaqualone M
80	6	2	6	3	Chlorphenesin Carbamate
81	0	0	7	3	Nordazepam/Diazepam M (Nor-)
81	0	0	9	3	Cocaine
81	0	0	9	3	Meprobamate
81	0	0	9	3	Nordazepam/Diazepam M (Nor-) L

Table 8.12. (Continued)

hRf	Color Code 1	2	3	4	Substance Name
81	0	0	9	3	Temazepam L
81	1	0	9	3	Temazepam
81	2	0	6	3	Tocainide M
81	2	0	9	3	Tocainide M L
81	4	4	9	3	Isoxsuprine
82	0	0	6	3	Metamfepramone
82	0	0	7	3	Carbamazepine M #1 L
82	1	0	7	3	Carbamazepine M #1
82	1	1	7	3	Carbamazepine M #1
82	3	0	9	3	Propoxyphene L/Dextropropoxyphene/Levopropoxyphene
82	3	3	9	3	Benzoctamine
82	6	3	9	3	Phenacetin
82	9	7	7	3	Propoxyphene /Dextropropoxyphene /Levopropoxyphene
83	0	0	1	3	Opipramol M #2
84	0	0	9	3	Ketamine
84	0	2	9	3	Oxypertine
85	0	0	6	3	Nitrazepam
85	0	0	9	3	Artifact (stopper)
85	0	0	9	3	Phencyclidine
85	0	6	7	3	Lofepramine
85	2	0	9	3	Salicylamide
85	3	0	6	3	Ethenzamide
85	3	0	6	9	Ethenzamide H
85	6	0	9	3	Carbromal
85	6	3	9	3	Nafcillin M
86	0	0	6	3	Camazepam
86	0	0	6	3	Clonazepam
86	0	0	9	3	Dextromoramide/Racemoramide/Levomoramide
86	0	0	9	3	Propyphenazone
86	3	0	6	3	Prenoxdiazine/Tibexin
86	3	0	9	3	Prenoxdiazine H/Tibexin H
87	0	0	6	3	Clorazepic Acid
87	2	3	3	3	Fluspirilene
87	4	4	3	3	Fluspirilene H
87	6	0	9	3	Clobazam
88	0	0	1	3	Clopentixol L
88	0	0	6	3	Flupirtine
88	0	0	6	3	Medazepam
88	0	0	7	3	Spironolactone L
88	0	0	9	3	Fluanisone
88	0	0	9	3	Fluanisone L
88	1	1	7	3	Spironolactone
88	1	9	4	3	Beclamide L
88	2	0	6	3	Lidocaine H
88	2	0	9	3	Lidocaine
88	2	3	7	3	Amiodarone L

Table 8.12. (Continued)

hR_f	Color Code 1	2	3	4	Substance Name
88	2	3	9	3	Amiodarone
88	3	2	9	3	Nifedipine H
88	4	0	1	3	Clopentixol
88	4	4	1	3	Clopenthixol/Zuclopenthixol/cis-Ordinol
88	4	4	1	3	Clopentixol H
88	4	4	6	3	Phenazopyridine
88	5	0	6	3	Pridinol
88	5	4	7	3	Ethaverine L
88	5	4	9	3	Ethaverine
88	6	2	7	3	Nifedipine
88	7	9	6	3	Beclamide
89	0	0	6	3	Ketazolam
89	3	3	6	3	Prenylamine L
89	3	3	9	3	Prenylamine
90	0	0	6	3	Flunitrazepam
90	0	0	7	3	Diazepam/Ketazolam M/Medazepam M
90	0	0	9	3	Clomethiazole
90	0	0	9	3	Methaqualone
90	1	0	6	3	Glutethimide
90	1	0	9	0	Phenobarbital
90	1	0	9	0	Phenytoin
90	1	0	9	3	Phenolphthalein L
90	1	0	9	3	Phenytoin H
90	1	4	9	3	Phenolphthalein
90	1	7	6	3	Glutethimide H
90	3	3	9	3	Ethchlorvynol M
90	4	4	6	3	Noscapine
90	4	4	9	3	Noscapine H
90	4	4	9	3	Phenolphthalein H
90	5	0	9	3	Artifact (syringe)
90	5	5	9	3	Bisacodyl
91	0	0	6	3	Tetrazepam
91	4	1	1	3	Clotiazepam
91	5	9	6	3	Flunarizine L
91	5	9	9	3	Flunarizine
92	0	0	7	3	Oxazolam
92	0	0	9	3	Clindamycin Palmitate L
92	0	0	9	3	Disulfiram
92	1	3	4	3	Selegiline
92	3	0	9	3	Fendiline
92	5	9	6	3	Cinnarizine L
92	5	9	9	3	Cinnarizine
92	6	0	6	3	Nimodipine
92	6	0	9	3	Clindamycin Palmitate
92	9	0	6	3	Fendiline L
93	0	0	6	3	Tilidine

Table 8.12. (Continued)

hR_f	Color Code 1	2	3	4	Substance Name
93	0	0	9	3	Meclozine
95	0	0	1	3	Opipramol M #1 L
95	0	0	6	3	Amfepramone/Diethylpropion
95	0	0	6	3	Halazepam
95	0	0	6	3	Prazepam
95	0	1	1	3	Opipramol M #1
95	1	7	0	3	Benzyl Alcohol L
95	3	0	6	3	Tilidine H
95	3	3	9	3	Benzyl Alcohol
99	3	3	7	3	Artifact (lipid)

8.13 hR_f Data in TLC System P1 (Pesticides) vs. Substance Names

P1	P2	UV	Color Code 1	2	3	4	5	6	Substance Name
0	0	+	0	0	0	0	0	0	Clopyralid
0	0	+	0	0	0	0	7	5	MCPA/(4-Chloro-o-tolyloxy)acetic acid
0	0	+	0	0	0	1	1	5	Picloram
0	0	–	0	0	0	0	3	6	Cyhexatin
0	0	–	0	1	0	0	0	6	Amitrole
0	0	–	6	1	1	0	0	6	Acephate
0	0	–	6	1	1	0	0	6	Omethoate
0	1	+	0	0	0	0	0	0	Chlorophacinon
0	2	+ –	0	0	0	0	7	6	(2,4-Dichlorophenoxy)acetic acid/2,4-D
0	13	–	0	0	1	0	0	9	Butocarboxim M (Sulfoxide)
1	0	–	0	0	0	0	0	6	Azocyclotin
1	0	–	6	1	1	0	0	5	Methamidophos
1	1	+	0	0	0	0	1	6	Fenoprop
1	1	+	0	6	0	3	1	6	Formetanate
1	1	+	6	0	0	0	0	6	Monocrotophos
1	3	–	6	4	1	0	1	0	Demeton-S-methylsulfone
2	2	–	0	0	0	0	0	6	Sulfaquinoxaline
3	0	+	6	4	1	0	0	5	Etrimfos
3	2	+	0	0	0	0	3	0	Dichlorprop
3	2	–	0	0	0	0	0	6	Butoxycarboxim
4	1	+	0	9	0	3	0	6	Nicotine
4	2	+	0	0	0	0	0	0	Hexazinone
4	2	–	6	0	0	3	9	9	Trichlorfon/Trichlorphos
4	4	–	6	4	1	1	0	0	Dimethoate
4	19	+	0	0	0	3	3	6	Ioxynil
5	2	+ –	0	0	0	0	7	0	Mecoprop/MCPP
5	2	+ –	0	9	0	3	1	6	Enilconazole/Imazalil
5	4	+	0	6	1	1	3	9	Metamitron
6	6	+	0	0	0	0	0	6	Methomyl
6	38	+	0	0	0	0	0	0	4,6-Dinitro-o-cresol/DNOC
7	2	+	6	0	1	0	0	0	Phosphamidon
7	3	–	0	0	0	0	0	6	Tiabendazole
7	4	+	0	0	0	3	1	6	Metoxuron
8	3	+	0	0	0	0	0	9	Benzthiazuron
8	5	+	0	0	0	1	9	6	Difenoxuron
8	6	+	0	0	0	3	1	1	MCPB
8	16	+	0	0	0	3	3	5	Bromoxynil
9	8	+	0	0	0	0	0	4	Fuberidazole
10	3	+	0	0	0	0	0	1	Bitertanol
10	4	+	0	0	0	0	0	6	Thiazafluron
10	7	+	0	0	0	3	9	6	Chloroxuron
11	3	–	0	0	0	0	0	0	Triadimenol
11	7	+	0	0	0	3	0	9	Fenuron
11	17	+	0	0	0	9	6	5	Phenmedipham

Table 8.13. (Continued)

P1	P2	UV	Color Code 1	2	3	4	5	6	Substance Name
12	10	+	6	0	0	0	0	0	Mevinphos/Phosdrin
12	11	+	0	0	0	0	0	6	Warfarin
12	16	+ –	0	0	0	0	9	0	Dichlorophen
13	5	–	0	0	0	3	0	9	Cycluron
13	7	+	0	0	0	3	9	6	Monuron
14	20	+	0	0	0	0	0	0	Nitrophenol,p-
14	30	+	0	0	0	0	0	6	Coumatetralyl
15	8	+	0	0	0	0	0	5	Lenacil
15	9	+	0	1	0	3	1	9	Chlortoluron
15	10	+	0	0	0	3	9	9	Diuron
15	11	+	0	9	1	0	1	6	Butocarboxim
16	9	+	0	0	0	0	0	9	Fluometuron
16	12	+	0	0	0	0	0	0	Fenarimol
16	12	+	0	0	0	0	0	6	Cyanazine
17	17	–	6	4	1	1	1	3	Demeton
17	20	+	0	0	0	9	6	6	Carbofuran
17	45	–	0	0	0	0	0	0	Pindone
18	0	–	6	4	1	0	0	0	Oxydemeton-methyl
18	11	–	0	0	0	3	0	0	Propiconazole
18	13	–	6	4	1	0	0	0	Demeton-S-methyl
18	22	+	0	0	0	0	1	5	Methabenzthiazuron
18	24	+	0	0	0	0	0	6	Diflubenzuron
18	25	+	0	0	0	0	6	6	Carbaryl
19	2	–	0	0	0	0	0	6	Triforine
19	11	+	0	0	0	0	1	9	Methoprotryne
19	51	–	0	1	1	0	0	0	Thiram
20	17	–	0	0	0	0	0	6	Thiofanox
20	18	–	6	0	0	0	1	0	Heptenophos
20	20	–	6	0	0	0	3	0	Dichlorvos/DDVP
20	21	+	0	0	0	0	3	6	Propoxur
20	29	+	0	0	1	1	9	9	Buturon
20	42	+	5	4	1	3	0	0	Azinphos-methyl
21	16	+	0	0	0	0	0	9	Desmetryn
21	25	+	0	0	1	3	1	6	Ethiofencarb
21	27	+	0	0	1	3	0	5	Metobromuron
21	38	+	6	4	1	0	0	6	Triazophos
21	52	+	0	0	0	0	0	6	Naptalam
22	18	+ –	0	0	0	0	0	6	Simazine
22	23	–	0	0	0	3	3	0	Triadimefon
22	29	+	0	0	0	0	3	0	Flamprop-methyl
22	30	+	0	0	0	3	9	5	Chlorbromuron
22	31	+	0	0	0	0	9	9	Linuron
22	32	+	0	0	0	1	9	6	Chlorphenol,p-
23	30	+	0	0	0	3	3	5	Monolinuron
23	75	+	6	4	1	3	0	0	Temephos
24	31	+	0	0	1	0	0	1	Metribuzin

Table 8.13. (Continued)

P1	P2	UV	Color Code						Substance Name
			1	2	3	4	5	6	
24	48	+	6	4	1	3	0	0	Azinphos-ethyl
25	29	+	6	4	0	3	7	6	Tetrachlorvinphos
25	50	+ –	0	0	0	0	5	0	Ethofumesate
26	11	–	0	0	1	0	0	6	Thiocyclam
26	17	+	0	0	0	0	0	6	Pirimicarb
26	23	+	0	0	0	0	0	9	Ametryne
26	24	+	0	0	0	3	9	9	Neburon
26	26	+	6	0	0	0	0	0	Chlorfenvinphos
27	61	+	0	0	1	0	0	0	Coumaphos
29	35	+	0	0	0	0	3	0	Flamprop-isopropyl
29	56	+	6	4	1	3	0	3	Methidathion
30	64	–	0	0	0	0	1	6	Dichlofluanid
30	73	+	6	4	1	1	1	0	Parathion-methyl
31	53	–	6	4	1	0	0	6	Malathion
31	67	+	6	4	1	0	0	0	Phosalone
31	71	+	6	4	1	1	1	1	Chlorthion
32	27	+	0	0	0	0	0	5	Terbutylazine
32	31	+	0	0	0	0	0	9	Prometryn
32	32	+	0	0	0	0	0	5	Terbutryn
32	47	+	0	4	1	0	0	0	Pyrazophos
32	76	+	6	4	1	1	1	0	Fenitrothion
33	28	–	6	0	1	0	0	5	Ethoprofos/Ethoprop
33	52	–	0	0	0	0	0	5	Propyzamide
33	67	–	0	0	0	0	6	6	Tolylfluanid
34	54	+	0	4	0	0	0	0	Iprodione
35	35	+	0	0	0	0	0	9	Dipropetryn
37	89	+	0	0	0	3	1	6	Cyfluthrin
39	57	+ –	0	0	0	0	0	6	Propham
40	45	+	0	0	0	0	0	0	Alachlor
40	77	–	0	0	0	0	1	0	Endosulfan
40	81	+	0	0	0	0	0	9	Dinoseb
40	90	+	0	0	0	3	1	0	Decamethrin/Deltamethrin
41	67	+	0	4	1	1	0	5	Isofenphos
41	81	+	5	0	1	0	0	0	Fenthion
41	84	+	1	4	1	1	1	0	Parathion-ethyl
41	91	+	0	0	0	0	3	0	Cypermethrin
41	92	+ –	0	0	0	0	0	0	Fenvalerate
42	50	+	0	0	0	0	0	6	Aziprotryne
42	76	–	0	0	0	0	1	6	Vinclozolin
42	86	+	6	4	1	1	0	0	Phoxim
43	42	+	0	0	0	0	0	5	Trietazine
43	76	+	0	0	0	3	7	0	Diclofop-methyl
43	80	+	0	0	0	3	3	0	Dicofol
43	84	+	0	0	0	3	1	0	Methoxychlor
46	66	+	0	0	0	0	0	0	Fluazifop-butyl
47	50	+	5	4	1	0	0	4	Dimpylate/Diazinon

Table 8.13. (Continued)

P1	P2	UV	Color Code 1	2	3	4	5	6	Substance Name
47	64	–	0	0	1	0	0	0	Pyridate
48	52	+	0	0	0	0	0	0	Bioallethrin
49	42	+	0	0	0	0	0	0	Furmecyclox
49	66	+ –	0	0	0	1	1	6	Carbosulfan
50	48	–	0	0	0	0	0	0	Sethoxydim
50	55	+	0	0	0	0	0	6	Isomethiozin
50	75	+	6	4	1	0	1	6	Pirimiphos-methyl
50	84	–	6	4	1	0	0	6	Sulfotep
50	88	+	0	0	0	0	0	0	Fenpropathrin
51	92	–	0	0	0	3	0	0	Lindane
52	67	+	6	4	1	0	0	5	Etrimfos
52	89	–	6	4	1	0	0	0	Tolclofos-methyl
55	91	+	6	4	1	1	0	0	Chlorthiophos
56	89	–	6	1	9	1	0	6	Chlorpyriphos-methyl
58	73	–	0	0	0	3	1	6	Trifenmorph
58	89	–	0	0	1	0	0	6	Disulfoton
58	92	+	6	4	1	3	9	0	Bromophos
59	89	+	6	4	1	1	0	6	Fonofos
60	92	+	0	0	0	0	0	0	Pendimethalin/Penoxalin
61	94	+	0	0	0	3	1	0	Permethrin
63	90	–	6	4	4	1	0	0	Terbufos
64	91	–	0	4	1	1	0	0	Chlormephos
64	95	+ –	6	4	1	3	0	0	Chlorpyriphos
65	87	–	0	0	0	3	1	0	Dieldrin
69	63	–	0	0	1	0	0	6	EPTC
69	93	+ –	6	4	1	1	9	0	Bromophos-ethyl
70	82	–	0	1	1	0	0	0	Quinomethionate
71	90	–	0	0	0	3	9	0	Endrin
72	97	+	0	0	0	0	0	0	Trifluralin
76	98	+ –	0	0	0	3	1	0	Clofenotane (p,p′-Isomer)/p,p′-DDT
77	34	–	0	0	0	0	0	6	Fenpropimorph
77	95	–	0	0	0	0	1	0	Endosulfan
78	97	–	0	0	0	3	0	0	S 421
82	90	+	0	1	1	1	0	5	Triallate
83	81	–	6	4	1	1	1	3	Demeton
84	97	–	0	0	0	3	1	0	Heptachlor
86	95	+	0	0	0	0	1	0	Quintozene/PCNB
89	98	+ –	0	0	0	3	3	0	Aldrin

8.14 hR$_f$ Data in TLC System P2 (Pesticides) vs. Substance Names

P2	P1	UV	Color Code 1	2	3	4	5	6	Substance Name
0	0	+	0	0	0	0	0	0	Clopyralid
0	0	+	0	0	0	0	7	5	MCPA/(4-Chloro-o-tolyloxy)acetic acid
0	0	+	0	0	0	1	1	5	Picloram
0	0	–	0	0	0	0	3	6	Cyhexatin
0	0	–	0	1	0	0	0	6	Amitrole
0	0	–	6	1	1	0	0	6	Acephate
0	0	–	6	1	1	0	0	6	Omethoate
0	1	–	0	0	0	0	0	6	Azocyclotin
0	1	–	6	1	1	0	0	5	Methamidophos
0	3	+	6	4	1	0	0	5	Etrimfos
0	18	–	6	4	1	0	0	0	Oxydemeton-methyl
1	0	+	0	0	0	0	0	0	Chlorophacinon
1	1	+	0	0	0	0	1	6	Fenoprop
1	1	+	0	6	0	3	1	6	Formetanate
1	1	+	6	0	0	0	0	6	Monocrotophos
1	4	+	0	9	0	3	0	6	Nicotine
2	0	+ –	0	0	0	0	7	6	(2,4-Dichlorophenoxy)acetic acid/2,4-D
2	2	–	0	0	0	0	0	6	Sulfaquinoxaline
2	3	+	0	0	0	0	3	0	Dichlorprop
2	3	–	0	0	0	0	0	6	Butoxycarboxim
2	4	+	0	0	0	0	0	0	Hexazinone
2	4	–	6	0	0	3	9	9	Trichlorfon/Trichlorphos
2	5	+ –	0	0	0	0	7	0	Mecoprop/MCPP
2	5	+ –	0	9	0	3	1	6	Enilconazole/Imazalil
2	7	+	6	0	1	0	0	0	Phosphamidon
2	19	–	0	0	0	0	0	6	Triforine
3	1	–	6	4	1	0	1	0	Demeton-S-methylsulfone
3	7	–	0	0	0	0	0	6	Tiabendazole
3	8	+	0	0	0	0	0	9	Benzthiazuron
3	10	+	0	0	0	0	0	1	Bitertanol
3	11	–	0	0	0	0	0	0	Triadimenol
4	4	–	6	4	1	1	0	0	Dimethoate
4	5	+	0	6	1	1	3	9	Metamitron
4	7	+	0	0	0	3	1	6	Metoxuron
4	10	+	0	0	0	0	0	6	Thiazafluron
5	8	+	0	0	0	1	9	6	Difenoxuron
5	13	–	0	0	0	3	0	9	Cycluron
6	6	+	0	0	0	0	0	6	Methomyl
6	8	+	0	0	0	3	1	1	MCPB
7	10	+	0	0	0	3	9	6	Chloroxuron
7	11	+	0	0	0	3	0	9	Fenuron
7	13	+	0	0	0	3	9	6	Monuron
8	9	+	0	0	0	0	0	4	Fuberidazole
8	15	+	0	0	0	0	0	5	Lenacil

Table 8.14. (Continued)

P2	P1	UV	Color Code						Substance Name
			1	2	3	4	5	6	
9	15	+	0	1	0	3	1	9	Chlortoluron
9	16	+	0	0	0	0	0	9	Fluometuron
10	12	+	6	0	0	0	0	0	Mevinphos/Phosdrin
10	15	+	0	0	0	3	9	9	Diuron
11	12	+	0	0	0	0	0	6	Warfarin
11	15	+	0	9	1	0	1	6	Butocarboxim
11	18	–	0	0	0	3	0	0	Propiconazole
11	19	+	0	0	0	0	1	9	Methoprotryne
11	26	–	0	0	1	0	0	6	Thiocyclam
12	16	+	0	0	0	0	0	0	Fenarimol
12	16	+	0	0	0	0	0	6	Cyanazine
13	0	–	0	0	1	0	0	9	Butocarboxim M (Sulfoxide)
13	18	–	6	4	1	0	0	0	Demeton-S-methyl
16	8	+	0	0	0	3	3	5	Bromoxynil
16	12	+ –	0	0	0	0	9	0	Dichlorophen
16	21	+	0	0	0	0	0	9	Desmetryn
17	11	+	0	0	0	9	6	5	Phenmedipham
17	17	–	6	4	1	1	1	3	Demeton
17	20	–	0	0	0	0	0	6	Thiofanox
17	26	+	0	0	0	0	0	6	Pirimicarb
18	20	–	6	0	0	0	1	0	Heptenophos
18	22	+ –	0	0	0	0	0	6	Simazine
19	4	+	0	0	0	3	3	6	Ioxynil
20	14	+	0	0	0	0	0	0	Nitrophenol,p-
20	17	+	0	0	0	9	6	6	Carbofuran
20	20	–	6	0	0	0	3	0	Dichlorvos/DDVP
21	20	+	0	0	0	0	3	6	Propoxur
22	18	+	0	0	0	0	1	5	Methabenzthiazuron
23	22	–	0	0	0	3	3	0	Triadimefon
23	26	+	0	0	0	0	0	9	Ametryne
24	18	+	0	0	0	0	0	6	Diflubenzuron
24	26	+	0	0	0	3	9	9	Neburon
25	18	+	0	0	0	0	6	6	Carbaryl
25	21	+	0	0	1	3	1	6	Ethiofencarb
26	26	+	6	0	0	0	0	0	Chlorfenvinphos
27	21	+	0	0	1	3	0	5	Metobromuron
27	32	+	0	0	0	0	0	5	Terbutylazine
28	33	–	6	0	1	0	0	5	Ethoprofos/Ethoprop
29	20	+	0	0	1	1	9	9	Buturon
29	22	+	0	0	0	0	3	0	Flamprop-methyl
29	25	+	6	4	0	3	7	6	Tetrachlorvinphos
30	14	+	0	0	0	0	0	6	Coumatetralyl
30	22	+	0	0	0	3	9	5	Chlorbromuron
30	23	+	0	0	0	3	3	5	Monolinuron
31	22	+	0	0	0	0	9	9	Linuron
31	24	+	0	0	1	0	0	1	Metribuzin

Table 8.14. (Continued)

P2	P1	UV	Color Code 1	2	3	4	5	6	Substance Name
31	32	+	0	0	0	0	0	9	Prometryn
32	22	+	0	0	0	1	9	6	Chlorphenol,p-
32	32	+	0	0	0	0	0	5	Terbutryn
34	77	–	0	0	0	0	0	6	Fenpropimorph
35	29	+	0	0	0	0	3	0	Flamprop-isopropyl
35	35	+	0	0	0	0	0	9	Dipropetryn
38	6	+	0	0	0	0	0	0	4,6-Dinitro-o-cresol/DNOC
38	21	+	6	4	1	0	0	6	Triazophos
42	20	+	5	4	1	3	0	0	Azinphos-methyl
42	43	+	0	0	0	0	0	5	Trietazine
42	49	+	0	0	0	0	0	0	Furmecyclox
45	17	–	0	0	0	0	0	0	Pindone
45	40	+	0	0	0	0	0	0	Alachlor
47	32	+	0	4	1	0	0	0	Pyrazophos
48	24	+	6	4	1	3	0	0	Azinphos-ethyl
48	50	–	0	0	0	0	0	0	Sethoxydim
50	25	+ –	0	0	0	0	5	0	Ethofumesate
50	42	+	0	0	0	0	0	6	Aziprotryne
50	47	+	5	4	1	0	0	4	Dimpylate/Diazinon
51	19	–	0	1	1	0	0	0	Thiram
52	21	+	0	0	0	0	0	6	Naptalam
52	33	–	0	0	0	0	0	5	Propyzamide
52	48	+	0	0	0	0	0	0	Bioallethrin
53	31	–	6	4	1	0	0	6	Malathion
54	34	+	0	4	0	0	0	0	Iprodione
55	50	+	0	0	0	0	0	6	Isomethiozin
56	29	+	6	4	1	3	0	3	Methidathion
57	39	+ –	0	0	0	0	0	6	Propham
61	27	+	0	0	1	0	0	0	Coumaphos
63	69	–	0	0	1	0	0	6	EPTC
64	30	–	0	0	0	0	1	6	Dichlofluanid
64	47	–	0	0	1	0	0	0	Pyridate
66	46	+	0	0	0	0	0	0	Fluazifop-butyl
66	49	+ –	0	0	0	1	1	6	Carbosulfan
67	31	+	6	4	1	0	0	0	Phosalone
67	33	–	0	0	0	0	6	6	Tolylfluanid
67	41	+	0	4	1	1	0	5	Isofenphos
67	52	+	6	4	1	0	0	5	Etrimfos
71	31	+	6	4	1	1	1	1	Chlorthion
73	30	+	6	4	1	1	1	0	Parathion-methyl
73	58	–	0	0	0	3	1	6	Trifenmorph
75	23	+	6	4	1	3	0	0	Temephos
75	50	+	6	4	1	0	1	6	Pirimiphos-methyl
76	32	+	6	4	1	1	1	0	Fenitrothion
76	42	–	0	0	0	0	1	6	Vinclozolin
76	43	+	0	0	0	3	7	0	Diclofop-methyl

Table 8.14. (Continued)

P2	P1	UV	Color Code 1	2	3	4	5	6	Substance Name
77	40	–	0	0	0	0	1	0	Endosulfan
80	43	+	0	0	0	3	3	0	Dicofol
81	40	+	0	0	0	0	0	9	Dinoseb
81	41	+	5	0	1	0	0	0	Fenthion
81	83	–	6	4	1	1	1	3	Demeton
82	70	–	0	1	1	0	0	0	Quinomethionate
84	41	+	1	4	1	1	1	0	Parathion-ethyl
84	43	+	0	0	0	3	1	0	Methoxychlor
84	50	–	6	4	1	0	0	6	Sulfotep
86	42	+	6	4	1	1	0	0	Phoxim
87	65	–	0	0	0	3	1	0	Dieldrin
88	50	+	0	0	0	0	0	0	Fenpropathrin
89	37	+	0	0	0	3	1	6	Cyfluthrin
89	52	–	6	4	1	0	0	0	Tolclofos-methyl
89	56	–	6	1	9	1	0	6	Chlorpyriphos-methyl
89	58	–	0	0	1	0	0	6	Disulfoton
89	59	+	6	4	1	1	0	6	Fonofos
90	40	+	0	0	0	3	1	0	Decamethrin/Deltamethrin
90	63	–	6	4	4	1	0	0	Terbufos
90	71	–	0	0	0	3	9	0	Endrin
90	82	+	0	1	1	1	0	5	Triallate
91	41	+	0	0	0	0	3	0	Cypermethrin
91	55	+	6	4	1	1	0	0	Chlorthiophos
91	64	–	0	4	1	1	0	0	Chlormephos
92	41	+ –	0	0	0	0	0	0	Fenvalerate
92	51	–	0	0	0	3	0	0	Lindane
92	58	+	6	4	1	3	9	0	Bromophos
92	60	+	0	0	0	0	0	0	Pendimethalin/Penoxalin
93	69	+ –	6	4	1	1	9	0	Bromophos-ethyl
94	61	+	0	0	0	3	1	0	Permethrin
95	64	+ –	6	4	1	3	0	0	Chlorpyriphos
95	77	–	0	0	0	0	1	0	Endosulfan
95	86	+	0	0	0	0	1	0	Quintozene/PCNB
97	72	+	0	0	0	0	0	0	Trifluralin
97	78	–	0	0	0	3	0	0	S 421
97	84	–	0	0	0	3	1	0	Heptachlor
98	76	+ –	0	0	0	3	1	0	Clofenotane (p,p′-Isomer)/p,p′-DDT
98	89	+ –	0	0	0	3	3	0	Aldrin

8.15 hR_f Data in TLC System B (Benzodiazepine Hydrolysis Products) vs. Substance Names

hR_f	Substance Name		
0	Clonazepam M Hy	DCB	2,5-Diamino-2′-chloro-benzophenone
0	Flurazepam Hy	DCFB	5-Chloro-2-(2-diethylamino)-ethylamino-2′-fluorobenzophenone
0	Nitrazepam M Hy	DAB	2,5-Diaminobenzophenone
2	Bromazepam Hy	ABP	2-(2-Amino-5-bromo-benzoyl)pyridine
2	Flurazepam M Hy	HCFB	5-Chloro-2′-fluoro-2-(2-hydroxyethylamino)-benzophenone
15	Nitrazepam Hy	ANB	2-Amino-5-nitrobenzophenone
16	Clonazepam Hy	ANCB	2-Amino-2-chloro-5-nitrobenzophenone
16	Flunitrazepam M Hy	ANFB	2-Amino-2′-fluoro-5-nitrobenzophenone
27	Oxazepam Hy	ACB	2-Amino-5-chlorobenzophenone
29	Flunitrazepam Hy	MNFB	2-Methylamino-2′-fluoro-5-nitrobenzophenone
29	Nimetazepam Hy	MNB	2-Methylamino-5-nitrobenzophenone
31	Flurazepam M Hy	ACFB	2-Amino-5-chloro-2′-fluorobenzophenone
32	Haloxazolam Hy	ABFB	2-Amino-5-bromo-2′-fluorobenzophenone
33	Lorazepam Hy	ADB	2-Amino-2′,5-dichlorobenzophenone
52	Diazepam Hy	MACB	2-Methylamino-5-chlorobenzophenone
55	Fludiazepam Hy	CFMB	5-Chloro-2′-fluoro-2-methylaminobenzophenone
57	Lormetazepam Hy	MDB	2′,5-Dichloro-2-methylaminobenzophenone
57	Pinazepam Hy	CPB	5-Chloro-2-(2-propinylamino)-benzophenone
68	Prazepam Hy	CCB	5-Chloro-2-cyclopropylmethylaminobenzophenone
74	Halazepam Hy	TCB	5-Chloro-2-(2,2,2-trifluoroethylamino)benzophenone
74	Quazepam Hy	CFTB	5-Chloro-2′-fluoro-2-(2,2,2-trifluoroethyl)aminobenzophenone

8.16 CAS Registry Numbers in Numerical Order vs. Substance Names

CAS No.	Substance Name
50-06-6	Phenobarbital
50-10-2	Oxyphenonium Bromide
50-11-3	Metharbital
50-12-4	Mephenytoin
50-18-0	Cyclophosphamide
50-23-7	Hydrocortisone
50-28-2	Estradiol
50-29-3	Clofenotane (p,p′-Isomer)/p,p′-DDT
50-33-9	Phenylbutazone
50-34-0	Propantheline Bromide
50-36-2	Cocaine
50-37-3	Lysergide/LSD
50-39-5	Protheobromine
50-44-2	Mercaptopurine,6-
50-47-5	Desipramine/Imipramine M
50-48-6	Amitriptyline
50-49-7	Imipramine
50-52-2	Thioridazine
50-53-3	Chlorpromazine
50-55-5	Reserpine
50-60-2	Phentolamine
50-65-7	Niclosamide
50-78-2	Acetylsalicylic Acid
51-06-9	Procainamide
51-12-7	Nialamide
51-34-3	Scopolamine
51-41-2	Levarterenol/Norepinephrine
51-43-4	Epinephrine
51-45-6	Histamine
51-55-8	Atropine
51-61-6	Dopamine
51-68-3	Meclofenoxate
51-71-8	Phenelzine
51-83-2	Carbachol
52-01-7	Spironolactone
52-24-4	Thiotepa
52-31-3	Cyclobarbital
52-43-7	Allobarbital
52-53-9	Verapamil
52-67-5	Penicillamine
52-68-6	Trichlorfon/Trichlorphos
52-86-8	Haloperidol
52-88-0	Atropine Methonitrate
53-41-8	Androsterone
53-46-3	Methanthelinium Bromide

Table 8.16. (Continued)

CAS No.	Substance Name
53-86-1	Indometacin
54-03-5	Hexobendine
54-04-6	Mescaline
54-05-7	Chloroquine
54-11-5	Nicotine
54-30-8	Camylofine
54-31-9	Furosemide
54-32-0	Moxisylyte
54-36-4	Metyrapone
54-49-9	Metaraminol
54-85-3	Isoniazid
54-91-1	Pipobroman
54-92-2	Iproniazid
54-95-5	Pentetrazol
55-38-9	Fenthion
55-56-1	Chlorhexidine
55-63-0	Nitroglycerin
55-65-2	Guanethidine
55-73-2	Betanidine
55-98-1	Busulfan
56-29-1	Hexobarbital
56-38-2	Parathion-ethyl
56-53-1	Diethylstilbestrol
56-54-2	Quinidine
56-72-4	Coumaphos
56-75-7	Chloramphenicol
57-24-9	Strychnine
57-27-2	Morphine
57-41-0	Phenytoin
57-42-1	Pethidine/Meperidine
57-43-2	Amobarbital
57-44-3	Barbital
57-47-6	Physostigmine
57-53-4	Meprobamate
57-63-6	Ethinylestradiol
57-66-9	Probenecid
57-67-0	Sulfaguanidine
57-68-1	Sulfadimidine
57-83-0	Progesterone
57-87-4	Ergosterol
57-88-5	Cholesterol
57-92-1	Streptomycin
57-95-4	Tubocurarine
57-96-5	Sulfinpyrazone
58-00-4	Apomorphine
58-08-2	Caffeine
58-14-0	Pyrimethamine

Table 8.16. (Continued)

CAS No.	Substance Name
58-15-1	Aminophenazone/Amidopyrine
58-18-4	Methyltestosterone,17-
58-22-0	Testosterone
58-25-3	Chlordiazepoxide
58-32-2	Dipyridamole
58-34-4	Thiazinamium Metilsulfate
58-37-7	Aminopromazine
58-38-8	Prochlorperazine
58-39-9	Perphenazine
58-40-2	Promazine
58-46-8	Tetrabenazine
58-54-8	Etacrynic Acid
58-55-9	Theophylline
58-56-0	Pyridoxine Hydrochloride
58-73-1	Diphenhydramine
58-74-2	Papaverine
58-89-9	Lindane
58-93-5	Hydrochlorothiazide
58-94-6	Chlorothiazide
59-26-7	Nikethamide
59-32-5	Chloropyramine
59-39-2	Piperoxan
59-40-5	Sulfaquinoxaline
59-42-7	Phenylephrine
59-46-1	Procaine
59-47-2	Mephenesin
59-63-2	Isocarboxazid
59-66-5	Acetazolamide
59-67-6	Nicotinic Acid
59-87-0	Nitrofural/Nitrofurazone
59-92-7	Levodopa
59-96-1	Phenoxybenzamine
59-98-3	Tolazoline
59-99-4	Neostigmine
60-26-4	Hexamethonium
60-40-2	Mecamylamine
60-44-6	Penthienate Bromide
60-45-7	Fenimide
60-51-5	Dimethoate
60-54-8	Tetracycline
60-56-0	Thiamazole
60-57-1	Dieldrin
60-79-7	Ergometrine
60-80-0	Phenazone/Antipyrine
60-87-7	Promethazine
60-89-9	Pecazine/Mepazine
60-91-3	Diethazine

Table 8.16. (Continued)

CAS No.	Substance Name
60-99-1	Levomepromazine/Methotrimeprazine
61-00-7	Acepromazine
61-01-8	Methopromazine
61-50-7	Dimethyltryptamine,N,N-/DMT
61-51-8	Diethyltryptamine,N,N-/DET
61-54-1	Tryptamine
61-56-3	Sultiame
61-57-4	Niridazole
61-68-7	Mefenamic Acid
61-82-5	Amitrole
62-44-2	Phenacetin
62-46-4	Thioctic Acid
62-67-9	Nalorphine
62-73-7	Dichlorvos/DDVP
63-12-7	Benzquinamide
63-25-2	Carbaryl
63-74-1	Sulfanilamide
63-75-2	Arecoline
63-98-9	Phenacemide
64-04-0	Phenethylamine
64-39-1	Trimeperidine
64-55-1	Mebutamate
64-65-3	Bemegride
64-77-7	Tolbutamide
64-86-8	Colchicine
64-95-9	Adiphenine
65-45-2	Salicylamide
65-49-6	Aminosalicylic Acid,p-
65-64-5	Mebenazine
65-85-0	Benzoic Acid
66-76-2	Dicoumarol
67-03-8	Thiamine Hydrochloride
67-20-9	Nitrofurantoin
67-45-8	Furazolidone
67-52-7	Barbituric Acid
68-35-9	Sulfadiazine
68-41-7	Cycloserine
68-88-2	Hydroxyzine
68-90-6	Benziodarone
69-23-8	Fluphenazine
69-72-7	Salicylic Acid
69-81-8	Carbazochrome
70-19-9	Thurfyl Nicotinate
70-30-4	Hexachlorophene
71-27-2	Suxamethonium Chloride
71-63-6	Digitoxin
71-81-8	Isopropamide Iodide

Table 8.16. (Continued)

CAS No.	Substance Name
72-14-0	Sulfathiazole
72-20-8	Endrin
72-43-5	Methoxychlor
72-44-6	Methaqualone
72-69-5	Amitriptyline M/Nortriptyline
73-09-6	Etozolin
73-48-3	Bendroflumethiazide/Bendrofluthiazide
73-49-4	Quinethazone
74-11-3	Chlorobenzoic Acid,p-
74-55-5	Ethambutol
75-80-9	Tribromoethanol
76-22-2	Camphor
76-23-3	Tetrabarbital
76-41-5	Oxymorphone
76-42-6	Oxycodone
76-44-8	Heptachlor
76-57-3	Codeine
76-58-4	Ethylmorphine
76-65-3	Amolanone
76-68-6	Cyclopentobarbital
76-73-3	Secobarbital
76-74-4	Pentobarbital
76-75-5	Thiopental
76-76-6	Probarbital
76-94-8	Methyl-5-phenylbarbituric Acid,5-
76-99-3	Methadone
77-01-0	Fenpipramide
77-02-1	Aprobarbital
77-04-3	Pyrithyldione
77-07-6	Levorphanol
77-09-8	Phenolphthalein
77-10-1	Phencyclidine
77-14-5	Proheptazine
77-15-6	Ethoheptazine
77-17-8	Pethidine M (Nor-)
77-19-0	Dicycloverine/Dicyclomine
77-20-3	Alphaprodine
77-21-4	Glutethimide
77-23-6	Pentoxyverine/Carbetapentane
77-26-9	Butalbital
77-28-1	Butobarbital/Butethal
77-30-5	Hexethal
77-32-7	Thiobarbital
77-36-1	Chlortalidone
77-37-2	Procyclidine
77-38-3	Chlorphenoxamine
77-39-4	Cycrimine

Table 8.16. (Continued)

CAS No.	Substance Name
77-41-8	Mesuximide
77-50-9	Propoxyphene
77-51-0	Isoaminile
77-65-6	Carbromal
77-66-7	Acecarbromal
77-67-8	Ethosuximide
78-11-5	Pentaerithritol Tetranitrate
78-44-4	Carisoprodol
79-55-0	Pempidine
79-57-2	Oxytetracycline
79-93-6	Phenaglycodol
80-35-3	Sulfamethoxypyridazine
80-49-9	Homatropine Methylbromide
80-77-3	Chlormezanone
81-07-2	Saccharin
81-81-2	Warfarin
81-82-3	Coumachlor
82-02-0	Khellin
82-54-2	Cotarnine
82-58-6	Lysergic Acid
82-66-6	Diphenadione
82-68-8	Quintozene/PCNB
82-88-2	Phenindamine
82-92-8	Cyclizine
82-93-9	Chlorcyclizine
82-95-1	Buclizine
82-98-4	Piperidolate
83-07-8	Ampyrone
83-12-5	Phenindione
83-26-1	Pindone
83-28-3	Isovaleryl-1,3-indanedione,2-
83-43-2	Methylprednisolon
83-67-0	Theobromine
83-73-8	Diiodohydroxyquinoline
83-74-9	Ibogaine/Rotenone
83-89-6	Mepacrine
83-98-7	Orphenadrine
84-04-8	Pipamazine
84-06-0	Thiopropazate
84-12-8	Phanquinone
84-22-0	Tetryzoline
84-55-9	Viquidil
84-96-8	Trimeprazine/Alimemazine
84-97-9	Perazine
85-00-7	Diquat Dibromide
85-01-8	Phenanthrene
85-18-7	Chlorotheophylline,8-

Table 8.16. (Continued)

CAS No.	Substance Name
85-73-4	Phthalylsulfathiazole
85-79-0	Cinchocaine/Dibucaine
86-12-4	Thenalidine
86-13-5	Benzatropine
86-14-6	Diethylthiambutene
86-21-5	Pheniramine
86-22-6	Brompheniramine
86-34-0	Phensuximide
86-35-1	Ethotoin
86-42-0	Amodiaquine
86-43-1	Propoxycaine
86-50-0	Azinphos-methyl
86-54-4	Hydralazine
86-78-2	Pentaquine
86-80-6	Quinisocaine
87-00-3	Homatropine
88-85-7	Dinoseb
89-83-8	Thymol
90-34-6	Primaquine
90-39-1	Sparteine
90-45-9	Aminoacridine
90-49-3	Pheneturide
90-54-0	Etafenone
90-69-7	Lobeline
90-82-4	Pseudoephedrine
90-84-6	Amfepramone/Diethylpropion
90-89-1	Diethylcarbamazine
91-33-8	Benzthiazide
91-64-5	Coumarine
91-75-8	Antazoline
91-79-2	Thenyldiamine
91-80-5	Methapyrilene
91-81-6	Tripelennamine
91-82-7	Pyrrobutamine
91-84-9	Mepyramine/Pyrilamine
91-85-0	Thonzylamine
92-12-6	Phenyltoloxamine
92-13-7	Pilocarpine
92-23-9	Leucinocaine
92-84-2	Phenothiazine
93-14-1	Guaifenesin
93-30-1	Methoxyphenamine
93-60-7	Methyl Nicotinate
93-72-1	Fenoprop
94-07-5	Synephrine/Oxedrine
94-09-7	Benzocaine
94-10-0	Ethoxazene

Table 8.16. (Continued)

CAS No.	Substance Name
94-13-3	Propyl p-Hydroxybenzoate
94-14-4	Isobutyl 4-Aminobenzoate
94-15-5	Dimethocaine
94-19-9	Sulfaethidole
94-20-2	Chlorpropamide
94-24-6	Tetracaine/Amethocaine
94-25-7	Butyl Aminobenzoate
94-35-9	Styramate
94-44-0	Benzyl Nicotinate
94-74-6	MCPA/(4-Chloro-o-tolyloxy)acetic Acid
94-75-7	Dichlorophenoxy)acetic Acid,(2,4-/2,4-D
94-78-0	Phenazopyridine
94-81-5	MCPB
95-04-5	Ectylurea
95-25-0	Chlorzoxazone
95-27-2	Dimazole/Diamthazole
96-88-8	Mepivacaine
97-23-4	Dichlorophen
97-24-5	Fenticlor
97-77-8	Disulfiram
98-92-0	Nicotinamide
98-96-4	Pyrazinamide
99-43-4	Oxybuprocaine
99-66-1	Valproic Acid
99-76-3	Methyl p-Hydroxybenzoate/Methyl Paraben
99-96-7	Hydroxybenzoic Acid, p-
100-02-7	Nitrophenol,p-
100-10-7	Dimethylaminobenzaldehyde,p-
100-33-4	Pentamidine
100-51-6	Benzyl Alcohol
100-55-0	Nicotinyl Alcohol
100-88-9	Cyclamic Acid
100-91-4	Eucatropine
100-92-5	Mephentermine
100-97-0	Methenamine
101-08-6	Diperodon
101-20-2	Triclocarban
101-31-5	Hyoscyamine
101-40-6	Propylhexedrine
101-42-8	Fenuron
102-45-4	Cyclopentamine
103-84-4	Acetanilide
103-90-2	Paracetamol
104-14-3	Octopamine
104-32-5	Propamidine
105-20-4	Betazole
106-48-9	Chlorphenol,p-

Table 8.16. (Continued)

CAS No.	Substance Name
107-43-7	Betaine
108-46-3	Resorcinol
110-85-0	Piperazine
110-89-4	Piperidine
111-30-8	Glutaral
112-38-9	Undecylenic Acid
113-15-5	Ergotamine
113-18-8	Ethchlorvynol
113-42-8	Methylergometrine
113-45-1	Methylphenidate
113-53-1	Dothiepin
113-59-7	Chlorprothixene
114-07-8	Erythromycin
114-26-1	Propoxur
114-86-3	Phenformin
114-90-9	Obidoxime Chloride
115-32-2	Dicofol
115-37-7	Thebaine
115-38-8	Methylphenobarbital
115-44-6	Talbutal
115-46-8	Azacyclonol
115-63-9	Hexocyclium Metilsulfate
115-67-3	Paramethadione
115-68-4	Sulfadicramide
116-42-7	Sulfaproxyline
116-43-8	Succinylsulfathiazole
117-10-2	Danthron
117-89-5	Trifluoperazine
118-08-1	Hydrastine
118-10-5	Cinchonine
118-23-0	Bromazine/Bromdiphenhydramine
118-42-3	Hydroxychloroquine
118-55-8	Phenyl Salicylate
118-91-2	Chlorobenzoic Acid,o-
119-36-8	Methyl Salicylate
120-20-7	Dimethoxyphenethylamine,3,4-
120-29-6	Tropine
120-36-5	Dichlorprop
120-47-8	Ethyl p-Hydroxybenzoate
120-57-0	Piperonal (precursor of MDA, MDE, MDMA)
120-97-8	Diclofenamide
121-75-5	Malathion
122-06-5	Stilbamidine
122-09-8	Phentermine
122-11-2	Sulfadimethoxine
122-14-5	Fenitrothion
122-34-9	Simazine

Table 8.16. (Continued)

CAS No.	Substance Name
122-39-4	Diphenylamine
122-42-9	Propham
123-30-8	Paracetamol M/p-Aminophenol
123-82-0	Tuaminoheptane
125-24-6	Pseudomorphine
125-28-0	Dihydrocodeine
125-29-1	Hydrocodone
125-33-7	Primidone
125-40-6	Secbutabarbital
125-42-8	Vinbarbital
125-53-1	Oxyphencyclimine
125-55-3	Narcobarbital
125-58-6	Levomethadone
125-60-0	Fenpiverinium
125-64-4	Methyprylon
125-70-2	Levomethorphan
125-71-3	Dextromethorphan
125-73-5	Dextrorphan
125-79-1	Methadone Intermediate/Pre-Methadone
125-84-8	Aminoglutethimide
126-07-8	Griseofulvin
126-27-2	Oxetacaine
126-52-3	Ethinamate
127-35-5	Phenazocine
127-69-5	Sulfafurazole
127-79-7	Sulfamerazine
127-90-2	S 421
128-46-1	Dihydrostreptomycin
128-62-1	Noscapine
129-03-3	Cyproheptadine
129-20-4	Oxyphenbutazone/Phenylbutazone M (Hydroxy-)
129-83-9	Phenampromide
130-26-7	Clioquinol
130-95-0	Quinine
131-01-1	Deserpidine
131-28-2	Narceine
131-69-1	Phthalylsulfacetamide
132-22-9	Chlorphenamine/Chlorpheniramine
132-60-5	Cinchophen
132-66-1	Naptalam
133-16-4	Chloroprocaine
133-67-5	Trichlormethiazide
134-49-6	Phenmetrazine/Phendimetrazine M (Nor-)
134-62-3	Diethyl-m-toluamide,N,N-/Deet
135-07-9	Methyclothiazide
135-09-1	Hydroflumethiazide
136-70-9	Protokylol

Table 8.16. (Continued)

CAS No.	Substance Name
136-82-3	Piperocaine
137-26-8	Thiram
137-58-6	Lidocaine
138-39-6	Mafenide
138-56-7	Trimethobenzamide
138-61-4	Nordefrin Hydrochloride
139-62-8	Cyclomethycaine
139-91-3	Furaltadone
140-65-8	Pramoxine
141-94-6	Hexetidine
143-52-2	Metopon
144-11-6	Trihexyphenidyl/Benzhexol
144-12-7	Tiemonium Iodide
144-14-9	Anileridine
144-44-5	Pentolinium
144-80-9	Sulfacetamid
144-82-1	Sulfamethizole
144-83-2	Sulfapyridine
146-22-5	Nitrazepam
146-36-1	Azapetine
146-48-5	Yohimbine
146-54-3	Triflupromazine
147-20-6	Diphenylpyraline
147-27-3	Dimoxyline
147-94-4	Cytarabine
148-32-3	Amprotropine
148-79-8	Tiabendazole
149-16-6	Butacaine
149-64-4	Butylscopolammonium Bromide,N-
150-13-0	Aminobenzoic Acid,p-
150-59-4	Alverine
150-68-5	Monuron
151-83-7	Methohexital
152-02-3	Levallorphan
152-47-6	Sulfalene
152-72-7	Acenocoumarol
153-00-4	Methenolone
153-76-4	Gallamine
153-87-7	Oxypertine
154-21-2	Lincomycin
155-09-9	Tranylcypromine
155-41-9	Methscopolamine Bromide
156-08-1	Benzphetamine
156-74-1	Decamethonium
244-63-3	Norharman
297-90-5	Racemorphan
298-00-0	Parathion-methyl

Table 8.16. (Continued)

CAS No.	Substance Name
298-04-4	Disulfoton
298-46-4	Carbamazepine
298-55-5	Clocinizine
298-57-7	Cinnarizine
299-42-3	Ephedrine
300-62-9	Amphetamine/Amphetaminil M artifact
301-12-2	Oxydemeton-methyl
302-27-2	Aconitine
302-33-0	Proadifen
302-40-9	Benactyzine
302-41-0	Piritramide
302-66-9	Methylpentynol Carbamate
303-49-1	Clomipramine
303-53-7	Cyclobenzaprine
303-69-5	Prothipendyl
304-84-7	Etamivan
305-03-3	Chlorambucil
309-00-2	Aldrin
309-29-5	Doxapram
314-35-2	Etamiphyllin
315-30-0	Allopurinol
315-72-0	Opipramol
316-81-4	Thioproperazine
317-34-0	Aminophylline
321-64-2	Tacrine
322-35-0	Benserazide
330-54-1	Diuron
330-55-2	Linuron
333-41-5	Dimpylate/Diazinon
339-43-5	Carbutamide
339-44-6	Glymidine
341-00-4	Etifelmine
344-80-9	Flunitrazepam M Hy/ANFB
346-18-9	Polythiazide
357-56-2	Dextromoramide
357-57-3	Brucine
358-52-1	Hexapropymate
359-83-1	Pentazocine
361-37-5	Methysergide
362-29-8	Propiomazine
364-62-5	Metoclopramide
364-98-7	Diazoxide
366-70-1	Procarbazine
370-14-9	Pholedrine
388-51-2	Metofenazate
389-08-2	Nalidixic Acid
390-28-3	Methoxamine

Table 8.16. (Continued)

CAS No.	Substance Name
390-64-7	Prenylamine
395-28-8	Isoxsuprine
396-01-0	Triamterene
427-00-9	Desomorphine
428-37-5	Profadol
434-43-5	Pentorex
435-97-2	Phenprocoumon
437-38-7	Fentanyl
437-74-1	Xanthinol Nicotinate
438-60-8	Protriptyline
439-14-5	Diazepam/Ketazolam M/Medazepam M
441-61-2	Ethylmethylthiambutene
442-16-0	Ethacridine
442-51-3	Harmine
442-52-4	Clemizole
443-48-1	Metronidazole
446-86-6	Azathioprine
447-41-6	Buphenine
452-35-7	Ethoxzolamide
456-59-7	Cyclandelate
457-87-4	Etilamfetamine
458-24-2	Fenfluramine
458-88-8	Coniine
461-78-9	Chlorphentermine
465-65-6	Naloxone
466-06-8	Proscillaridine
466-40-0	Isomethadone
466-90-0	Thebacon
466-97-7	Morphine M (Nor-)
466-99-9	Hydromorphone
467-14-1	Neopine
467-15-2	Codeine M (Nor-)
467-18-5	Myrophine
467-36-7	Thialbarbital
467-60-7	Pipradrol
467-65-2	Clidinium
467-83-4	Dipipanone
467-84-5	Phenadoxone
467-85-6	Methadone M (Nor-)
467-86-7	Dioxaphetyl Butyrate
468-07-5	Phenomorphan
468-50-8	Betameprodine
468-51-9	Alphameprodine
468-59-7	Betaprodine
468-61-1	Oxeladine
469-21-6	Doxylamine
469-62-5	Dextropropoxyphene

Table 8.16. (Continued)

CAS No.	Substance Name
469-79-4	Ketobemidone
469-81-8	Morpheridine
469-82-9	Etoxeridine
470-68-8	Eucaine,α-
470-90-6	Chlorfenvinphos
471-53-4	Enoxolone
477-30-5	Demecolcine
477-32-7	Visnadin
477-93-0	Dimethoxanate
478-43-3	Rhein
478-84-2	Bromolysergide,2-
478-94-4	Lysergamide
479-18-5	Diprophylline/Propyphylline
479-92-5	Propyphenazone
481-06-1	Santonin
481-37-8	Cocaine M/Ecgonine
481-72-1	Aloe-emodin
481-74-3	Chrysophanol
482-15-5	Isothipendyl
482-74-6	Cryptopine
483-04-5	Ajmalicine/Raubasine
483-17-0	Cephaeline
483-18-1	Emetine
483-63-6	Crotamiton
484-23-1	Dihydralazine
485-35-8	Cytisine
485-71-2	Cinchonidine
486-12-4	Triprolidine
486-16-8	Carbinoxamine
486-17-9	Captodiame
486-47-5	Ethaverine
486-56-6	Cotinine
486-84-0	Harman
487-54-7	Hydroxyhippuric Acid,o-
487-93-4	Bufotenine
490-55-1	Amiphenazole
490-79-9	Gentisic Acid
492-41-1	Ephedrine M/Phenylpropanolamine
493-75-4	Bialamicol
493-78-7	Methaphenilene
493-80-1	Histapyrrodine
493-92-5	Prolintane
495-83-0	Tigloidine
495-84-1	Salinazid
495-99-8	Hydroxystilbamidine
496-67-3	Bromisoval
499-67-2	Proxymetacaine

Table 8.16. (Continued)

CAS No.	Substance Name
500-28-7	Chlorthion
500-34-5	Eucaine,β-
500-42-5	Chlorazanil
500-55-0	Apoatropine
500-64-1	Kavain
500-89-0	Thiambutosine
500-92-5	Proguanil/Chlorguanide
501-68-8	Beclamide
502-59-0	Octamylamine
503-01-5	Isometheptene
504-17-6	Thiobarbituric Acid
509-60-4	Dihydromorphine
509-67-1	Pholcodine
509-86-4	Heptabarbital
510-53-2	Racemethorphan
511-08-0	Ergocristine
511-09-1	Ergocryptine
511-12-6	Dihydroergotamine
511-45-5	Pridinol
512-15-2	Cyclopentolate
514-65-8	Biperiden
515-49-1	Sulfathiourea
515-64-0	Sulfisomidine
519-09-5	Cocaine M/N-Benzoylecgonine
519-37-9	Etofylline
519-88-0	Ambucetamide
520-52-5	Psilocybine
520-53-6	Psilocin
521-67-5	Cinnamoylcocaine
521-74-4	Broxyquinoline
522-00-9	Profenamine/Ethopropazine
522-18-9	Chlorbenzoxamine
522-66-7	Hydroquinine
523-54-6	Etymemazine
523-87-5	Dimenhydrinate
523-87-5	Dimenhydrinate
524-81-2	Mebhydrolin
524-84-5	Dimethylthiambutene
524-99-2	Medrylamine
525-66-6	Propranolol
526-08-9	Sulfaphenazole
526-36-3	Xylometazoline
528-92-7	Apronal
530-08-5	Isoetarine
530-78-9	Flufenamic Acid
531-76-0	Sarcolysis/Merphalan
532-03-6	Methocarbamol

Table 8.16. (Continued)

CAS No.	Substance Name
533-06-2	Mephenesin Carbamate
533-45-9	Clomethiazole
534-52-1	Dinitro-o-cresol, 4,6-/DNOC
535-80-8	Chlorobenzoic Acid,m-
536-21-0	Norfenefrine
536-24-3	Ethylnorepinephrine
536-25-4	Orthocaine
536-33-4	Ethionamide
536-69-6	Fusaric Acid
537-21-3	Chlorproguanil
537-26-8	Tropacocaine
537-46-2	Methamfetamine
539-15-1	Hordenine
539-21-9	Ambazone
543-15-7	Heptaminol
545-59-5	Racemoramide
545-93-7	Ibomal/Propallylonal
546-06-5	Conessine
547-44-4	Sulfacarbamid
548-00-5	Ethyl Biscoumacetate
548-73-2	Droperidol
549-68-8	Octaverine
550-81-2	Amopyroquine
552-25-0	Diampromide
552-79-4	Methylephedrine,N-
553-69-5	Fenyramidol
555-30-6	Methyldopa
555-37-3	Neburon
555-57-7	Pargyline
555-65-7	Brocresine
561-27-3	Diamorphine
561-48-8	Norpipanone
561-76-2	Properidine
561-83-1	Nealbarbital
561-86-4	Brallobarbital
561-94-4	Ergosine
562-26-5	Phenoperidine
568-21-8	Isothebaine
569-31-3	Meconin
569-65-3	Meclozine
575-74-6	Buclosamide
579-38-4	Diloxanide
584-79-2	Bioallethrin
586-06-1	Orciprenaline
586-60-7	Dyclonine
596-51-0	Glycopyrronium Bromide
599-79-1	Sulfasalazine

Table 8.16. (Continued)

CAS No.	Substance Name
599-88-2	Sulfaperine
603-00-9	Proxyphylline
603-50-9	Bisacodyl
604-51-3	Deptropine
604-75-1	Oxazepam/M of Chlordiazepoxide, Clorazepic Acid, Camazepam, Diazepam, Halazepam, Ketazolam, Medazepam, Oxazolam, Prazepam
606-90-6	Piprinhydrinate
608-08-2	Indolyl Acetate,3-
632-00-8	Sulfasomizole
633-47-6	Cropropamide
634-03-7	Phendimetrazine
635-41-6	Trimetozine
636-54-4	Clopamide
637-07-0	Clofibrate
639-48-5	Nicomorphine
642-72-8	Benzydamine
644-26-8	Amylocaine
644-62-2	Meclofenamic Acid
651-06-9	Sulfametoxydiazine
653-03-2	Butaperazine
655-05-0	Tozalinone
657-24-9	Metformin
657-24-9	Metformin
673-31-4	Phenprobamate
692-13-7	Buformin
709-55-7	Etilefrine
719-59-5	Oxazepam Hy/ACB
721-50-6	Prilocaine
723-46-6	Sulfamethoxazole
731-27-1	Tolylfluanid
735-06-8	Flunitrazepam Hy/MNFB
738-70-5	Trimethoprim
739-71-9	Trimipramine
742-20-1	Cyclopenthiazide
749-02-0	Spiperone
749-13-3	Trifluperidol
751-97-3	Rolitetracycline
759-94-4	EPTC
768-94-5	Amantadine
784-38-3	Flurazepam M Hy/ACFB
789-02-6	Clofenotane (o,p′-Isomer)/o,p′-DDT
791-35-5	Clofedanol
804-10-4	Carbocromen
834-12-8	Ametryne
835-31-4	Naphazoline
841-06-5	Methoprotryne
846-49-1	Lorazepam/Lormetazepam M

Table 8.16. (Continued)

CAS No.	Substance Name
846-50-4	Temazepam/M of Camazepam, Diazepam, Ketazolam, Medazepam
848-75-9	Lormetazepam
853-34-9	Kebuzone
865-04-3	Methoserpidine
865-21-4	Vinblastine
886-50-0	Terbutryn
886-74-8	Chlorphenesin Carbamate
894-76-8	Flunitrazepam M (7-Aminonor-)
908-54-3	Diminazene Aceturate
910-86-1	Tiocarlide
911-45-5	Clomifene
911-65-9	Etonitazene
915-30-0	Diphenoxylate
919-86-8	Demeton-S-methyl
938-73-8	Ethenzamide
944-22-9	Fonofos
950-37-8	Methidathion
952-54-5	Morinamide
959-98-8	Endosulfan
959-98-8	Endosulfan
963-39-3	Demoxepam/Chlordiazepoxide M
968-63-8	Butinoline
968-81-0	Acetohexamide
972-02-1	Difenidol
982-24-1	Clopenthixol/Zuclopenthixol/cis-Ordinol
982-43-4	Prenoxdiazine/Tibexin
985-16-0	Nafcillin
991-42-4	Norbormide
1014-69-3	Desmetryn
1022-13-5	Diazepam Hy/MACB
1028-33-7	Pentifylline
1050-79-9	Moperone
1082-57-1	Tramazoline
1082-88-8	Trimethoxyamphetamine,3,4,5-
1085-98-9	Dichlofluanid
1088-11-5	Nordazepam/M of Chlordiazepoxide, Clorazepic Acid, Diazepam, Halazepam, Ketazolam, Medazepam, Oxazolam, Prazepam
1098-97-1	Pyritinol
1113-02-6	Omethoate
1131-64-2	Debrisoquine
1134-47-0	Baclofen
1142-70-7	Butallylonal
1154-09-2	Dothiepin M (Nor-)
1156-05-4	Phenglutarimide
1156-19-0	Tolazamide
1165-48-6	Dimefline
1209-98-9	Fencamfamine

Table 8.16. (Continued)

CAS No.	Substance Name
1216-40-6	Sigmodal (5-(2-Bromoallyl)-5-(1-methylbutyl)barbituric Acid)
1223-36-5	Clofexamide
1260-17-9	Carminic Acid
1403-66-3	Gentamycin
1404-04-2	Neomycin
1420-06-0	Trifenmorph
1420-55-9	Thiethylperazine
1421-14-3	Propanidid
1435-55-8	Hydroquinidine
1477-39-0	Noracymethadol
1480-19-9	Fluanisone
1491-59-4	Oxymetazoline
1506-86-1	Fluorofentanyl,p-
1508-75-4	Tropicamide
1518-86-1	Hydroxyamphetamine/Amphetamine M
1531-12-0	Levorphanol M (Nor-) → Dextrorphan M
1548-36-3	Fludiazepam Hy/CFMB
1563-56-0	Bromazepam Hy/ABP
1563-66-2	Carbofuran
1580-83-2	Paraflutizide
1582-09-8	Trifluralin
1617-90-9	Vincamine
1622-61-3	Clonazepam
1622-62-4	Flunitrazepam
1649-18-9	Azaperone
1668-19-5	Doxepin
1679-76-1	Drofenine
1689-83-4	Ioxynil
1689-84-5	Bromoxynil
1694-78-6	Medazepam M (Nor-)
1702-17-6	Clopyralid
1746-81-2	Monolinuron
1764-85-8	Epithiazide
1775-95-7	Nitrazepam Hy/ANB
1812-30-2	Bromazepam
1824-58-4	Ethiazide
1830-32-6	Azintamid
1841-19-6	Fluspirilene
1861-21-8	Enallylpropymal
1861-40-1	Benfluralin
1882-26-4	Pyricarbate
1893-33-0	Pipamperone
1910-68-5	Metisazone
1912-26-1	Trietazine
1918-02-1	Picloram
1929-88-0	Benzthiazuron
1951-25-3	Amiodarone

Table 8.16. (Continued)

CAS No.	Substance Name
1952-67-6	Crotylbarbital
1977-10-2	Loxapine
1982-37-2	Methdilazine
1982-47-4	Chloroxuron
2011-66-7	Clonazepam Hy/ANCB
2011-67-8	Nimetazepam
2016-63-9	Bamifylline
2030-63-9	Clofazimine
2058-52-8	Clotiapine
2062-78-4	Pimozide
2062-84-2	Benperidol
2086-83-1	Berberine
2090-89-3	Butethamine
2104-96-3	Bromophos
2127-01-7	Clorexolone
2139-47-1	Nifenazone
2152-34-3	Pemoline
2156-27-6	Benproperine
2163-69-1	Cycluron
2164-08-1	Lenacil
2164-17-2	Fluometuron
2165-19-7	Guanoxan
2167-85-3	Pipazetate
2179-37-5	Bencyclane
2180-92-9	Bupivacaine
2201-15-2	Eticyclidine/PCE/N-Ethyl-1-phenylcyclohexylamine
2201-24-3	Phenylcyclohexylamine,1-
2201-39-0	Rolicyclidine/PHP/1-(1-phenylcyclohexyl)pyrrolidine
2201-40-3	Phenylcyclohexyl)morpholine,4-(1-/PCM
2210-63-1	Mofebutazone
2235-90-7	Etryptamin
2259-96-3	Cyclothiazide
2303-17-5	Triallate
2310-17-0	Phosalone
2338-37-6	Levopropoxyphene
2373-84-4	Allyl-5-ethylbarbituric Acid,5-
2385-81-1	Furethidine
2430-49-1	Vinylbital
2438-72-4	Bufexamac
2439-01-2	Quinomethionate
2447-57-6	Sulfadoxine
2470-73-7	Dixyracine
2541-01-6	Terpin hydrate
2545-39-3	Clamoxyquine
2558-30-7	Flunitrazepam M (Nor-)
2565-43-7	Allylbarbituric Acid,5-
2574-78-9	Orazamide

Table 8.16. (Continued)

CAS No.	Substance Name
2589-47-1	Prajmalium Bitartrate
2609-46-3	Amiloride
2622-26-6	Periciazine
2622-30-2	Carfenazine
2622-37-9	Trifluomeprazine
2642-71-9	Azinphos-ethyl/Ethyl Guthion
2667-89-2	Bisbentiamin
2709-56-0	Flupentixol
2751-68-0	Acetophenazine
2784-73-8	Morphine-6-acetate/Diamorphine M
2801-68-5	Dimethoxyamphetaminc,2,5-
2829-19-8	Rolicyprine
2886-65-9	Flurazepam M (N(1)-Dealkyl-)
2886-65-9	Quazepam M (N-Dealkyl-2-oxo-)
2894-67-9	Delorazepam
2897-00-9	Prazepam Hy/CCB
2898-12-6	Medazepam
2898-13-7	Sulazepam
2921-88-2	Chlorpyrifos
2922-20-5	Butaxamine
2933-94-0	Toliprolol
2955-38-6	Prazepam
2958-30-3	Lorazepam Hy/ADB
3060-89-7	Metobromuron
3099-52-3	Nicametate
3146-66-5	Idobutal (5-Allyl-5-butylbarbituric Acid)
3215-70-1	Hexoprenaline
3321-80-0	Methylpiperidyl Benzilate/JB-336
3354-67-4	Amidefrine
3362-45-6	Noxiptiline
3383-96-8	Temephos
3416-26-0	Lidoflazine
3426-08-2	Prozapine
3562-84-3	Benzbromarone
3563-01-7	Diethylaminoethyl Diphenylpropionate
3565-72-8	Embramine
3567-12-2	Ethylpiperidyl Benzilate/JB-318
3572-43-8	Bromhexine
3572-80-3	Cyclazocine
3575-80-2	Melperone
3576-64-5	Clefamide
3583-64-0	Bumadizon
3605-01-4	Piribedil
3614-30-0	Emepronium Bromide
3615-24-5	Isopropylaminophenazone/Isopyrin
3625-06-7	Mebeverine
3625-25-0	Reposal

Table 8.16. (Continued)

CAS No.	Substance Name
3627-48-3	Pethidine Intermediate C
3627-62-1	Pethidine Intermediate A
3647-71-0	Benethamine
3684-46-6	Broxaldine
3686-58-6	Tolycaine
3688-65-1	Codeine M (N-Oxide)
3688-66-2	Nicocodine
3689-24-5	Sulfotep
3689-50-7	Oxomemazine
3691-21-2	Buzepide
3691-35-8	Chlorophacinon
3691-78-9	Benzethidine
3703-79-5	Bamethan
3731-59-7	Moroxydine
3734-52-9	Metazocine
3736-08-1	Fenetylline
3737-09-5	Disopyramide
3766-60-7	Buturon
3785-21-5	Butanilicaine
3811-25-4	Clorprenaline
3819-00-9	Piperacetazine
3820-67-5	Glafenine
3833-99-6	Homofenazine
3861-76-5	Clonitazene
3867-15-0	Phencyclidine intermediate/1-Piperidino-1-cyclohexanecarbonitrile/PCC
3878-19-1	Fuberidazole
3900-31-0	Fludiazepam
3930-20-9	Sotalol
3964-81-6	Azatadine
4093-35-0	Bromopride
4147-51-7	Dipropetryn
4171-13-5	Valnoctamine
4205-90-7	Clonidine
4255-23-6	Aletamine
4268-36-4	Tybamate
4360-12-7	Ajmaline
4378-36-3	Fenbutrazate
4394-00-7	Niflumic Acid
4406-22-8	Cyprenorphine
4498-32-2	Dibenzepin
4658-28-0	Aziprotryne
4685-14-7	Paraquat
4757-55-5	Dimetacrine
4764-17-4	Methylenedioxyamphetamine,3,4-/MDA
4824-78-6	Bromophos-ethyl
4846-07-5	Deoxyephedrine,(+/-)-
4914-30-1	Dehydroemetine

Table 8.16. (Continued)

CAS No.	Substance Name
4928-02-3	Nitrazepam M (7-Amino-)
4928-03-4	Nitrazepam M (7-Acetamido-)
4945-47-5	Bamipine
4958-56-9	Nimetazepam Hy/MNB
4959-17-5	Clonazepam M (7-Amino-)
4969-02-2	Metixene
5001-32-1	Guanoclor
5003-48-5	Benorilate
5011-34-7	Trimetazidine
5036-02-2	Tetramisole
5104-49-4	Flurbiprofen
5118-29-6	Melitracene
5140-28-3	Morphine-3-acetate
5377-20-8	Metomidate
5560-72-5	Iprindole/Glycophene
5585-64-8	Aminoxytriphene
5585-93-3	Oxypendyl
5588-33-0	Mesoridazine
5591-45-7	Tiotixene
5598-13-0	Chlorpyriphos-methyl
5626-86-3	Lormetazepam Hy/MDB
5632-44-0	Tolpropamine
5636-83-9	Dimetindene
5636-92-0	Picloxydine
5638-76-6	Betahistine
5666-11-3	Levomoramide
5668-06-4	Mecloxamine
5677-84-9	Methylphenazone,4-
5697-56-3	Carbenoxolone
5786-21-0	Clozapine
5836-29-3	Coumatetralyl
5915-41-3	Terbutylazine
5966-41-6	Diisopromine
6028-35-9	Thibenzazoline
6055-90-9	Narcophin
6168-76-9	Crotethamide
6376-26-7	Salverine
6452-71-7	Oxprenolol
6493-05-6	Pentoxifylline
6506-37-2	Nimorazole
6536-18-1	Morazone
6556-11-2	Inositol Nicotinate
6592-85-4	Hydrastinine
6621-47-2	Perhexiline
6673-35-4	Practolol
6703-27-1	Acetylcodeine
6703-39-5	Diphenazoline

Table 8.16. (Continued)

CAS No.	Substance Name
6740-88-1	Ketamine
6923-22-4	Monocrotophos
7009-54-3	Pentapiperide
7085-19-0	Mecoprop/MCPP
7195-27-9	Mefruside
7224-08-0	Imiclopazine/Chlorimiphenine
7248-28-4	Strychnine M (N(6)-Oxide)
7261-97-4	Dantrolene
7287-19-6	Prometryn
7413-36-7	Nifenalol
7416-34-4	Molindone
7456-24-8	Dimetiotazine
7683-59-2	Isoprenaline
7722-15-8	Chlordiazepoxide M (Nor-)
7728-40-7	Lidocaine M/MEGX
7773-52-6	Cetylpyridinium
7786-34-7	Mevinphos/Phosdrin
8001-54-5	Benzalkonium Chloride
8006-25-5	Ergotoxine
8025-81-8	Spiramycin
8044-71-1	Cetrimide
8051-02-3	Veratrine
8063-07-8	Kanamycin
8065-48-3	Demeton
8065-48-3	Demeton
9014-67-9	Aloxiprin
10061-32-2	Levophenacylmorphan
10118-90-8	Minocycline
10176-39-3	Gitoformate
10238-21-8	Glibenclamide/Glyburide
10262-69-8	Maprotiline
10265-92-6	Methamidophos
10310-32-4	Tribenoside
10379-11-0	Tetrazepam M (Nor-)
10379-14-3	Tetrazepam
10402-90-1	Eprazinone
10405-02-4	Trospium Chloride
10457-90-6	Bromperidol
10539-19-2	Moxaverine
11005-63-3	Strophanthin
11032-41-0	Ergoloid
13042-18-7	Fendiline
13071-79-9	Terbufos
13121-70-5	Cyhexatin
13132-73-5	Bromazepam M (3-Hydroxy-)
13171-21-6	Phosphamidon
13194-48-4	Ethoprofos/Ethoprop

Table 8.16. (Continued)

CAS No.	Substance Name
13292-46-1	Rifampin
13360-45-7	Chlorbromuron
13392-18-2	Fenoterol
13392-28-4	Rimantadine
13457-18-6	Pyrazophos
13460-98-5	Theodrenaline
13461-01-3	Aceprometazine
13495-09-5	Piminodine
13523-86-9	Pindolol
13539-59-8	Azapropazone
13655-52-2	Alprenolol
13669-70-0	Nefopam
13674-05-0	Methoxy-4,5-methylenedioxyamphetamine,3-/MMDA
13684-63-4	Phenmedipham
13912-77-1	Octacaine
13912-80-6	Butoxyethyl Nicotinate,2-
14007-64-8	Butetamate
14007-67-1	Dimethylphenylthiazolanimin (Dimethyl-5-phenyl-2-thiazolidinimine,3,4-
14008-44-7	Metopimazine
14028-44-5	Amoxapine
14051-33-3	Benzetimide
14088-71-2	Proclonol
14176-49-9	Tiletamine
14214-32-5	Difenoxuron
14222-60-7	Protionamide
14293-44-8	Xipamide
14297-87-1	Benzylmorphine
14334-40-8	Pramiverine
14376-16-0	Sulfaloxic Acid
14504-73-5	Tritoqualine
14521-96-1	Etorphine
14556-46-8	Bupranolol
14611-51-9	Selegiline
14759-06-9	Sulforidazine
14769-73-4	Levamisole
14816-18-3	Phoxim
14860-49-2	Clobutinol
15301-48-1	Bezitramide
15301-69-6	Flavoxate
15301-93-6	Tofenacin
15307-86-5	Diclofenac
15351-09-4	Metamfepramone
15351-13-0	Nicofuranose
15500-66-0	Pancuronium Bromide
15545-48-9	Chlortoluron
15574-96-6	Pizotifen
15588-95-1	Dimethoxy-4-methylamphetamine,2,5-/STP/DOM

Table 8.16. (Continued)

CAS No.	Substance Name
15676-16-1	Sulpiride
15686-51-8	Clemastine
15686-61-0	Fenproporex
15686-68-7	Volazocine
15687-07-7	Cyprazepam
15687-16-8	Carbifene
15687-27-1	Ibuprofen
15687-37-3	Naftazone
15687-41-9	Oxyfedrine
15793-40-5	Terodiline
15876-67-2	Distigmine Bromide
15972-60-8	Alachlor
16008-36-9	Methyldesorphine
16110-51-3	Cromoglicic Acid
16509-23-2	Ethomoxane
16662-47-8	Gallopamil
16752-77-5	Methomyl
16773-42-5	Ornidazole
17086-28-1	Doxycycline
17199-54-1	Alphamethadol
17199-58-5	Alphacetylmethadol
17199-59-6	Betacetylmethadol
17243-39-9	Benzoctamine
17243-57-1	Mefenorex
17365-01-4	Etiroxate
17479-19-5	Dihydroergocristine
17560-51-9	Metolazone
17575-22-3	Lanatoside C
17617-23-1	Flurazepam
17617-59-3	Flurazepam M (Dideethyl-)
17656-74-5	Flurazepam M (Monodeethyl-)
17692-31-8	Dropropizine
17692-34-1	Etodroxizine
17780-72-2	Clorgiline
18053-31-1	Fominoben
18109-80-3	Butamirate
18323-44-9	Clindamycin
18330-94-4	Nitrazepam M Hy/DAB
18559-94-9	Salbutamol/Albuterol
18683-91-5	Bromhexine M/Ambroxol
18691-97-9	Methabenzthiazuron
18717-72-1	Cocaine M (Nor-)
18818-61-6	Prazepam M (3-Hydroxy-)
18833-13-1	Acefylline Piperazine
19216-56-9	Prazosin
19420-52-1	Methyl-1-phenylcyclohexyl)piperidine,1-(4-
19794-93-5	Trazodone

Table 8.16. (Continued)

CAS No.	Substance Name
19937-59-8	Metoxuron
20290-09-9	Morphine M (-3-glucuronide)
20380-58-9	Tilidine
20448-86-6	Bornaprine
20736-11-2	Codeine M (-6-glucuronide)
20830-75-5	Digoxin
20971-53-3	Flurazepam M (N(1)-Hydroxyethyl-)
21087-64-9	Metribuzin
21187-98-4	Gliclazide
21363-18-8	Viminol
21500-98-1	Tenocyclidine/TCP/1-[1-(2 Thienyl)cyclohexyl]piperidine
21725-46-2	Cyanazine
21829-25-4	Nifedipine
21888-48-2	Dexetimide
22004-32-6	Dimethoxy-4-ethylamphetamine,2,5-/DOET
22071-15-4	Ketoprofen
22089-22-1	Trofosfamide
22131-35-7	Butalamine
22131-79-9	Alclofenac
22204-53-1	Naproxen
22204-89-3	Psilocin-(eth)/CZ-74
22232-54-8	Carbimazole
22232-71-9	Mazindol
22248-79-9	Tetrachlorvinphos
22259-30-9	Formetanate
22316-47-8	Clobazam
22316-55-8	Clobazam M (Nor-)
22345-47-7	Tofisopam
22365-40-8	Triflubazam
22494-42-4	Diflunisal
22664-55-7	Metipranolol
22881-35-2	Famprofazone
22916-47-8	Miconazole
23031-25-6	Terbutaline
23047-25-8	Lofepramine
23092-17-3	Halazepam
23103-98-2	Pirimicarb
23155-02-4	Fosfomycin
23239-32-9	Methoxyamphetamine,4-/PMA
23271-74-1	Fedrilate
23465-76-1	Caroverine
23560-59-0	Heptenophos
23593-75-1	Clotrimazole
23694-81-7	Mepindolol
23744-24-3	Pipoxolan
23779-99-9	Floctafenine
23887-31-2	Clorazepic Acid

Table 8.16. (Continued)

CAS No.	Substance Name
23930-37-2	Alfadolone Acetate
23943-90-0	Orientalidine
23950-58-5	Propyzamide
23964-58-1	Articaine
24017-47-8	Triazophos
24143-17-7	Oxazolam
24166-13-0	Cloxazolam
24219-97-4	Mianserin
24526-64-5	Nomifensine
24815-24-5	Rescinnamine
24934-91-6	Chlormephos
25046-79-1	Glisoxepide
25311-71-1	Isofenphos
25322-68-3	Polyethyleneglycol
25366-23-8	Thiazafluron
25384-17-2	Allylprodine
25614-03-3	Bromocriptine
25717-80-0	Molsidomine
26171-23-3	Tolmetin
26225-79-6	Ethofumesate
26644-46-2	Triforine
26807-65-8	Indapamide
26839-75-8	Timolol
26864-56-2	Penfluridol
26944-48-9	Glibornuride
27031-08-9	Sulfaguanole
27060-91-9	Flutazolam
27112-37-4	Diamocaine
27203-92-5	Tramadol
27220-47-9	Econazole
27223-35-4	Ketazolam
27848-84-6	Nicergoline
28395-03-1	Bumetanide
28782-42-5	Difenoxine
28797-61-7	Pirenzepine
28860-95-9	Carbidopa
28911-01-5	Triazolam
28981-97-7	Alprazolam
29094-61-9	Glipizide
29122-68-7	Atenolol
29177-84-2	Ethyl Loflazepate
29216-28-2	Mequitazine
29232-93-7	Pirimiphos-methyl
29973-13-5	Ethiofencarb
29975-16-4	Estazolam
30544-47-9	Etofenamate
30560-19-1	Acephate

Table 8.16. (Continued)

CAS No.	Substance Name
30748-29-9	Feprazone
30896-57-2	Alprazolam M (4-Hydroxy-)
31036-80-3	Lofexidine
31329-57-4	Naftidrofuryl
31430-15-6	Flubendazole
31431-39-7	Mebendazole
31793-07-4	Pirprofen
31828-71-4	Mexiletine
31868-18-5	Mexazolam
31879-05-7	Fenoprofen
31895-21-3	Thiocyclam
32156-26-6	Bromo-2,5-dimethoxyamphetamine,4-/DOB
32385-11-8	Sisomicin
32986-56-4	Tobramycin
33005-95-7	Tiaprofenic Acid
33125-97-2	Etomidate
33342-05-1	Gliquidone
33369-31-2	Zomepirac
33396-37-1	Meproscillarine
33671-46-4	Clotiazepam
34084-50-9	Flunitrazepam M (7-Amino-)
34368-04-2	Dobutamine
34580-13-7	Ketotifen
34681-10-2	Butocarboxim
34681-23-7	Butoxycarboxim
34784-64-0	Tertatolol
34915-68-9	Bunitrolol
35080-11-6	Prajmaline
35231-38-0	Flurazepam M Hy/HCFB
35322-07-7	Fosazepam
35367-38-5	Diflubenzuron
35554-44-0	Enilconazole/Imazalil
35941-65-2	Butriptyline
36104-80-0	Camazepam
36105-18-7	Flurazepam Hy/DCFB
36322-90-4	Piroxicam
36330-85-5	Fenbufen
36505-84-7	Buspirone
36637-18-0	Etidocaine
36688-78-5	Clindamycin Palmitate
36734-19-7	Iprodione
36735-22-5	Quazepam
36894-69-6	Labetalol
37115-32-5	Adinazolam
37115-33-6	Adinazolam M (Nor-)
37115-43-8	Alprazolam M (α-Hydroxy-)
37115-45-0	Triazolam M (α-Hydroxy-)

Table 8.16. (Continued)

CAS No.	Substance Name
37148-27-9	Clenbuterol
37350-58-6	Metoprolol
37517-28-5	Amikacin
37517-30-9	Acebutolol
37640-71-4	Aprindine
38194-50-2	Sulindac
38260-54-7	Etrimfos
38260-54-7	Etrimfos
38304-91-5	Minoxidil
38363-40-5	Penbutolol
39196-18-4	Thiofanox
39235-63-7	Methylenedioxyphentermin,3,4-
39393-56-3	Pseudoephedrine M (Nor-)/Cathine
39562-70-4	Nitrendipine
39715-02-1	Endralazine Mesilate
39860-99-6	Pipotiazine
40054-69-1	Etizolam
40487-42-1	Pendimethalin/Penoxalin
40951-53-9	Amino-5-bromo-3-hydroxyphenyl)(2-pyridyl)ketone,(2-
41083-11-8	Azocyclotin
41394-05-2	Metamitron
41708-72-9	Tocainide
41859-67-0	Bezafibrate
41993-29-7	Clonazepam M (7-Amino-3-hydroxy-)
41993-30-0	Clonazepam M (7-Acetamido-)
42116-76-7	Carnidazole
42200-33-9	Nadolol
42399-41-7	Diltiazem
42542-10-9	Methylenedioxymethamphetamine,3,4-/MDMA/XTC
42794-76-3	Midodrine
43121-43-3	Triadimefon
43200-80-2	Zopiclone
46817-91-8	Viloxazine
48141-64-6	Etafedrine
50275-61-1	Glutethimide M (Amino-)
50471-44-8	Vinclozolin
50567-35-6	Metamizol/Dipyrone
50679-08-8	Terfenadine
50993-68-5	Propyphenazone M (Nor-)
51012-32-9	Tiapride
51235-04-2	Hexazinone
51322-75-9	Tizanidine
51333-22-3	Budesonide
51338-27-3	Diclofop-methyl
51481-61-9	Cimetidine
51630-58-1	Fenvalerate
51781-06-7	Carteolol

Table 8.16. (Continued)

CAS No.	Substance Name
51940-44-4	Pipemidic Acid
52315-07-8	Cypermethrin
52463-83-9	Pinazepam
52468-60-7	Flunarizine
52485-79-7	Buprenorphine
52645-53-1	Permethrin
52756-22-6	Flamprop-isopropyl
52756-25-9	Flamprop-methyl
52918-63-5	Decamethrin/Deltamethrin
53008-88-1	Lonazolac
53164-05-9	Acemetacine
53179-11-6	Loperamide
53716-49-7	Carprofen
54063-52-4	Pitofenone
54063-53-5	Propafenone
54063-56-8	Suloctidil
54143-55-4	Flecainide
54739-18-3	Fluvoxamine
54910-89-3	Fluoxetine
55179-31-2	Bitertanol
55219-65-3	Triadimenol
55285-14-8	Carbosulfan
55512-33-9	Pyridate
55985-32-5	Nicardipine
56775-88-3	Zimeldine
56980-93-9	Celiprolol
56995-20-1	Flupirtine
57018-04-9	Tolclofos-methyl
57052-04-7	Isomethiozin
57132-53-3	Proglumetacin
57775-29-8	Carazolol
57801-81-7	Brotizolam
57808-65-8	Closantel
57808-66-9	Domperidone
58166-83-9	Cafedrine
58479-51-9	Clonazepam M Hy/DCB
59026-32-3	Methylprimidone,4
59128-97-1	Haloxazolam
59467-70-8	Midazolam
59468-85-8	Midazolam M (4-Hydroxy-)
59468-90-5	Midazolam M (α-Hydroxy-)
59729-31-6	Lorcainide
60168-88-9	Fenarimol
60207-90-1	Propiconazole
60238-56-4	Chlorthiophos
60568-05-0	Furmecyclox
60607-34-3	Oxatomide

Table 8.16. (Continued)

CAS No.	Substance Name
61197-73-7	Loprazolam
62551-41-1	Brotizolam M (α-Hydroxy-)
62571-86-2	Captopril
63659-18-7	Betaxolol
64257-84-7	Fenpropathrin
64638-05-7	Methoxy-4,5-methylenedioxyamphetamine,2-/MMDA-2
64740-68-7	Midazolam M (α,4-Dihydroxy-)
65277-42-1	Ketoconazole
65686-11-5	Triazolam M (4-Hydroxy-)
66085-59-4	Nimodipine
66357-35-5	Ranitidine
66722-44-9	Bisoprolol
66778-36-7	Encainide
66796-40-5	Propoxyphene M (Nor-)
67306-03-0	Fenpropimorph
67664-86-2	Bromazepam M (N(py)-Oxide)
67739-71-3	Flunitrazepam M (3-Hydroxy-)
67739-72-4	Flunitrazepam M (7-Acetamido-)
68359-37-5	Cyfluthrin
68692-83-1	Ethyl-2-(p-tolyl)malonamide,2-
68844-77-9	Astemizole
68883-08-9	Melperone M (FG 5155)
69806-50-4	Fluazifop-butyl
74050-98-9	Ketanserin
74051-80-2	Sethoxydim
74191-85-8	Doxazosin
75847-73-3	Enalapril
76330-73-9	Ketanserin M/Ketanserinol
78755-81-4	Flumazenil/RO 15-1788
79704-88-4	Alphamethylfentanyl
80866-90-6	Primidone M/PEMA
81098-60-4	Cisapride
83915-83-7	Lisinopril
84031-17-4	Metaclazepam
84378-44-9	Flumazenil M/RO 15-3890
86298-26-2	Metaclazepam M (N-Nor-)
86298-28-4	Metaclazepam M (Dinor-)
87112-45-6	Bromo-STP/Bromo-2,5-dimethoxy-4-methylamphetamine
88883-43-6	Brotizolam M (6-Hydroxy-)

9 References

Akkerboom, J.C., Schepers, P., Werff, J. v.d.: Thin-layer chromatography, a case study. Statist. Neerl. **34** (1980) 173-187.

Bogusz, M., Klys, M., Wijsbeek, J., Franke. J.P., de Zeeuw, R.A.: Impact of biological matrix and isolation methods on detectability and interlaboratory variations of TLC R_f values in systematic toxicological analysis. J.Anal.Toxicol. **8** (1984) 149-154.

Bogusz, M., Gierz, J., de Zeeuw, R.A., Franke, J.P.: Influence of the biological matrix on retention behaviour in thin-layer chromatography: Evidence of systematic differences between pure and extracted drugs. J. Chromatogr. **342** (1985a) 241-244.

Bogusz, M., Bialka, J., de Zeeuw, R.A., Franke, J.P.: Erratic data in thin-layer chromatography due to anomalous behaviour of correction standards. J. Anal. Toxicol. **9** (1985b) 139-140.

Brose, C., Rochholz, G., Erdmann, F., Schütz, H.: Ein DC-Detektionsprogramm für 178 Pestizide. Beiträge Ger. Med., in press.

De Clerq, H., Massart, D.L.: Evaluation and selection of optimal solvents and solvent combinations in thin-layer chromatography. Application of the method to basic drugs. J. Chromatogr. **115** (1975) 1-7.

De Zeeuw, R.A., Schepers, P., Greving, J.E., Franke, J.P.: A new approach to the optimization of chromatographic systems and the use of a generally accessible data bank in systematic toxicological analysis. In: Instrumental Applications in Forensic Drug Chemistry. M. Klein, A.V. Kruegel, S.P. Sobol, (Eds.): US Government Printing Office, Washington DC 1978, pp. 167-179.

Degel, F; Paulus, N., Toxi-Lab International Drug Compendium, Vol. 1. Toxi-Lab, Inc., Irvine, CA 1991.

Degel, F., Schulzki, T., Weideman, G.: Integration of thin-layer chromatographic color reactions in a computerized identification of drugs in clinical toxicological analysis. Mitt. Deutsch. Ges. Klin. Chemie **21** (1990) 251.

Erdmann, F., Brose, C., Schütz, H.: TLC screening program for 170 commonly used pesticides using the corrected R_f value. Int. J. Legal Med. **104** (1990) 25-31.

Franke, J.P., Schepers, P., Bosman, J., de Zeeuw, R.A.: Optimization of thin-layer chromatography for toxicological screening: Applicability of shorter retention distances. J. Anal. Toxicol. **6** (1982) 131-134.

Franke, J.P., Kruyt, W., de Zeeuw, R.A.: Influence of developing distance on resolution in thin-layer chromatography. J. High Resol. Chromatogr. Chromatogr. Commun. **6** (1983) 82-88.

Franke, J.P., de Zeeuw, R.A., Schepers, P.: Retrieval of analytical data and substance identification by the Mean List Length approach. J. Forens. Sci. **30** (1985) 1074-1081.

Galanos, D.S., Kapoulas, V.M.: The paper chromatographic identification of compounds using two reference compounds. J. Chromatogr. **13** (1964) 128-138.

Gill, R., Law, B., Brown, C., Moffat, A.C.: A computer search for the identification of drugs using a combination of thin-layer chromatographic, gas-liquid chromatographic and ultraviolet spectroscopic data. Analyst **110** (1985) 1059-1065.

Hegge, H.F.J., Franke, J.P., de Zeeuw, R.A.: Combined information from retardation factor (R_f) values and color reactions on the plate greatly enhances the identification power of thin-layer chromatography in systematic toxicological analysis. J. Forensic Sci. **36** (1991) 1094-1101.

Martindale: The Extra Pharmacopoeia, 29th edn. Pharmaceutical Press, London 1989.

Massart, D.L.: The use of information theory for evaluating the quality of thin-layer chromatographic separations. J. Chromatogr. **79** (1973) 157-163.

Merck Index: 11th edn., Merck and Co., Rahway, New Jersey 1989.

Moffat, A.C.: The standardisation of thin-layer chromatographic systems for the identification of basic drugs. J. Chromatogr. **110** (1975) 341-347.

Moffat, A.C.: Thin-layer chromatography. In: Clarke's Isolation and Identification of Drugs. A.C. Moffat (Ed.): Pharmaceutical Society, London 1986, pp. 160-177.

Moffat, A.C., Clare, B.: The choice of paper and thin-layer chromatographic systems for the analysis of basic drugs. J. Pharm. Pharmac. **26** (1974) 665-670.

Moffat, A.C., Smalldon, K.W.: Optimum use of paper, thin-layer and gas-liquid chromatography for the identification of basic drugs. 2. Paper and thin-layer chromatography. J. Chromatogr. **90** (1974) 9-17.

Moffat, A.C., Smalldon, K.W., Brown, C.: Optimum use of paper, thin-layer and gas-liquid chromatography for the identification of basic drugs. 1. Determination of effectiveness for a series of chromatographic systems. J. Chromatogr. **90** (1974) 1-7.

Müller, R.K.: Selection of methods for systematic toxicological analysis. In: Forensic Toxicology, Proc. TIAFT Meeting Glasgow. J.S. Oliver, Editor. Croom Helm, London 1980, pp. 26-33.

Müller, R.K. (Ed.): Toxicological Analysis. Ullstein Mosby, Berlin 1991.

Müller, R.K., Mückel, W., Wallenborn, H., Weihermüller, A., Weihermüller, C., Lauermann, I.: Objektive Kriterien zur Auswahl optimaler chromatographischer Systeme. 1. Mitteilung: Optimierung von DC-Fließmitteln auf Grund der R_f Verteilung größerer Stoffgruppen. Beitr. Gerichtl. Med. **34** (1976) 265-269.

Musumarra, G., Scarlata, G., Romano, G., Clementi, S.: Identification of drugs by principal components analysis of R_f data obtained by TLC in different eluent systems. J. Anal. Toxicol. 7 (1983) 286-292.

Musumarra, G., Scarlata, G., Cirma, G., Romano, G., Palazzo, S., Clementi, S., Giulietti, G.: Application of principal components analysis to the evaluation and selection of eluent systems for the thin-layer chromatography of basic and neutral drugs. J. Chromatogr. **295** (1984) 31-47.

Nanogen Index: Nanogen International. Freedom, California 1975.

Owen, P., Pendlebury, A., Moffat, A.C.: Choice of thin-layer chromatographic systems for the routine screening for neutral drugs during toxicological analyses. J. Chromatogr. **161** (1978a) 187-193.

Owen, P., Pendlebury, A., Moffat, A.C.: Choice of thin-layer chromatographic systems for the routine screening for acidic drugs during toxicological analyses. J. Chromatogr. **161** (1978b) 195-203.

Schepers, P., Franke, J.P., de Zeeuw, R.A.: System evaluation and substance identification in systematic toxicological analysis by the Mean List Length approach. J. Anal. Toxicol. **7** (1983) 272-278.

Schütz, H.: Benzodiazepines II. Springer, Berlin-Heidelberg-New York 1989.

Smalldon, K.W., Moffat, A.C.: The calculation of discriminating power for a series of correlated attributes. J. Forens. Sci. Soc. **13** (1973) 291-295.

Stead, A.H., Gill, R., Wright, T., Gibbs, J.P., Moffat, A.C.: Standardized thin-layer chromatographic systems for the identification of drugs and poisons. Analyst **107** (1982) 1106-1168.

10 Addresses of the Authors

Prof.Dr. R.A. de Zeeuw	Universitair Centrum voor Farmacie Universiteit Groningen A. Deusinglaan 2 NL-9713 AW Groningen The Netherlands
Dr. J.P. Franke	Universitair Centrum voor Farmacie Universiteit Groningen A. Deusinglaan 2 NL-9713 AW Groningen The Netherlands
Dr. F. Degel	Klinikum Nürnberg Institut für Klinische Chemie Flurstrasse 17 D-8500 Nürnberg 90 Germany
Prof.Dr. G. Machbert	Institut für Rechtsmedizin Universitätsstrasse 22 D-8520 Erlangen Germany
Prof.Dr. H. Schütz	Institut für Rechtsmedizin Frankfurter Strasse 58 D-6300 Giessen Germany
Ing. J. Wijsbeek	Universitair Centrum voor Farmacie Universiteit Groningen A. Deusinglaan 2 NL-9713 AW Groningen The Netherlands

11 Members of the TIAFT Committee for Systematic Toxicological Analysis

Prof.Dr. B.S. Finkle	Center for Human Toxicology University of Utah Salt Lake City, UT 84018 USA
Dr. J.P. Franke	Universitair Centrum voor Farmacie Universiteit Groningen A. Deusinglaan 2 NL-9713 AW Groningen The Netherlands
Dr. G.R. Jones	Office of the Chief Medical Examiner Edmonton, Alberta T5J 2P4 Canada
Prof.Dr. M.R. Möller	Institut für Rechtsmedizin Universität des Saarlandes D-6650 Homburg/Saar Germany
Prof.Dr. R.K. Müller (Chairman)	Institut für Gerichtliche Medizin Universität Leipzig Johannisallee 28 D-7010 Leipzig Germany
Dr. M.D. Osselton	Central Research and Support Establishment Home Office Forensic Science Service Aldermaston, Reading, Berkshire RG7 4PN, Great Britain
Prof.Dr. R.A. de Zeeuw	Universitair Centrum voor Farmacie Universiteit Groningen A. Deusinglaan 2 NL-9713 AW Groningen The Netherlands
Prof.Dr.R. Wennig	Laboratoire National de Santé Boîte Postale 11012 1A, Rue Auguste Lumière L-1011 Luxembourg Luxembourg

12 Members of the DFG Commission for Clinical-Toxicological Analysis

(Mandate finished per 1990)

Dr. J. Bäumler	Sonnmattstrasse 20 CH-4142 Münchenstein Switzerland
Prof.Dr. H. Brandenberger (Member until Summer 1987, thereafter Guest)	Lindenhofrain 8 CH-8708 Männedorf Switzerland
Prof.Dr.Dr. J. Büttner	Institut für Klinische Chemie I Zentrum Laboratoriumsmedizin Medizinische Hochschule Hannover Konstanty-Gutschow-Strasse 8 D-3000 Hannover 61 Germany
Prof.Dr. M. von Clarmann	Toxikologische Abteilung II. Medizinische Klinik u. Poliklinik Technische Universität Ismaninger Strasse 22 D-8000 München 80 Germany
Prof.Dr.Dr. M. Geldmacher-von Mallinckrodt (Chairperson)	Institut für Rechtsmedizin Universitätsstrasse 22 D-8520 Erlangen Germany
Prim.Dr. H.J. Gibitz	Chemische Zentrallaboratorien Landeskrankenhaus Salzburg Müllner Hauptstrasse 48 A-5020 Salzburg Austria
Prof.Dr. A.N.P. van Heijst (Member until Summer 1987, thereafter Guest)	Baarnseweg 42-A NL-3735 MJ Bosch en Duin The Netherlands

Prof.Dr. K. Ibe † (Deceased, 1991)	Freie Universität Berlin Universitätsklinikum Rudolf Virchow Standort Charlottenburg Spandauer Damm 130 D-1000 Berlin 19 Germany
Prof.Dr. G. Machata	Chemische Abteilung Institut für Gerichtliche Medizin Sensengasse 2 A-1090 Wien IX Austria
Prof.Dr. R.A.A. Maes	Faculteit Farmacie Sorbonnelaan 16 NL-3584 CA Utrecht The Netherlands
Dr. H. Moll	Kinderklinik des Marienhospitals Hauptkanal 75 D-2990 Papenburg Germany
Prof.Dr. M. Oellerich	Abteilung Klinische Chemie Georg August Universität Robert Koch Strasse 40 D-3400 Göttingen Germany
Prof.Dr. H. Schütz	Institut für Rechtsmedizin Frankfurter Strasse 58 D-6300 Giessen Germany
Prof.Dr.Dr. D. Stamm	Friedrich-Rein-Weg 21 D-8000 München 16 Germany
Dr. M. Stoeppler	Mariengartenstrasse 1a D-5170 Jülich 1 Germany

Prof.Dr.R. Wennig	Laboratoire National de Santé Boîte Postale 11012 1A, Rue Auguste Lumière L-1011 Luxembourg Louxembourg
Prof.Dr.Dr.H. Wisser	Abteilung für Klinische Chemie Robert-Bosch-Krankenhaus Auerbachstrasse 10 D-7000 Stuttgart 50 Germany
Prof.Dr. R.A. de Zeeuw	Universitair Centrum voor Farmacie Universiteit Groningen A. Deusinglaan 2 NL-9713 AW Groningen The Netherlands

Permanent Guest:

Dr. W. Fabricius	Bundesgesundheitsamt Thielallee 88-92 D-1000 Berlin 33 Germany

Program Director in charge of the
Deutsche Forschungsgemeinschaft:

Dr. Christoph Schneider	Deutsche Forschungsgemeinschaft Kennedyallee 40 D-5300 Bonn 2 Germany

13 Publications of the DFG Commission for Clinical-Toxicological Analysis

Mitteilung I	Gaschromatographische Retentionsindices toxikologisch relevanter Verbindungen auf SE-30 oder OV-1	1982
Report II	Gas-Chromatographic Retention Indices of Toxicologically Relevant Substances on SE-30 or OV-1 (Second, Revised and Enlarged Edition)	1985
Mitteilung III	Empfehlungen zum Nachweis von Suchtmitteln im Urin	1985
Report IV	Pharmacokinetics: Classic and Modern – Application to Pharmacology and Toxicology	1985
Report V	Theophylline Profile	1985
Mitteilung VI	Dünnschichtchromatographische Suchanalyse für 1,4-Benzodiazepine in Harn, Blut und Mageninhalt	1986
Report VII	Thin-Layer Chromatographic Rf-Values of Toxicologically Relevant Substances on Standardized Systems	1987
Mitteilung VIII	Photometrische Bestimmung von Carboxy-Hämoglobin (CO-Hb) im Blut	1988
Report IX	Tobramycin Profile	1988
Mitteilung X	Empfehlungen zur klinisch-toxikologischen Analytik, Folge 1: Einsatz von immuno-chemischen Testen in der Suchtmittelanalytik	1988
Mitteilung XI	Empfehlungen zur klinisch-toxikologischen Analytik, Folge 2: Einsatz der Gaschromatographie in der klinisch-toxikologischen Analytik	1988
Mitteilung XII	Empfehlungen zur klinisch-toxikologischen Analytik, Folge 3: Einsatz der Hochleistungsflüssigchromatographie in der klinisch-toxikologischen Analytik	1989
Report XIII	Amiodarone Profile	1990

Mitteilung XIV	Empfehlungen zur klinisch-toxikologischen Analytik, Folge 4: Einsatz elektrochemischer Techniken in der klinisch-toxikologischen Analytik	1990
Mitteilung XV	Orientierende Angaben zu therapeutischen und toxischen Konzentrationen von Arzneimitteln und Giften in Blut, Serum oder Urin	1990
Mitteilung XVI	Empfehlungen zur klinisch-toxikologischen Analytik, Folge 5: Empfehlungen zur Dünnschichtchromatographie	1991
Report XVII	Thin-Layer Chromatographic R_f Values of Toxicologically Relevant Substances on Standardized Systems (Second, Revised and Enlarged Edition)	1992
Report XVIII	Gas Chromatographic Retention Indices of Toxicologically Relevant Substances on Packed or Capillary Columns with Dimethylsilicone Stationary Phases (Third, Revised and Enlarged Edition)	1992
Report XIX	Gas Chromatographic Retention Indices of Solvents and Other Volatile Substances for Use in Toxicological Analysis	1992
Denkschrift	Klinisch-toxikologische Analytik	1983
Denkschrift	Dokumentation und Information in der Klinisch-toxikologischen Analytik	1987
Rundgespräche und Kolloquien	Klinisch-toxikologische Analytik – Gegenwärtiger Stand und Forderungen für die Zukunft	1987
Rundgespräche und Kolloquien	Klinisch-toxikologische Analytik bei akuten Vergiftungen und Drogenmißbrauch	1989
Rundgespräche und Kolloquien	Analytik für Mensch und Umwelt (unter Mitwirkung anderer Senatskommissionen)	1990

Appendix A

Computer Program in BASIC for the Calculation of Corrected R_f Values.

```
10 REM Program for the correction of Rf values
20 REM maximum 4 reference substances
30 GOTO 200
40 REM erase part of the screen
50 L=0
60 L=L+1: LOCATE (L+11),24:PRINT” “
70 IF L<4 THEN GOTO 60
80 RETURN
100 REM subroutine for reading a number (NR)
110 NR=-1
120 INPUT NR$
130 IF NR$=”” THEN NR=-1 :GOTO 180
140 FOR L = 1 TO LEN(NR$)
150 IF (ASC(MID$(NR$,L,1))>45) AND (ASC(MID$(NR$,L,1))<58) THEN NEXT L
160 IF L-1=LEN(NR$) THEN NR=VAL(NR$)
170 IF (NR<0) OR (NR>100) THEN GOTO 100
180 RETURN
200 CLS
210 PRINT “ Four reference substances must be used for correction”
220 PRINT “ First the program requests input of the reference/data base values”
230 PRINT “ Then the measured values for the reference substances has to be given”
240 PRINT “ Note: Input of hRf-values (Rf x 100); the substance with the lowest”
250 PRINT “ migration first and then with increasing hRF”
260 PRINT: PRINT” Input can be stopped by the input of ’0’”
400 LOCATE 10,10: PRINT”reference hRF and measured hRF”
410 LOCATE 12,10: PRINT”substance 1 : “;:GOSUB 100
420 IF NR=0 THEN CLS: END
430 RFC(1)=NR: I=1 440 I=I+1
450 LOCATE (I+11),10: PRINT “substance “;I;”: “;: GOSUB 100
460 RFC(I)=NR
470 IF I<4 THEN GOTO 440
500 I=0
510 I=I+1
520 LOCATE (I+11),34):GOSUB 100
530 IF NR=0 THEN GOSUB 100
540 RF(I)=NR
550 IF I<4 THEN GOTO 510
600 REM Input of unknown hRf
610 LOCATE 18,32:PRINT” “
620 LOCATE 18,10: PRINT “hRf of unknown spot : “;:GOSUB 100
```

```
630 IF NR=0 THEN GOSUB 50: GOTO 400
640 RFX=NR
700 REM Rf correction subroutine
710 IF RFX< RF(1) THEN RFCA=0:RFCB=RFC(1):RFA=0:RFB=RF(1)
720 IF RFX>RF(1) THEN RFCA=RFC(1):RFCB=RFC(2):RFA=RF(1):RFB=RF(2)
730 IF RFX>RF(2) THEN RFCA=RFC(2):RFCB=RFC(3):RFA=RF(2):RFB=RF(3)
740 IF RFX>RF(3) THEN RFCA=RFC(3):RFCB=RFC(4):RFA=RF(3):RFB=RF(4)
750 IF RFX>RF(4) THEN RFCA=RFC(4):RFCB=100:RFA=RF(4):RFB=100
760 RFCX=RFCA+(RFCB-RFCA)*(RFX-RFA)/(RFB-RFA)
770 LOCATE 20,10: PRINT "Corrected hRf value of unknown:";:PRINT
USING"###.#";RFCX
780 GOTO 600
```

Appendix B

Computer Program in TURBO PASCAL for the Calculation of Corrected R_f Values.

```
Program Rf_correction;

Uses Crt;

Var i,nr:integer;
 rfa,rfb,rfca,rfcb,rfcx,rfx:real;
 rfc,rf:array[1..4] of real; {hRf values of reference substances}
 number:string;

Procedure readnumber;
var code:integer;
begin nr:=-1;
 number:=''; readln(number);
 if number='' then nr:=-1 else val(number,nr,code);
 if code<>0 then nr:=-1
end;

Procedure input; {input of hRf data reference substances}
begin clrscr;
 writeln('Four reference substances must be used for correction.');
 writeln('First the program requests input of the reference/data base values');
 writeln('Then the measured values for the reference substances has to be given');
 writeln('Note: Input of hRf-values (Rf x 100); the substance with the lowest');
 writeln('migration first and then with increasing hRf');
 writeln;writeln('Input can be stopped by the input of "0"');
 gotoxy(10,10); write('reference hRf and measured hRf of reference substances');
 For i:=1 to 4 do
 begin
 gotoxy(10,i+11);
 write('substance ',I:1,': ');
 repeat gotoxy(24,i+11); readnumber until (nr>-1) and (nr<=100);
 if nr=0 then halt;
 rfc[i]:=nr;
 end;
 For i:=1 to 4 do
 begin
 repeat gotoxy(32,i+11); readnumber until (nr>0) and (nr<=100);
 rf[i]:=nr
 end
end;
```

```
Procedure unknown; {input hrf value unknown spot}
begin
 repeat gotoxy(32,18); write(' ');
 gotoxy(10,18); Write('hRf of unknown spot: ');
 readnumber;
 until (nr>=0) and (nr<=100);
 rfx:=nr
 end;

Procedure correction;
begin
 if rfx<rf[1] then begin rfca:=0; rfcb:=rfc[1]; rfa:=0; rfb:=rf[1] end;
 if rfx>rf[1] then begin rfca:=rfc[1]; rfcb:=rfc[2]; rfa:=rf[1]; rfb:=rf[2] end; if
rfx>rf[2] then begin rfca:=rfc[2]; rfcb:=rfc[3]; rfa:=rf[2]; rfb:=rf[3] end;
 if rfx>rf[3] then begin rfca:=rfc[3]; rfcb:=rfc[4]; rfa:=rf[3]; rfb:=rf[4] end;
 if rfx>rf[4] then begin rfca:=rfc[4]; rfcb:=100; rfa:=rf[4]; rfb:=100; end;
 rfcx:=rfca + (rfcb-rfca) * (rfx-rfa)/(rfb-rfa);
 gotoxy(10,20); write('Corrected hRf value of unknown: ',rfcx:4:1);
end;

{********** main program ************}
Begin input;
 repeat unknown; if nr<>0 then correction else input; until nr=0
end.
```

Appendix C

Details for the Detection of Pesticides Using TLC Systems P1 and P2.

After development, dry the plate in a stream of warm air for about 3 min and then apply one of the following six detection reagents, abbreviated RG1 through RG6.

RG1: 4-(4-Nitrobenzyl)-pyridine/tetraethylenepentamine

Preparation

a. 4-(4-nitrobenzyl)-pyridine (NBP): Dissolve 5 g of the substance in 100 ml of acetone.
b. tetraethylenepentamine (Tetren): Dilute 20 ml of a Tetren solution (Riedel-de Haën, AG, D-3016 Seelze, FRG) with acetone to 100 ml.

Procedure

Dip the plate in the NBP solution for 3 sec and dry in an oven for 30 min at 110 °C. After cooling, dip the plate for 3 sec in the diluted Tetren solution and observe the blue to violet spots on a white background. The colors are not stable.

RG2: 2,6-Dibromoquinone-4-chlorimide (Gibbs reagent)

Preparation

Two grams of 2,6-dibromoquinone-4-chlorimide (DBQ) are dissolved in 100 ml acetic acid. The solution is stable for two weeks.

Procedure

Dip the plate in DBQ solution for 2 sec and dry in an oven for 5 min at 110 °C. Observe the spots against a lightly yellow background. This reagent is able to detect sulfur-containing substances.

RG3: Palladium chloride

Preparation

Dissolve 0.5 g $PdCl_2$ in 2.5 ml HCl (32%) and carefully dilute with ethanol to 100 ml.

Procedure

Dip the plate in the reagent for 3 sec, dry for about 1 min in air and then dry in an oven for 20 min at 110 °C. Observe the spots against a lightly brown background.

RG4: Silver nitrate

Preparation
Dissolve 1 g $AgNO_3$ in 5 ml distilled water and 2.5 ml ammonia (25%). Dilute with acetone to 100 ml.

Procedure
Dip the plate in the reagent for about 3 min and dry for 15 min at 150 °C. A variety of colors may be seen against a brown background.

RG5: N,N-dimethyl-p-phenylenediamine hydrochloride

Preparation
Dissolve 0.1 g N,N-dimethyl-p-phenylenediamine hydrochloride in 10 ml ethanol. Mix with 10 ml sodium ethylate just before spraying.

Procedure
Irradiate the plate with UV light of 366 nm for 10 min and spray homogeneously with the reagent. Irradiate again with UV light of 366 nm for 10 min and observe the spots under the UV light against a gray to black background.

RG6: o-Tolidine

Preparation
Dissolve 1 g o-tolidine in 10 ml anhydrous acetic acid and 4 g potassium iodide in 10 ml distilled water. Mix the two solutions and dilute with distilled water to 1 L.

Procedure
Put the plate in a closed tank with chlorine gas (1 g potassium permanganate + 2 ml HCl (32%)) for 1 min. Remove excess of chlorine from the plate with a stream of cold air for about 30 sec. Dip the plate in the reagent for about 3 sec and observe the spots against a white to blueish background.

The resulting colors can be encoded using the Toxi-Lab Color Wheel and the observed R_f values and color codes can then be compared with the reference data in Tables 8.13 and 8.14.

Appendix D

Details for the Thin-Layer Chromatographic Analysis of Benzodiazepine Hydrolysis Products Using System B.

Thin-layer chromatography (TLC) is the preferred method for screening benzodiazepines and their metabolites. The procedure involves hydrolysis to yield aminobenzophenone derivatives, which are then extracted, separated by TLC and photolytically dealkylated. The products are diazotized and coupled with azo-dyes (e.g. the Bratton-Marshall reagent).

Experimental

Recommended reference substances for commonly used benzodiazepines:

ABFB	2-amino-5-bromo-2'-fluorobenzophenone (e.g. from haloxazolam)
ABP	(2-amino-5-bromophenyl)(2-pyridyl)methanone (e.g. from bromazepam)
ACB	2-amino-5-chlorobenzophenone (e.g. from oxazepam)
ACFB	2-amino-5-chloro-2'-fluorobenzophenone (e.g. from desalkylflurazepam)
ADB	2-amino-2',5-dichlorobenzophenone (e.g. from lorazepam)
ANB	2-amino-5-nitrobenzophenone (e.g. from nitrazepam)
ANCB	2-amino-2'-chloro-5-nitrobenzophenone (e.g. from clonazepam)
ANFB	2-amino-2'-fluoro-5-nitrobenzophenone (e.g. from 1-desmethylflunitrazepam)
CCB	5-chloro-2-[(cyclopropylmethyl)amino]benzophenone (e.g. from prazepam)
CFMB	5-chloro-2'-fluoro-2-(methylamino)benzophenone (e.g. from fludiazepam)
CFTB	5-chloro-2'-fluoro-2-(2,2,2-trifluoroethylamino)-benzophenone (e.g. from quazepam)
CPB	5-chloro-2-(2-propinylamino)-benzophenone (e.g. from pinazepam)
DAB	2,5-diaminobenzophenone (e.g. from aminonitrazepam)
DCB	2'-chloro-2,5-diaminobenzophenone (e.g. from aminoclonazepam)
DCFB	5-chloro-2-[2-(diethylamino)-ethylamino]-2'-fluorobenzophenone (e.g. from flurazepam)
HCFB	5-chloro-2'-fluoro-2-(2-hydroxyethylamino)-benzophenone (e.g. from hydroxyethylflurazepam)
MACB	5-chloro-2-(methylamino)benzophenone (e.g. from diazepam)
MDB	2',5-dichloro-2-(methylamino)benzophenone (e.g. from lormetazepam)
MNB	2-(methylamino)-5-nitrobenzophenone (e.g. from nimetazepam)
MNFB	2'-fluoro-2-(methylamino)-5-nitrobenzophenone (e.g. from flunitrazepam)
TCB	5-chloro-2-(2,2,2-trifluoroethylamino)benzophenone (e.g. from halazepam).

Spray Solution (Bratton-Marshall Reagent)

Dissolve 1g of N-(1-naphthyl)ethylenediamine in a mixture of 50 ml of dimethylformamide and 50 ml of 4 M hydrochloric acid, with warming if necessary. Filter the cooled solution if it is not clear. A slight violet color does not affect its use. If kept in the refrigerator the solution is stable for about a year.

Standard Solution for TLC

Dissolve 1 mg each of the reference substances in 2 ml of methanol. Note that for screening not all the reference substances are absolutely necessary, but ANB, ACB, MACB and CCB should all be used for comparison purposes and for calculation of the corrected R_f-value. If stored in glass bottles in the refrigerator (4 °C) and protected from light, the solution is stable for several months.
To avoid interference in the TLC, no other substances should be present that give a color with the Bratton-Marshall reagent.

Hydrolysis

Place 100 ml of the urine sample in a 500-ml Erlenmeyer flask and add 50 ml of concentrated hydrochloric acid. Heat the mixture for 30 min under a reflux condenser, in a boiling water-bath, and if necessary rinse the condensate from the condenser into the flask with a little concentrated hydrochloric acid.

Neutralization and Extraction

After the hydrolysis, cool the solution to room temperature, and then, with further cooling, adjust the pH to between 8 and 9 (universal indicator paper) by addition of 8 M sodium hydroxide (about 5 ml or so will be needed). Wear safety goggles during this operation, which should be conducted under an efficient fumehood on account of the very unpleasant smell. Extract the aminobenzophenone derivatives with about 200 ml of diethyl ether. Note that the acid hydrolysis of bromazepam and its metabolites does not yield benzophenone derivatives, but benzoylpyridine derivatives. As the latter are also primary aromatic amines, they can be detected with Bratton-Marshall reagent as well. To increase the yield, the extraction can be repeated with 100 ml of diethyl ether, at pH 11. Reduce the combined extracts to a volume of about 3 ml in a rotary evaporator, and transfer this concentrate to a glass-stoppered centrifuge tube and carefully evaporate it to dryness (at about 30-40 °C; it is not necessary to use reduced pressure). Cool the residue to 4 °C and reserve it for analysis; for this dissolve it in 0.1 ml of methanol.

Thin-Layer Chromatography

Use 20x20 cm silica gel 60 F_{254} TLC plates (Merck), layer thickness 0.25 mm. Apply the sample and standard spots 1.5 cm from the lower edge of the plate, with 2 μl capillaries. For each sample use three capillary-loads overlapped to give an approximately straight line of sample. To avoid any cross-contamination apply the test solutions before the standards. Run the chromatogram until the solvent front has travelled 10 cm. Use the ascending method, without chamber saturation. No special activation of the plates is needed, and would not improve the results anyway. Use about 100 ml of toluene as the mobile phase.

Photolytic Dealkylation

After the development of the chromatogram (which takes 30-40 min), leave the plate to drip in the development tank for a short time, then dry it in a cold air-stream under the fumehood. Expose the dried plate to a suitable ultraviolet source (e.g. a sun-lamp) at a distance of 30-40 cm for about 20 min. For rapid analysis a 6-min exposure is sufficient. Immediately cool the plate to room-temperature, or the yield in the diazotization step will be impaired.
Note that the dealkylation step is only necessary when testing for diazepam, camazepam, temazepam, ketazolam, prazepam, flurazepam, flunitrazepam, lormetazepam, fludiazepam, nimetazepam, pinazepam, quazepam and halazepam.

Diazotization

Place the cool dry plate in an empty chromatographic tank, on the bottom of which is a small beaker (20-50 ml) containing 10 ml of 20% sodium nitrite solution. Pipette 5 ml of 25% v/v hydrochloric acid into the beaker as fast as possible, to liberate nitrogen oxides, and seal the tank with its lid. Leave the plate in the tank for 3-5 min, which is sufficient time for diazotization of primary aromatic amine groups. Remove the lid, and when most of the nitrous gases have dispersed take out the plate and leave it under the fumehood for 20-30 min in a stream of cold air (e.g. from a fan heater set at "cold").
For rapid work it is sufficient to air the plate for only 5 min to remove the nitrous gases and then to spray it gently with a 1% aqueous solution of urea. (Alternatively, put the plate in a vacuum desiccator and draw vacuum for about 2 min to remove excess nitrous gases). Finally, spray the plate thinly and uniformly (meander pattern) with Bratton-Marshall reagent at 4 °C to couple the diazonium salts. The result: violet and redviolet colors.

N.B. As nitrous gases are poisonous, work in a fume hood and wear gloves!